Clone Being

Clone Being

Exploring the Psychological and Social Dimensions

Stephen E. Levick

ROWMAN & LITTLEFIELD PUBLISHERS, INC.
Lanham • Boulder • New York • Toronto • Oxford

ROWMAN & LITTLEFIELD PUBLISHERS, INC.

Published in the United States of America
by Rowman & Littlefield Publishers, Inc.
A wholly owned subsidary of The Rowman & Littlefield Publishing Group, Inc.
4501 Forbes Boulevard, Suite 200, Lanham, Maryland 20706
www.rowmanlittlefield.com

PO Box 317
Oxford
OX2 9RU, UK

British Library Cataloguing in Publication Information Available

Library of Congress Cataloging-in-Publication Data

Levick, Stephen E., 1951–
  Clone being : exploring the psychological and social dimensions /
Stephen E. Levick.
    p. cm.
Includes bibliographical references and index.
  ISBN 0-7425-2989-4 (alk. paper) — ISBN 0-7425-2990-8 (pbk. : alk.
paper).
  1. Human cloning—Social aspects. 2. Human cloning—Psychological
aspects. I. Title.
  QH442.2.L485 2004
  176—dc22

                                                      2003015575

Printed in the United States of America

♾ ™ The paper used in this publication meets the minimum requirements of the
American National Standard for Information Sciences—Permanence of Paper for
Printed Library Materials, ANSI/NISO Z39.48-1992.

Contents

Acknowledgments   vii

Preface   ix

Introduction   xiii

1. The Identical Twin Model   1
2. The Assisted Reproductive Technologies and Arrangements Model   23
3. The Stepchild Model   41
4. The Adoption Model   49
5. The Parent–Child Resemblance Model   71
6. The Child of the Famous Model   97
7. The Replacement Child Model   111
8. The Namesake Model   133
9. The Models Integrated   161
10. Wider Social and Cultural Implications of Cloning   183
11. Intimacy, Sex, and Sexuality   207
12. Implications for Cloning Ethics and Policy   235

Notes   259

Index   303

About the Author   319

Acknowledgments

Over the course of writing this book, I discussed many of the ideas with friends and colleagues. B. Hibbs, Rob Fuentes, Martha Webster, Bob Dworkin, Jamie Walkup, Alan Schwartz, and Elizabeth Gosch responded to my earliest formulations, and Elio Frataroli's encouraging words to my editor about the project were a boon.

As the project progressed, I bounced selected ideas off John Steidl, Syd Pulver, Eric Lager, Vic Malatesta, Jim Hoyme, Mike McCarthy, and Andrew Smolar. Wolf Rieger provided insightful German translation. The reviewers for the publisher, Rogers Smith and Alex Burland, offered crucial insights and advice, and discussions with Homer Curtis were always illuminating. Nancy Segal and Richard Alford made me aware of some highly relevant research in their respective areas of expertise.

Librarians at the University of Pennsylvania, the College of Physicians of Philadelphia, and Pennsylvania Hospital were quite helpful, especially Donna Quinn and Mary McMullen at Pennsylvania Hospital. Jessica Murakami, then an undergraduate at the University of Pennsylvania, read the manuscript and usefully responded as a lay reader, and Nolan Shenai provided library research assistance at a critical time.

Karen Shirley introduced me to my original editor at Rowman and Littlefield, Dean Birkenkamp. Dean provided critical guidance as I developed and executed this project. As the new sponsoring editor, Eve DeVaro has been a delight, and Lynn Weber, as production editor, and her staff have been most helpful.

It would have been truly impossible to write this book without the support of my wife, Judy Saltzberg. I also wish to thank my son, Noah, for

vii

teaching me invaluable things about fatherhood, the development of individual identity, and, though it is not a subject of this book, basketball.

Too numerous to acknowledge are the many great teachers and supervisors in my medical and psychiatry training who helped me develop the knowledge and understanding to engage this project. Among them, Behnaz Jalali and Paul Fleischman stand out from my residency at Yale. In more recent years, my colleagues at Pennsylvania Hospital, Rick Kluft and Catherine Fine, have helped open new psychotherapeutic vistas for me.

Making invaluable contributions are many people I cannot acknowledge by name for reasons of doctor–patient confidentiality. Their disguised cases and case composites illustrate the various models of cloning, the essential substance of this book.

Preface

Sometimes we can be shown a way of seeing that makes us feel more favorably disposed to something that had been distasteful or frightening. But we can also be alerted to menacing implications of something that we had previously thought harmless or frivolously amusing. Cloning provides a case study in the power of scientific thinking to change our minds in both directions.

—Richard Dawkins

What might be the psychological and social consequences of cloning people? Curious to see how experts on the ethics of cloning were considering this question, I took a day off from my psychiatric practice to attend a session of the American Association for the Advancement of Science Annual Meeting devoted to the ethics of cloning.[1] There I found leaders in biology, biomedical ethics, theology, and the law alluding to possible psychological and social implications of human cloning in their discussion of its ethics. Although their concerns seemed legitimate, they did not appear to be grounded in psychological theory, research, or clinical experience that might be relevant by analogy to aspects of cloning.

At the conference I was fortunate to be sitting next to Farida Shamali. After sharing some of my views with her, she invited me to write something for a conference in the United Arab Emirates. Not quite believing that considerable efforts to try to understand cloning from a psychological and social perspective had not already been made, I called Ruth Macklin. She had been my favorite professor in college and is a renowned bioethicist. Ruth agreed that the voices in the cloning debate had been pretty much confined to individuals in biology, biomedical ethics, theology, the law, and politics. She also pointed out that the then recently completed National Bioethics Advisory Commission report on human cloning did not reflect ideas and opinions from psychology and social science.[2] Since that conversation, the President's Commission on Bioethics issued its report in 2002, and although it does consider possible psychological and social consequences of cloning, it does not much tap into relevant theory and research.[3]

Ruth then asked the obvious question: "What do you want to say?" I honestly replied that, while I had misgivings about cloning, I did not yet know but that I hoped to systematically analyze the issue. She wished me luck, and I began the project, which eventually evolved into this book.

"Nothing that is human is alien to me."[4] This sentiment of the ancient philosopher Terence became the light of open-mindedness, helping to illuminate my way through this topic. With cloning such a controversial topic, I felt that this was the only way to speak to everyone and have a chance of being heard. In the beginning, staying open-minded was relatively easy, but, as the book progressed, it sometimes became a challenge. In addition to trying to remain scientifically dispassionate, I was guided by Terrence's attitude on many occasions. It helped me to assume a variety of perspectives on cloning, and I sincerely hope that it is reflected throughout the book.

But I should tell you that there is an issue related to reproductive cloning about which I have some strong biases. I favor embryonic stem cell research, which utilizes cloning not to produce a baby but to provide tissues for medical research. That research could lead to treatments with the potential to cure previously incurable diseases, and some people I love dearly could benefit. Because some might perceive my personal investment in stem cell research as interfering with my objectivity, I did not want to deal with it in this book. Specifically, I did not want to try to counter the claim made by many people opposed to abortion—that all cloning is reproductive cloning.[5] Not wanting to risk alienating readers with strong views on either side of the abortion and reproductive rights debate, I did not want to have to explain why we ought to separate the cloning of embryonic stem cell research from cloning for reproduction. But by the final chapter, on ethics and policy, the evolving logic of the book compelled me, most reluctantly, to confront that issue.

In this book, we'll try to draw logical inferences from psychological and social theory and research that seem relevant by analogy to key dimensions of cloning. If you wish, challenge my conclusions; and if you find the analogies unconvincing or feel that they are misapplied, then pinpoint where and how. And if you feel that reasoning by analogy is itself a mistake, consider that while a human clone or two could even have been born by the time this book is published, we have nothing else to serve as a basis for serious and systematic examination of the topic. There is already more than enough dogma, ungrounded speculation, and overheated rhetoric.

Jargon pervades psychiatry, psychology, and the social sciences, as it does any specialized field, but insider shorthand can make what ought to be clear ideas esoteric and opaque to everyone on the outside. But insiders, too, may befuddle themselves with their own private language. Being able to use jargon does not mean that you know what you are talking about. This haze must be avoided, and so I have tried to define technical terms in ordinary language and concretely illustrate the abstract discussion. Making research findings accessible and illustrating points with clinical examples vivify and humanize this project.

Assisting me in the writing of this book were groups of imagined readers. I have tried to speak to each in a way that all may understand: First, and foremost, is the general public, people without advanced education in any area this book touches on but no less able to reason than an expert, if things are laid out clearly. I hope that this book facilitates wider participation in the ethical debate on cloning and helps move that debate beyond the "fairly predictable lines" it has followed until now.[6]

Mental health professionals will find much that is familiar here—but applied to a most unfamiliar domain. Clinical practice of psychiatry and psychology in the twenty-first century may well encounter clones and cloning, and this book, in part, is an attempt to prepare for that eventuality. In addition, perhaps the very attempt to apply clinical theory to cloning may have some unforeseen general implications for those concepts and theories. By the same token, perhaps researchers in psychology and the social sciences, in finding their research applied to cloning, may have insights that never would have occurred to me.

Reproductive scientists and physicians will make cloning people a reality, if it has not already become one by the time this book appears. I have tried to be sensitive to their motivations to help the infertile, and I hope that this book will help to deepen their understanding and broaden their perspective on reproductive human cloning.

Advocates of, and personal aspirants for, human cloning have been much maligned and misunderstood. Figuring prominently among them, the infertile couple, gay and lesbian people desiring parenthood, the grieving parent, and even the narcissistic individual all deserve a compassionate hearing. I have tried to imagine some of their possible motivations, both conscious and unconscious. Whether or not they proceed with cloning, this book will have succeeded if it invites them to introspect about their own unique motives and helps them to empathize with the many others who may be directly or indirectly affected by their decision.

I have also written this book for a group people who, as far as we know, do not yet exist—individuals cloned from an already or previously existing person. I hope that, should they come to be, this book will prove useful to them.

SOME NOTES ON LANGUAGE

For brevity's sake, when I use the word *cloning* without elaboration I mean reproductive cloning by somatic cell nuclear transfer. I have not tried to achieve gender neutrality or equality in pronoun usage. Some things we will be examining are very gender specific, and many are not. The context will make it clear when the gender of the pronoun is not changeable.

Introduction

If, by the time this book is published, a human clone has not been born, one inevitably will be. After all, some people are personally motivated to be parents of a clone, and some scientists and clinicians with the knowledge and skills to reproductively clone humans have announced their intentions to try to do so.[1] Most scientists and fertility experts who might be able to attempt to clone a human being have been deterred by the low survival rate of implanted animal embryo clones and the high degree of medical risk run by those successfully implanted.[2] But what if cloning techniques improved sufficiently so that those risks were no longer major or could be reduced to a level now associated with currently practiced forms of medically assisted sexual reproduction? Would cloning people then become ethically acceptable, and permissible, with or without restriction? In pondering these ethical and policy questions, we need to ask whether cloning might carry other risks, in particular, psychological and social ones.

We need not wait for cloning techniques to be sufficiently perfected to try to anticipate possible psychological and social consequences. The first people to consider the possibility of psychological and social harm from cloning did so from ethical and legal perspectives.[3] It is primarily from those perspectives that such questions continue to be framed, with some input from pioneers in the biology of cloning, whose understandable focus is that of medical risk.[4]

Why have only a few people in the field thought that the psychological and social sciences might be able to contribute to cloning ethics and policy?[5] Undoubtedly, there has been a reluctance to pose and try to answer questions, given that research directly bearing on the subject is impossible. Without cloned people to study, wouldn't any attempt to apply psychological and social sciences to cloning be speculative? Certainly, but there is a difference between correctly derided "nebulous speculation" and systematic and thoughtful attempts at anticipation.[6] Hopefully, this book falls into the latter category and will further the kind of "informed insight" that Nobel Prize–winning biologist Joshua Lederberg believes can guide ethical judgment and social policy on human cloning.[7]

In 1938, Hans Spemann conceived his *gedanken,* or thought experiment, that became the basis for asexual reproduction or cloning by transferring the nucleus from an adult somatic cell into an egg whose nucleus had been removed. What now goes by the name "somatic cell nuclear transfer" seemed "fantastical," even to him.[8] This book is my thought experiment on the possible psychological and social consequences of the likely realization in humans of Spemann's *gedanken.* I hope that my approach makes my conjectures seem less fantastical than they might otherwise appear.

A few years before Spemann's thought experiment, the psychiatrist Paul Schilder mused about biological phenomena as analogues to psychoanalytic concepts.[9] One phenomenon he examined was asexual reproduction in primitive organisms. As far as I can tell, this was the first attempt to apply psychology to thinking about asexual reproduction, if only as a metaphor. Now, with the cloning of people soon a reproductive option, and a new reality of the human condition, we need to move beyond metaphor.

Science both grounds and transcends metaphor with models. While a metaphor suggests comparison between one concept and another, a scientific model seeks to represent aspects of one phenomenon or concept by another that is better understood. With the right model, scientists can make specific predictions about a phenomenon that is not well understood, but they may find that certain aspects of that phenomenon might be better represented by another model.[10] That is exactly what we will find here with these models of clones and cloning.

We will systematically analyze possible psychological and social consequences of human reproductive cloning through eight models of situations relevant by analogy to it. After noting biological, psychological, and social conceptual parallels and differences of each model with cloning, we will marshal theory and data relevant to each model and, with appropriate caveats, apply them to cloning.[11] As a result, we will see what psychological and social science theory and data imply for clones and cloning.

I want to emphasize that the psychological and social issues likely to be faced by clones and those involved with their upbringing are not unique to them. Cloning simply presents these issues to us in an unfamiliar but by no means alien context. Indeed, we can all reflect on aspects of our personal history and experience that may help us to empathically understand clones and cloning. The conceptual models and the scientific and clinical theory and data presented here are nothing more than tools to inform and thereby enhance our natural abilities as people to try to understand other people and ourselves.

Trying to understand cloning does not necessarily mean condoning it, but if we prejudge it, then we will hamper our ability to understand it. Because we best understand human phenomena when we try to do so from a range of perspectives, this will also be my approach before coming to ethical conclusions and policy recommendations.

In the interest of avoiding lengthy preliminaries to orient the reader, chapter 1, "The Identical Twin Model," also functions as an extended introduction. One way it does this is by presenting a number of fundamental psychological concepts that readers will also need at various other points in the book. Another introductory feature of the first chapter arises from the fact that identical twins are themselves clones, though of a different kind than we wish to understand here. Understanding how a clone by somatic cell nuclear transfer may differ psychologically and socially from a naturally occurring identical twin requires us to digest a minimal account of the biology of cloning.

Yet another key introductory aspect of chapter 1 arises from the fact that it demonstrates the need to supplement the Identical Twin Model with other models. The models presented subsequently are relevant by analogy to human reproductive cloning in some crucial ways in which identical twins are not.

In examining each model, we will try to understand not only the clone but also the progenitor (the person who was cloned) and both of the clone's rearing parents (only one of whom may be the progenitor). In doing so, we will also focus on a variety of possible origins of the wish to clone, as well as attend to the clone's larger family and his social and cultural worlds. Most of our focus will be on the clone's individual psychological and interpersonal development throughout his life, but especially in his formative early years. In each model, we will examine salient psychological, interpersonal, and social issues from the perspectives that would appear to be most relevant to cloning. In doing so, we will look for parallels and differences between the model and cloning.

Psychotherapy case material helps to illustrate the models, with several of the cases illustrating more than one model. Cases are drawn primarily from published clinical psychiatric literature (primarily psychoanalytic) and from my own clinical practice. Though some individuals might believe that they recognize themselves in these examples, I have disguised details that might have made them identifiable to others and in some instances have created composites of several actual cases to assure confidentiality.

No single model fully reflects all the psychological and social aspects of cloning. That is why we will look at eight. Some illuminate certain dimensions, issues, and perspectives better than others, and they also vary in both how and the degree to which they are relevant to cloning. After examining each model, we will note where they converge and complement one another and integrate them into a composite. We will then focus on wider social consequences, and in the penultimate chapter, specifically on the issues of intimacy, sex, and sexuality. In the process, a ninth model will emerge. The final chapter focuses on ethics and policy recommendations.

Although it is important to be fair in our assessments, we need not strive for perfect balance. From both medical and social policy perspectives, we must examine any new, allegedly beneficial practice for what harm it might cause. As he tries to heal, every physician is obligated by the Hippocratic oath to try even harder to do no harm. And so we will focus much more on possible risks than on the conceivable benefits of reproductive cloning.

Utilizing the methods I have outlined, let us begin where I started—uncertain of what we might conclude. Trusting in a systematic process of free inquiry, we may expect some surprises in the process.

1

The Identical Twin Model

A CLONE "ON THE COUCH"

Jerry came for his first appointment when I was already well into writing this book. As always in my clinical work, and especially in a first session, my mind was neither blank nor naively open. Instead, I listened with several clinical conceptual schemes of my profession in mind.[1] But with Jerry and some other patients, certain aspects of their cases also began to resonate with one or more models of cloning starting to take shape in my mind. While my primary agenda was always the patient, sometimes I could attend to his or her concerns and also seize the opportunity to better understand a cloning model.[2] This occurred with Jerry halfway through his first appointment when he told me that he was an identical twin. As one, Jerry truly was a clone on my couch—sitting, as do all my patients, not lying on it, as is the custom in psychoanalysis.

Identical twins are clones by a natural mechanism, and though not the sort we want to understand, they are clones nonetheless. As such, they might help us understand some things about the sort of clone that is our focus. Jerry and future cases illustrate a variety of cloning models—putting clones and cloning, as it were, "on the couch."

In the process of exploring the current model, we will familiarize ourselves with some basic facts about cloning, as well as some psychological concepts and issues also key to other models. Meeting Jerry will help us get started.

JERRY

Jerry, a retired businessman in his mid-fifties, had been having panic attacks for years. He had become too familiar with the racing heartbeat, the profuse sweating, and the feeling of slow suffocation, which no amount of rapid breathing could alleviate. All were part of a nameless dread that would come over him, seemingly without cause or explanation. These attacks had been worsening since his wife had almost died from complications from elective surgery, and he set his first appointment with me almost a year to the day from that event. Recently, she had begun talking about having the same procedure done again, and he had been trying to dissuade her, to that point unsuccessfully. She did not remember how sick she had been. Jerry could never forget.

He had gone several times to the hospital emergency room fearing that he was having a heart attack, but all testing showed nothing wrong physically. He finally was forced to agree with doctors who told him that it must be "in his head," but it sure did not feel that way.

Jerry was determined not to "hit the bottle" again to quell the terror that he had been feeling. He had almost lost his marriage over it and had alienated his children before the past decade of sobriety helped him slowly regain their trust. He knew that his alcoholism had left a scar on the people that meant the most to him, his wife, four sons, and one daughter.

Jerry also described feelings of anxiety that would well up when he was away from home for long. In these agoraphobic states, he would worry about his wife, who in a severe depression had several times tried to commit suicide years earlier. He also recalled the night he had been out drinking when his daughter had been in a serious automobile accident. He worried: "If I'm not at home something bad might happen."

Born a half hour after his brother, George, Jerry has always been the shorter of the two. To stand next to six-foot, four-inch, and 220-pound Jerry, it is hard to believe that he could be the smaller one. He certainly towers over me.

Jerry and George talk almost daily. They spend winters in the southeastern United States, where they live in the same community, and the rest of the year they reside not far from each other in New Jersey. Family is important to Jerry. He talks frequently with his other siblings—but with no one as much as his twin.

Jerry and George went to the same schools, including college, and were sometimes mistaken for one another by teachers and other students. They were tempted to capitalize on the confusion and sometimes did so in humorous, though innocent, ways.

In Jerry's opinion, being an identical twin is an asset. He values the special bond between George and himself. Years earlier, Jerry had been instrumental in helping George to get help with his alcoholism, another disorder they have in common, and since then they have supported each other's sobriety. In his treatment with me, Jerry described how hearing his twin describe his own panic symptoms helped him accept his own illness. When Jerry found relief through psychotherapy and a serotonin reuptake inhibitor medication, he hoped that George would also seek psychiatric help.[3]

Jerry and I discussed the genetics and biology of Panic Disorder, including how people with it often turn to alcohol as self-medication. He hoped that as a recovering alcoholic, George would understand that. In addition, he planned to explain to George that because they are identical genetically, there was a good chance that the same medicine that was helping him would help George. That Panic Disorder also has a strong genetic and biological basis would seem to make this even more likely. Unfortunately, despite his best efforts to convince his twin to seek help, George continued to delay.

George finally asked for psychiatric help for the first time in the wake of the awful events of September 11, 2001. George's panic attacks worsened following the terrorist attacks, and Jerry called to tell me that his brother desperately needed psychiatric help and was now willing to accept it. I agreed to see him the next day but said that both should understand that I could only provide psychiatric "first aid" while I worked to arrange a referral to a colleague, so as to avoid potential conflicts in their therapy.

I met with George only once, but differences between the twins were quickly apparent. Jerry joined George in my office for the first few minutes of George's one and only meeting with me, the two clones (by embryo splitting) sitting next to each other on the couch. One physical difference was not immediately obvious—Jerry is right-handed while George is left-handed, a difference that implies underlying differences in the right–left organization of the brain.[4]

The differences between the twins emotionally and interpersonally were even more striking and relevant to our purposes here. In contrast to naturally cheerful and gregarious Jerry, George was reserved, even withdrawn. George seemed bitter and sad.

As the session progressed, it became apparent that, in addition to Panic Disorder, George was also suffering from a major depressive episode layered on top of a relatively chronic unremitting, less severe depression. The chronic depression was often punctuated by brief periods of feeling good and powerful, requiring little sleep, and behaving with irritable impatience toward others. All this led me to suspect that George might also have Bipolar Affective Disorder or manic-depressive illness. He was probably going to need a mood-stabilizing medication like lithium or one of the several anticonvulsant medications also effective for this purpose.

I started George on a potent antianxiety medication. It would provide immediate relief until he could see a psychiatrist colleague. My colleague could evaluate him more carefully from diagnostic and psychodynamic perspectives and decide on other medications and a psychotherapeutic approach.

Jerry had told me that he believed that being an identical twin had been less important than other factors in shaping his life, and George concurred. For them both, their relationship with their father seemed more crucial to their problems. Jerry admired him as a great but flawed man who seldom showed affection to his children. George was bitter about that lack of affection and could say nothing positive about him.

Their father, Sam, had been a well-known and beloved politician in the small southern town where they were raised. Sam did not pay as much attention to his family as he did to his friends and acquaintances. He was comfortable hugging and kissing the babies of his constituents, but Sam's most intimate physical contact with Jerry was the handshake that he gave him when Jerry graduated high school. Sam was uninvolved and was only vaguely aware of what his seven children were doing or even what grades they were in. Jerry's mother raised the children without complaint, pretty much on her own. She adored Sam and fostered respect and admiration of him in the children.

At Sam's urgings, Jerry moved north as a young man to become part of a business venture run by an old friend of Sam's. Jerry commented on several occasions that this was the best thing his father ever did for him, as it allowed him to perceive himself and be perceived by others as his own person and not just as his father's son. Resembling his father more than his other brothers did only added to the difficulties he encountered establishing his own identity. The resemblance was both physical and behavioral. Jerry's easy, gregarious, boisterous style resembled that of his father much more than the often morose style of Jerry's twin, George, did.

A key feature to this case is the strong positive and helping bond between the twin brothers. Granted, each has his problems, but I had to agree with both Jerry and George that these were *unrelated to their status as identical twins.* Nothing has emerged in Jerry's psychotherapy to indicate otherwise.

Alcoholism and Panic Disorder, conditions afflicting both twins, have a strong genetic basis. For these conditions, the Identical Twin Model makes a contribution to understanding Jerry and George. But, even in the realm of disorders with strong biological determinants, the Identical Twin Model has its limitations.

Twins, by sharing the same genes, equally share some but not all biological, including psychobiological, vulnerabilities. That is because genes alone do not determine biological vulnerability. Environment also plays a role. In some important ways, Jerry and George had different experiences of the family they shared, and those experiences may even have been partially determined by their prenatal environment. The same womb had not been the same environment for them both. Recall that George weighs more, was born first, and is left-handed, while Jerry is smaller, was born second, and is right-handed.[5] Psychological evidence of biological differences between them are George's withdrawn and irritable temperament, detectable at a very early age, and the fact that one twin apparently has Bipolar Disorder while the other does not.

Jerry found having an identical twin advantageous in dealing with the disorders he and George share. Often, it is easier to recognize a problem in someone else before one recognizes it in oneself. But, when that someone else is one's identical twin, the implications for oneself become harder to deny. In addition, once one twin, Jerry, got help for himself, he was highly motivated to try to help his twin, feeling sure that what worked for him biologically would work for George.

The Identical Twin Model is most biologically relevant to cloning, but it fails to illuminate Jerry's problems. Being an identical twin does not appear to have contributed to psychological problems for either Jerry or George. My psychotherapeutic work with Jerry has pointed to other factors, instead.

Jerry's experience of his daughter's car accident and the approaching anniversary of almost losing his wife had nothing to do with being an identical twin. These appear to be the most potent psychological factors in his Panic Disorder. His psychotherapy has focused on trying to help him reprocess the first experience and give him a chance to discuss the trauma of the second. Regarding his wife, it has also been important for him to discuss how he has coped with her recurrent depressions and suicidal tendencies. The most effective psychotherapeutic interventions have been to help him set gentle but firm limits with his wife, especially regarding her desire to again have elective surgery. From what he tells me, it also appears that standing up for himself in the relationship has also helped her emotionally, and their relationship has benefited. As a result, Jerry is less anxious and has less to be worried about.

When it comes to Jerry's other problems, *his status as a clone of his twin brother does not appear to be a factor; in a psychological and social sense, but his status as being as if a "clone" of his father is quite relevant.* In this regard, Jerry's case can help to illustrate several other models we will be examining—the Parent–Child Resemblance Model and the Child of the Famous Model. I focus on these other features later.

THE IDENTICAL TWIN MODEL

As a model relevant by analogy to reproductive cloning, the experience of identical twins is superficially the most obvious and, biologically, the most parallel. Dr. Nancy Segal, an authority on twin psychology, makes it clear that although identical twins are clones, the sort we are considering here are not identical twins.[6] Such clones do not fulfill her three "twinship criteria": simultaneous conception, shared prenatal environment, and common birth. Keeping these differences in mind, let us look at the biological formation of identical twins and compare it with cloning.

Biology of Identical Twins versus that of the Clones That Are Our Focus

Identical twinning occurs naturally by a phenomenon known as "embryo splitting" whereby the fertilized egg or a very early embryo, known as a "blastocyst," splits in two.[7] Identical twins are also known as "monozygotic" to describe the fact that they both come from the same zygote—that is, from the same fertilized egg. Genetically, they are completely identical.

The body's genetic information is coded in DNA. Most DNA is found in the cell's "nucleus." The nucleus is the command center within the cell, directing its activity and function. Every somatic or nonreproductive cell contains the same DNA as every other somatic cell, even though during development cells come to "differentiate" into different tissues and form the various organs of the body. At the earliest stages, cells are "totipotent," that is, able to form any tissue.[8]

A small fraction of a person's DNA is found outside the nucleus but within the cell. That DNA is not floating free in the "cytoplasm," or complex biological soup, but, rather, is contained within particular structures within it called "mitochondria." Mitochondria are energy-generation machines found in every cell of the body. "Mitochondrial DNA" differs in some important respects from "nuclear DNA" in structure and function. For our purposes, the most important difference is that mitochondrial DNA is inherited only from the mother. Identical twins are identical both in their nuclear and in their mitochondrial DNA.

Genetic identity does not, however, make for biological identity in structure, function, and physical appearance—the distinction between what biologists call "genotype," or the genetic makeup of an individual, and "phenotype," which is their actual appearance and behavior. The identical genotypes of identical twins do not predestine them to identical phenotypes. Phenotypic differences can start at the earliest stages, and later environmental influences and life experiences may magnify or minimize them. While Jerry and George are identical twins, their height and weight were never the same. While both have Panic Disorder, only

George is possibly manic-depressive. Both are alcoholic, and both have been able to abstain from alcohol for years. Perhaps their common childhood experiences with a loving and supportive mother helped form their personalities in ways that give them the strength to resist their addiction.

Even twins reared together do not have exactly the same environment. Recall that George experienced his father, and most other things, more negatively than Jerry did. It is possible that his attitude also shaped how his parents related to him. Recall that it is likely that his prenatal environment was probably significantly different from George's even though they shared the same womb.

In some unusual circumstances, identical twins have been reared apart from one another and so lack the common family environment most twins share. Comparing twins reared together and reared apart has helped us learn a lot about the relative influence of "nature" versus "nurture" in the development of various psychiatric disorders.[9]

Although identical twins are clones of each other, they are clones that arise from the same egg and not from a more mature organism. Asexual reproduction from mature organisms is the rule for single-cell organisms supplemented by their version of sex, "conjugation." It is common in plants, as any gardener who has grown a new plant from a cutting of another can tell you. It may occur naturally for a variety of nonmammalian species of animals, but it does not occur naturally in mammals.[10]

The term *intergenerational cloning* describes the result of asexual reproduction through the process originally conceived by Spemann in 1937, cloning by somatic cell nuclear transfer (SCNT).[11] In such cloning, the nucleus from a somatic cell, or nonreproductive cell, of an already existing organism would become the source of genetic information to produce a new life, his or her clone. Perhaps *intergenerational cloning* is too restrictive a term for what would be possible with SCNT.[12] That is because a child could serve as the source of the nuclear DNA. In such cases, although there would be an age difference between clone and "progenitor," it would be less than a generation.

To understand the basics of cloning by SCNT, we should know that "germ" cells produce "gametes" or the cells involved in sexual reproduction—eggs and sperm. Individual gametes have half the number of chromosomes found in somatic cells. Germ cells have what is called a "haploid" number of chromosomes (twenty-three in humans). One of those is a sex-determining chromosome, either an X or a Y. Eggs always carry an X, but sperm have either an X or a Y. Nonreproductive or somatic cells have double that number, and so are called "diploid."

After fertilization of the "ovum," and before the first cell begins to divide, an additional process ensues. Matching chromosomes pair up with one another and "cross over." During the process of "recombination" they exchange many segments that serve the same functions. All pairs of chromosomes, except for the sex-determining X and Y chromosomes, engage in this exchange of homologous segments. Thus, while each parent contributes 50 percent of the nuclear DNA in the fertilized egg, recombination gives each offspring unique chromosomes.

The egg or ovum is an amazing and mysterious cell. If the cycle of cell division or the "cell cycle" of a somatic cell is put into a resting phase by depriving it of nutrients, it approximates the state of the nucleus of an unfertilized egg. If that nucleus is then inserted into an egg whose nucleus has been extracted, then the somatic cell nucleus may sometimes begin to direct cell division. That is how a clone is "conceived." When this is successful, cell division may proceed to form an embryo that can be implanted in the uterus of an animal, usually but not always of the same species.[13] A fetus may then develop, eventually to be born as a clone of the "donor" or "progenitor" animal.

So far, the success rate has been abysmal, and abnormalities in development and pre- and postnatal complications have been common. Even among otherwise apparently normal clones, large fetal size and respiratory distress in the newborn are very frequent. The large size of the fetus may create physical problems for the "mother" animal gestating and giving birth to it, and premature births are common.[14] These problems are cause for serious concern in applying SCNT to humans, but some renegade physicians and scientists are undeterred by these risks.

In this book, I use the term *progenitor* to refer to the person who is the source of the nucleus, which when transferred into an enucleated egg develops into a clone of that person. I restrict the term *clone* to refer to the nuclear genetic replica of the progenitor. The progenitor may or may not function as a rearing "parent" to the clone. Genetically, the clone's parents are the same as those of their progenitor.

Of course, only a woman is capable of becoming an egg mother to a clone. We might also refer to an egg mother as a "mitochondrial mother." A recently developed infertility technique for a rare condition involves injecting egg cytoplasm from one woman's eggs into another's fertilized eggs. These offspring have one nuclear DNA mother and another mitochondrial DNA mother. Unless or until a fetus can develop outside the body (a process known as "ectogenesis"), a woman would also be required to serve as a "gestational" and "birth mother" of a clone, regardless of who the progenitor might be.

In the biological dimension, there are two ways in which identical twins appear to be an appropriate model of cloning by SCNT. One is genotypic: Identical twins have identical nuclear DNA. The other is phenotypic: Like identical twin pairs, clone and progenitor would likely be highly similar in physical appearance—but in a peculiar way. The similarity may not be apparent at any given point in time. Instead, the similarity would be greatest when the clone is retrospectively compared with his or her progenitor when at the same age or stage of development or maturation.

A medical advantage of being an identical twin derives from each twin carrying the same inherited disease predispositions. Identical twins who understand this fact could use it to their advantage.[15] An unaffected twin can purposely avoid known environmental triggers of illnesses the other twin has manifested, and doctors presume, often correctly, that medication that helped one twin would be likely to help the other. Witnessing the negative effects of his brother's continued drinking after Jerry had stopped reinforced Jerry's decision to remain abstinent. A clone would also have the same inherited vulnerabilities as his progenitor. Although the situation is complicated by the risk of knowing too much too soon, a clone might also benefit from knowing the health profile of his or her progenitor.

Biologically, an identical twin and a clone created by SCNT would be different in several ways. The latter would be completely identical genetically to her progenitor if and only if the progenitor was also the "egg" or "mitochondrial" mother, whereas a male clone would necessarily have mitochondrial DNA different from that of his progenitor. At this point, we do not know how and in what way that difference would show up phenotypically—that is, in bodily structure, function, and appearance.

What the Psychology of Identical Twins Does and Does Not Suggest about Clones by Somatic Cell Nuclear Transfer

While genes and biology powerfully determine psychology, environment, including life experience, does as well, making it possible for identical twins to be as different psychologically as they are the same genetically. Just as environmental differences temper the effects of genetic identity and biological resemblance between identical twins, it also does so for psychological resemblance. An additional factor for clone and progenitor would be that of a difference in age. The progenitor could be a child, an adult, or even a dying or dead person.

Even if clone and progenitor shared the same family environment, they would not experience it as the same. Many patients have described to me that the family environment they experienced as children was much different from that experienced by siblings born before or after them. Those who were born either first or last in large families have often remarked on this. This level of dissimilarity in family environment would presumably be matched, if not magnified, if the progenitor were an older sibling, a dead loved one, a person admired by the family (living or dead), or, perhaps most especially, a rearing parent to the clone.[16]

The early psychological research on identical twins suggests that mutual attachment and dependency may conflict with the wish of twins to be autonomous and independent of each other, despite their strong physical resemblance and the reaction it engenders in others.[17] This makes intuitive sense, but science demands that we test our intuitions, if we can.

Nancy Segal challenges the earlier view that identical twins have trouble becoming separate individuals and are more likely to have psychiatrically significant problems. From her many interviews of identical twins, she also believes that the twinship bond brings with it many benefits, such as special lifelong companionship, mutual assistance, and sharing.[18] Jerry and George have certainly found this to be true of their experience as twins. And, while very attached to each other, they do not appear to be unusually dependent on each other. Although that may be the case for them, it might not be so for others. Perhaps it might account for a statistically significant finding in a very large study of twins: Compared with their non-twin relatives, the twins scored higher on the "panic-phobia" factor.[19]

In clinical medicine, psychology, and psychiatry, observing a phenomenon in a single case or a small number of cases may lead to generating novel hypotheses to explain it. The hypotheses need to be scientifically tested utilizing samples large enough to have the statistical power to support or disprove them. For this reason, Segal understandably gives the most credence to the larger studies on twins, research that found either no or a small effect of twin status on mental health. But what one gains with the statistical power of larger studies, one may pay for with a diminished ability to appreciate the complexity of the individual twin and twin pairs. Those older studies and case reports might still be of value to us.

There may be another reason not to dismiss those older, smaller research and case studies that originally raised mental health concerns for clones by embryo splitting, that is, identical twins. The issues they raise, though perhaps not often valid for identical twins, can remain part of our Identical Twin Model *to help us think about cloning by SCNT*. Although not of equal weight scientifically with large, carefully controlled research studies on twins, in some dimensions they are more relevant to cloning conceptually. At the same time, some of the larger more recent studies, while reassuring for identical twins, may not be so applicable and hence reassuring for clones. Let us group psychological theory and data on identical twins into a number of categories relevant to understanding clones.

Confusing Resemblance with Sameness

Jerry told me, "Just because you're an identical twin doesn't mean you're the same." While twins know this, it is sometimes harder for others to realize it, and when the twins' parents do not, we might expect the children to be affected emotionally. It all can occur quite innocently, as some parents just cannot tell their twins apart. Perhaps his experience may not fairly represent how commonly that occurs, but Dr. René Zazzo found that the parents of 10 percent of the identical twins he studied could not tell who was who.[20] These parents resorted to using distinctive marks such as bracelets to identify the children. Sometimes the bracelets were lost, and name attribution came to be made at random. The case of Annie and Genevieve is a case in point illustrating the confusion engendered in the children. One of them said: "People called me by my name, others by my sister's name, at the end I just didn't know anymore."

Innocent parental befuddlement is one thing, but it is quite another to expect the twins to be the same and even demand it. How often are parents of identical twins guilty of "psychic child abuse"?[21] Is the biologist Richard Lewontin correct in saying that many parents of twins try to create "an inhuman identity" between the twins that threatens the "individuality" of these "genetically identical individuals"?[22] Does psychological research support his assertions?

Certainly there must be parents who abuse their identical twins in the way Lewontin suggests, just as there are parents who abuse their non-twin children in a way that undermines their identity. However, it is important to note that parental expectations or demands that the twins be similar might not be the only or even the main reason that twins so often dress and act the same. Making similar choices might be caused instead by the mutual dependency of identical twins, especially female twins, and an identical genetic predisposition.[23] Jerry told me that he and George sometimes surprise each other by finding that they independently purchased identical items of clothing.

The twin model would predict that clones would have similar preferences and often make choices similar to those made by their progenitors. Simply based on genetic factors, this might be the case even if someone other than the progenitor reared the clone. Making similar choices as one's co-twin does not necessarily imply a deficiency in a twin's sense of individual autonomy or personal identity, and the same could be said of a clone making choices similar to those of his progenitor. However, if the rearing parents expect the clone to be like his or her progenitor, those expectations could powerfully influence the clone to conform and comply.

A rearing parent of a clone, like all rearing parents, has tremendous advantages in the relationship with the child. These include precedence, greater physical size and strength, and superior cognitive and language skills. Additional parental advantages derive from the child's unavoidable dependency on and natural tendency to want to please and identify with the parent, as well as the parent's societally sanctioned license to direct and discipline the child. We will see that models other than this one better illustrate these and a range of other factors with likely psychological and social significance for clones and cloning.

Attachment

Identical twins may be especially empathic with one another.[24] It must feel wonderful to feel as connected to another person as identical twins can be with one another. The attachment between identical twins may even grow over time, especially with girl twins.[25] Jerry told me that he and George have diverged.

Whatever benefits there may be to the attachment between identical twins, we must at least consider the assertion that the presence of a twin as a "replica" of oneself may complicate the development of an individual twin's self-concept and retard his differentiation into a separate person.[26] An intense and prolonged mutual dependency may sometimes occur between identical twins.[27] We will return in later chapters to bonding and attachment, a process best described for mother and infant.[28]

A significant component contributing to the attachment twins feel for one another is based on how much they believe that they physically resemble one another. As one would expect, both physical resemblance and a sense of closeness are greater between identical twins than between fraternal twins.[29] Progenitor and clone are likely to believe that they strongly resemble each other, and, indeed, they probably would resemble one another. However, any resemblance should be less than that seen in identical twins simply because of the age difference between progenitor and clone. Perhaps as important as actual or perceived resemblance is the expectation of resemblance. We'll delve more deeply into the issue of resemblance in chapter 5, "The Parent–Child Resemblance Model."

Much of the twinship bond is surely related to their shared environment and common age, factors that may be very different for clone and progenitor. However, we might expect that the bond could be even greater for the progenitor when he or she is a rearing parent, and especially the mother, of a clone.

It must be marvelous to be as empathetically connected as identical twins can be with each other, but only being in love can come close to competing with the bliss experienced by an infant in the arms of an empathic mother. Unfortunately, parental precedence and dominance make it possible for a parent to empathetically overwhelm a child. A clone of a rearing parent would likely be at great risk for this.

Erica.   As a case in point, let us take Erica, a college student who came to me in a major depression. Whenever Erica seemed even slightly upset, her mother would promptly try to spare her daughter (and perhaps herself) emotional pain and discomfort. Because her mother was sure she knew what Erica felt and told her, she could not help Erica discover her own feelings. In addition, there was little room or privacy for Erica to experience feelings on her own.

We have no idea how accurate Erica's mother was about what Erica felt. Even if she was always on the mark, telling Erica what she was sure she felt and why interfered with Erica coming to know herself. Her mother's overeager "empathic" responsiveness had the paradoxical effect of contributing to Erica feeling cut off from her inner life. As a result, she became more dependent on her mother and prone to depression, with or without her. *If Erica were a clone of her mother, then might her mother have been even more likely to feel sure that she "knew" what her daughter was feeling and feel even more entitled to intervene with that "knowledge"?*[30]

Separation-Individuation and the Development of a Separate Self and Identity

Some authors believe that identical twins face special hurdles during a process known as "separation-individuation," Margaret Mahler's term for the process emerging out of the state of "mother–infant symbiosis."[31] Separation-individuation extends from about six weeks to the end of the first year, when the infant experiences the mother mainly as a "need-satisfying quasi-extension of his or her own self."[32] Beginning around five–six months of age to about ten months, the infant shows a growing awareness of and interest in the world around him and behaves as if he is "hatching" out of the symbiotic unity with the mother. This phase is termed "differentiation."

"Separation" is "the cognitive awareness of our own existential separation, the discovery that we exist in our own skin." "Individuation" is "the psychological development of a sense of oneself as a capable, self-reliant individual." The process of separation-individuation peaks between sixteen and twenty-four months during the subphase termed "rapprochement." It is "the process and phase during which the child must resolve the intrapsychic crisis between the wish to remain with the mother (in symbiotic union) and the wish for autonomy that accompanies the awareness of the self as a separate individual." The child during this time intensely experiences the "excitement of autonomy and the fear of separateness from the protective and nurturing mother of infancy."[33]

Could the normal mother–child symbiosis be diminished or replaced by a symbiosis between twins?[34] A sign of this may be what has been called a "twinning reaction" in which one twin's sense of self fuses with his or her twin. A diffuse boundary between self and other can lead to a variety of psychologically problematic manifestations. One of these is a tendency toward "projection." Projection is a mental process "whereby a personally unacceptable impulse or idea is attributed to the external world," including other people.[35] Excessive introjection and overidentification may also occur. "Introjection" is the taking in of aspects of another person, and identification is a more developmentally advanced version of the same. Specifically, identification is the process by which one comes to selectively incorporate aspects of another person and integrate them into one's own separate identity.[36] Some believe that an identical twin might be less likely to identify with people other than their co-twin, resulting in greater risk of forming a "narcissistic" character.[37] Later on, we will examine narcissism in detail. For now, it is sufficient to understand the term *narcissistic* as it is used colloquially.

Some case studies point to problems with the sense of self and identity in which twinship status apparently played a role. One report suggests that, at least for the twins described, an awareness of the twin pair preceded awareness of separate selves.[38] These twins named their twinship pair "Gaga" before they referred to each of themselves or the other by name. Twins may also have a private language ("cryptophasia").[39] Of course, because clone and progenitor would be of different ages, any problems with self and identity, including manifestations in peculiarity of language, would not arise in the same way for a clone as it might possibly for some identical twins.

Identical twins, René Zazzo claims, are "restrained in their self-identification . . . living for a long time as an echo of the other."[40] The word *echo* implies a time delay that he probably did not intend, but that metaphor might better apply to the psychology of the relationship of clone and progenitor. A clone might be at risk to develop a self that is but an "echo" of his progenitor's, whereas a deeply insecure progenitor might be quite distressed to perceive his clone as an "echoic" distortion of self.

The Death of a Twin

While identical twins come into the world almost simultaneously, they die separately, except for inseparable conjoined twins. The special bond of attachment between twins does not protect the individual twin from leaving life on his own. The impact of losing a co-twin appears to be even more devastating for the surviving twin than losing a spouse, a parent, a non-twin brother or sister, or a friend.[41]

The psychiatrist George Engel, an identical twin, wrote of his experiences surviving his twin brother. He saw the twin pair as a "dual unit."[42] Hence, mourning a dead twin is partially mourning a loss of the self.

In an interview-based study of monozygotic and dizygotic twins, Joan Woodward discerned some recurrent themes.[43] Those who were teenagers or younger when they lost their co-twins often perceived their mourning parents as overprotective or severely rejecting of them, as the surviving twin. A clone who loses his parent/progenitor at an early age might be likely to experience similar reactions by the surviving spouse of the deceased progenitor/parent.

Another theme that emerged from Woodward's interviews of surviving twins was guilt for surviving. Some of the surviving twins spoke of feeling that they had to "'live for two' as if they had to justify being alive."[44] This might also be likely if a clone's progenitor were a child. We will focus more on guilt in the Replacement Child Model.

Some twins voiced and usually quickly denied a sense of relief at their twin's death. One said, "I suddenly realized I would no longer be described as one of the boys." It is not hard to imagine the analogous situation for a clone when his progenitor dies, especially if that progenitor were famous, something we'll examine in the Child of the Famous Model.

We have already seen other studies of identical twins that highlight the theme of "closeness," and Woodward's does as well. One man said that the loss of his identical twin was like watching himself die. To be predeceased by the other might have a similar emotional impact for either clone or progenitor.

Another theme Woodward identifies is that of "polarization." She sees it as deriving from the closeness of twins. That closeness may lead some twins to feel a strong need to be different from each other. Polarization resulted from the twins trying to differentiate themselves in complementary fashion to make a good team. After one twin in such a polarized pair died, the surviving twin was prone to feel like "only a half."[45] We are reminded of Engel's phrase, "dual unit," to describe the twin pair.

The theme most reported by surviving twins is that of being afraid to seek closeness with another out of fear of loss accompanied by the fear of being alone. An essential feature of the cloning situation is like that of identical twins—but with a twist. Although both identical twins start life genetically identical to another person, in the progenitor–clone pair, only the clone would have started life that way. Just as there are instances of identical twins having been separated at birth who are determined to find each other and when they do, marvel at their similarities, a clone who does not start out living with his progenitor might feel similarly driven to seek him out.[46]

We might suppose that most commonly one of a clone's rearing parents would be his or her progenitor. The young child naturally fears losing a parent, particularly the mother. Regardless of a clone's age, we might expect a clone's grief at losing a parent to be intensified when that parent is also the progenitor. Likewise, the parent might feel even more bereft at losing a child who is also that parent's clone.

Reinforcing this concern for clones is Woodward's finding that the worst aspect of losing a twin for the surviving twin is fearing that the parents would separate from each other or die. By extension, losing a progenitor parent would be a double loss for a clone.

Other surviving twins felt that their parents or the spouse of the deceased twin minimized or denied the importance of their loss. By analogy, perhaps a clone might encounter a similar sort of reaction when his progenitor parent dies. His surviving rearing parent, the progenitor's spouse, might not realize how extremely profound, and possibly conflictual, that loss might be for the clone.

One factor that ameliorated the loss for the surviving member of these twin pairs was to name one of their children after the deceased. We will explore the namesake as a model later. Furthermore, one twin said that he had "absorbed the spirit of his dead brother." Later, we will see how the theme of "the double" might be important to understand that sort of reaction.

Losing an identical twin also brings fears of impending death for the surviving twin, and the bereaved identical twin can become "a living reminder of the deceased, experiencing the confusion of others."[47] One might expect a similar kind of social response to the surviving clone of a progenitor.

The psychological consequences for a twin surviving a twin sibling who died before birth or in infancy might be quite relevant to some cloning scenarios, specifically the cloning of the dead or dying. Some twins express sadness or anger over this event.[48] This has not been systematically studied, but a biography of Elvis Presley speculates that his survivor guilt and drive toward individuality might be explainable on the basis of the early loss of his twin.[49] The connection between clone and progenitor, especially when the progenitor is the rearing parent, might be even more intense than either the bond between identical twins or the bond between parent and noncloned child. As a result, the emotional consequences of a clone losing a progenitor or of a progenitor losing her clone might be more devastating than losing a spouse, parent, or sexually reproduced child.

The Special Bond between Twins Might Diminish a Twin's Ability to Bond with Others

Having an identical twin may have effects on one's other relationships in and outside of the family. Comparing identical with fraternal twins in their experience of loss, Nancy Segal found that grief for identical twins consequent to losing a co-twin appeared to be more enduring than their grief at losing a spouse or a parent. She concludes that "identical twins may invest their emotions more exclusively in their co-twins than in their spouses."[50] The intense nature of bonds between twins is surely a factor in and of itself, but might there also be a reciprocally diminished ability to bond with others? Might the special bond of attachment between co-twins interfere, to a degree, with deeply attaching to others?[51]

A diminished ability to attach to others should be reflected in lower rates of marriage, and on that possibility, reports are contradictory.[52] One might also expect that once an identical twin forms an attachment, he would be less likely to maintain it. On this possibility, one would expect higher rates of divorce, something not found.[53] If the attachment of clone and progenitor parent were to be more intense and exclusive than normal, then we might well expect the clone to be less able to attach normally to others.

Dominance and Submission in the Twin Pair

Twins raised together are, in fact, a special sort of "couple"; and, as in couples, one member is typically more dominant than the other.[54] One might imagine that later on the dominated twin could have difficulty relating to others as an adult.[55] To study the dominance dimension, Pekka Tienari interviewed identical twin pairs to document and find predictors of different aspects of dominance.[56]

His large and careful research study showed that psychic dominance correlates with the role of being the spokesperson for the pair and that physical dominance correlates with psychic dominance. The firstborn twin is typically heavier at birth, more likely to be the physically and psychically dominating one of the pair, and less likely to be emotionally disturbed in childhood. There does not appear to be a lasting impact of dominance status in childhood on the twins' psychological development. Birth order does not predict socioeconomic achievement in adulthood, a rough indicator that birth order effects are also transient. In addition, later achievement of independence by twins is independent of dimensions of childhood dominance.

These findings, while reassuring for identical twins, are of limited applicability for our purposes because birth order for twins reflects only a small temporal difference. It is many times less than that of non-twin siblings and many orders of magnitude less than the generational difference between parent and child.[57] For this reason, a major deficiency of the Identical Twin Model of cloning is that it cannot reflect a birth order difference so large as that between progenitor and clone, as well as its psychological impact.

Unlike the reassuring results with identical twins showing a transient effect of dimensions of dominance, we might expect profound and lasting effects of the dominance of progenitor over clone. We will find other models of cloning superior in helping us to understand these dimensions of cloning.

Perhaps the most major deficiency of the twin model is that it does not reflect the likely striking differences in other dimensions of dominance. These include size, strength, cognitive and language abilities, psychic dominance, and dependency between clone and progenitor, especially when the progenitor is a rearing parent. The dominance of one twin over his counterpart, based in part on physical size and strength, would pale in comparison to the magnitude of dominance between progenitor and clone, even when the progenitor is a living child.

The physical size difference of an adult to the child clone is enormous, and that alone would create an immense differential in dominance and power. Add to that a parent–child relationship, and the dominance increases further with the child's dependency on the adult for care and nurturance, the adult's superior intellect, and his or her social license as a rearing parent to regulate, control, and discipline the child.[58]

As to a supposed advantage, the possibility of enhanced empathy of progenitor for his or her clone, the dominance of the progenitor parent over the clone might increase the risk that a progenitor parent might, quite innocently, misuse a real or imagined twinship-type empathy for the clone. The relationship of Erica and her mother is relevant by analogy.

To summarize, the Identical Twin Model fails to reflect vast differences in sources of power, dominance, and dependency between the generations. Other models of cloning will better reflect those dimensions.

The Social Experience of Identical Twins

Socially, identical twins are expected to be like one another, and their similar appearance may sometimes make others uncomfortable and cause social problems for the twins. Let me tell you about my experience of Kenny and Larry, my identical twin first cousins, eight years my senior. As a child, I anxiously expected that I ought to be able to easily and confidently identify the individuals constituting the twin unit, "Kenny n' Larry." In addition to being nearly identical physically, they had many of the same mannerisms, interests, and abilities. Both were excellent chess players and pianists.

I recall quizzing my parents before our infrequent visits to try to learn how they could tell who was who, but nothing they said helped. When I was with just one of the twins, I would intentionally avoid using either's name, not wanting to use the wrong one. If one twin referred by name to the other, then I would be relieved that he had made the identification. But then I was reluctant to leave the room, fearing that unless they stayed put, I would again not know who was who when I returned. Sometimes to try to hide my embarrassment and confusion, I would try to address them both as a unit with the name "Kenny n' Larry," being careful not to look at both and never only one.

It was not until I grew older and more aware that I could confidently identify my cousins as individuals. At the same time, they also differentiated themselves more from each other. Now I have no doubt who of my terrific cousins is who. Perhaps my experience of Kenny and Larry reveals more about me as a shy and anxious child than anything general about how most people respond to twins, but I suspect that some elements of my reaction might be universal, if perhaps less pronounced.

Psychological and social research on those in the lives of identical twins is essentially nonexistent. That is a shame because it might help us understand attitudes toward cloning. Part of the negative reaction to cloning is a reaction against creating a replica human being, though some would argue that opposing cloning for that reason "affronts" the dignity of identical twins.[59] But perhaps there is a deeper logic underlying these polemics deriving from the confused and anxious reactions of a child encountering identical twins for the first time.

Reactions like mine are tempered by the fact that identical twins are a given to which society has acclimated. Individually, we all go through acculturation and psychological maturation. Perhaps one component has to do with accepting the presence of identical twins in the universe of our experience. Because each individual must acclimate and mature, the raw responses of children to identical twins might tell us a lot about psychological and social reactions to cloning.

To their emotional and social detriment, conjoined twins and children of multiple births often become social curiosities, and so might the first human clones.[60] While legitimate scientific and medical interest would lead to demands for multiple tests and studies, social curiosity could lead to unwanted public scrutiny. That said, novelty effects diminish once a practice becomes more common, something we'll look at in the chapters "The Assisted Reproductive Technologies and Arrangements Model" and "Wider Social and Cultural Implications of Cloning."

Social, cultural, and historical contexts for the clone and progenitor would surely be discordant, but the degree of skew would vary on a case-by-case basis. Identical twins reared apart can only very partially model these.[61] The temporal and historical context for a pair of identical twins is indistinguishable: They start life outside the womb at the same time, give or take some minutes or hours. However, with adoption of foreign babies becoming more common in Western countries, it may now be feasible to study identical twins raised in very different cultures or socioeconomic strata.

CONCLUSION

Despite its limitations, the Identical Twin Model powerfully illustrates the strong influence and limitation of identical genes in a person's development.

Jerry and George are cases in point. As an expert on twin psychology, Nancy Segal believes that a twin model of cloning predicts no particular problems and some advantages for reproductive cloning. While I greatly respect her research on twins, we differ in the implications we find in it for cloning. I agree with her "that the experiences of identical twins, both reared apart and together, offer a stringent test of the concerns that identical genes might imply loss of self." Research, she says, shows that "this does not occur in identical twins born in the same generation." Even if we accept that identical twins run minimal or no risk of reduced autonomy, that should not reassure us when it comes to clones. Segal asserts that the fact that twins share an environment from the very beginning of life makes problems with developing a separate self less likely for a clone than for a twin. I believe that we should come to the opposite conclusion.

Here is how: Most problems might arise from knowing that the genes of clone and progenitor are identical, with the progenitor having had a head start on life and perhaps functioning as a rearing "parent" to the clone. The crucial question is how that knowledge might affect the attitudes and behavior of the rearing "parents" and others toward the clone. My analysis in this chapter does not point to a rosy conclusion for clones but, rather, suggests that the risk of reduced autonomy would be *greater* for intergenerational clones than it may be for identical twins. Even more generally, an age difference and, most importantly, a rearing relationship between progenitor and clone have the potential to make other risks of possible psychological and social harm raised by the Identical Twin Model more problematic for clones than they apparently are for twins.

The differences in age and maturity and power and dominance in identical twins appear to be minimal, and the lasting effects are negligible. However, the differential would be vastly magnified were the progenitor also the rearing "parent" of a clone. For this reason, we cannot really conclude from the generally reassuring data on the development of autonomy and individual identity in identical twins that clones would not have problems in these areas. Likewise, we should not be reassured by the research that finds no deficit for identical twins in forming and maintaining relationships with others.

The Identical Twin Model is strongest psychologically in demonstrating the special bond of attachment between twins. It implies that the already powerful bonds between parent and child might be even greater when a progenitor is also the rearing parent of a self-clone. Whether or not the strong bond between twins at all diminishes the capacity of each individual twin to bond as strongly with others, an intensified bond between a parent progenitor and his or her clone might reduce that capacity.

Segal acknowledges that the twin model of intergenerational cloning has serious limitations and suggests that other models be explored. Despite this, she thinks that "twin research findings may dispel some psychological concerns raised by human cloning, but may not justify all aspects of the procedure."[62] I believe that it is premature to hold this view before following her suggestion to focus on "naturally occurring models that mimic essential aspects of intergenerational cloning."[63] That is exactly why we will examine other models of cloning. Taken in combination with this one, these other models can illuminate important psychological and social issues for clones and cloning.

Jerry and George helped us understand that the biological fact of being that kind of clone can be of minor psychological and social importance compared with other life circumstances and experiences. By analogy, to focus primarily on the genetic identity of clone and progenitor and its possible consequences is to miss a lot. The Identical Twin Model of cloning, while very relevant biologically, has serious limitations in modeling psychological and social aspects of cloning. In particular, it fails to reflect important aspects of the parent–child relationship that would also be intrinsic to the relationship of rearing parent to clone.

Several later chapters model these dimensions. For Jerry, we see that being the son of a locally famous father and highly resembling him have been more important to his psychological development and social experience than being a twin. None of the cases we will examine in future models is in any sense biologically a clone, but those models will help us see how the cases are "clone-like" psychologically or socially. Although the Identical Twin Model is unrivaled in helping us to understand the relative contributions of genes and environment, we will see that other models better reflect many other important psychological and social dimensions for clones and cloning.

2

∞

The Assisted Reproductive Technologies and Arrangements Model

MARIA

Maria loved her husband, Jack, and really wanted to have a baby with him. Both in their late thirties, and childless after three years of marriage, they had finally sought medical help. Jack's sperm count was normal, as were Maria's sex hormone levels. Further tests showed that her fallopian tubes were scarred, the unfortunate consequence of a seemingly insignificant infection she had years before. Those narrow roadways had become impassable, preventing sperm from fertilizing an egg, and if one had been able to make the journey, the fertilized egg would not have been able to complete its trip in the opposite direction—to implant in the uterus.

When she learned she was infertile, Maria was glad that she was in psychiatric treatment and taking an antidepressant. Prior to starting her on the medicine, we had discussed what is and is not known about antidepressant use and pregnancy. We agreed that the risk of not treating her depression outweighed other considerations. To try to prevent a relapse of recurrent major depression, while she was trying to become pregnant we agreed to continue the antidepressant. If her obstetrician concurred, then we might even continue the medicine throughout most of a pregnancy. Severe depression would be detrimental not only to her but to the fetus and might impair her ability to mother the baby she hoped to have.

Upon learning that she was infertile, Maria embarked on the expensive and physically demanding route of the assisted reproductive technique of in vitro fertilization. After several cycles, the resulting embryos were just not implanting in her uterus, an essential requirement for further development. She and her husband, middle-class people, considered the costly prospect of enlisting another woman to carry one of the embryos but decided against it. They were deterred not only by the considerable expense of hiring a "surrogate mother" but also by the wish to not involve another person so intimately in their lives and in the life of the child they hoped to have. Now they are considering adoption.

Being clear that I was not recommending it but simply inquiring so that we might learn more about her thinking, I asked Maria, if she had the money for it, whether she would consider a new assisted reproductive technology. Specifically, if it were available and safe, would she consider having either herself or her husband cloned? Without hesitation she replied that she feared the child might face some insurmountable emotional obstacles. Maria knew more than most people about emotional pain and would not want to impose what she thought might be such a burden on her child. Cloning was not for her. Other parents in a similar situation might feel differently and be in a financial position to be able to make a different decision about cloning.

Reproductive cloning might become one more option among other assisted reproductive technologies that have made it possible for infertile people to have genetic children of their own. In this chapter we will examine the currently employed assisted reproductive technologies and arrangements as a model of cloning. We will look for parallels and differences between them and cloning and at their psychological and social consequences to try to anticipate possible consequences of cloning.

INFERTILITY

Medically, infertility is defined as the inability to achieve conception or carry a baby to term after a year or more of trying to conceive. Fifteen percent of couples in Western countries are infertile trying for their first child.[1] Of course, that statistic comes from heterosexual couples. Homosexual couples are an intrinsically infertile pairing in the realm of the only naturally possible mode of reproduction for human beings, sexual reproduction.[2] The main medical causes of infertility for women are scarred fallopian tubes from prior infection or a hormonal status that does not allow ovulation. Surgical attempts to establish a patent pathway for the eggs to traverse to the uterus are sometimes successful.

A low sperm count is the cause for infertility in men, but that condition can have a number of causes, including anything that can adversely affect the testes or the production of sperm. Some of those conditions are correctable, and others are not.

Because fertility occupies a significant place in human psychological development and mental life, it may have profound individual psychological and psychosocial consequences. This is especially true for women.[3] The psychodynamic orientation has been the most useful psychological perspective on pregnancy and reproduction.[4] It is a way of understanding mental life and behavior as the result of interacting and often opposing goals and motives, including unconscious ones.

Learning that one is infertile may cause a person to grieve as profoundly as losing a loved one, with the same stages of disbelief, anger, guilt, and hopefully, resolution.[5] Others may not recognize that person's loss and grief because no one actually died. However, the potential of having a child did expire. If the person fails to grieve the loss, then she may develop a clinical depression.[6] In addition, people with fertility problems frequently fear social stigma, and they may suffer decreased self-esteem and marital and sexual difficulties.[7]

Infertility for men means losing the capacity to procreate and parent children who are genetically one's own, but some men experience it as an essential loss of manhood and may become impotent after learning that their fertility is impaired. The factually incorrect linking of infertility with virility probably also explains why infertility is seen as less socially acceptable in men.[8] It is probably also the underlying reason why a significant proportion of women who themselves have no fertility problem sometimes lie to friends and take responsibility for the infertility of the couple.[9]

THE WISH TO PARENT A CHILD WHO IS GENETICALLY ONE'S OWN

Beyond wanting to be fertile to feel "whole" as individuals, people frequently wish to rear children who are their genetic offspring. In several later chapters we will focus on why people want children and, specifically, why people want children who are their genetic offspring. We will examine possible answers provided by psychoanalysis and the newer psychological framework of sociobiology and evolutionary psychology.

Psychological research on motivations for parenthood has centered mainly on people having difficulty achieving it. We might wish to look beyond their socially acceptable explanations, such as wanting "to give and receive love."[10] Whether or not a person is conscious of them, some motives may not be so noble.

Alice Miller warns that "the urgent wish for a child . . . may express among other things the wish to have an available mother. Unfortunately, children are too often wished for only as symbols to meet repressed needs."[11] Among the more remarkable cases one may encounter as a psychiatrist are patients with "pseudocyesis." This psychotic delusion that one is pregnant may be understood as a way to compensate for early losses and deprivation that have left the woman despairing of feeling complete and generative.[12]

A woman need not be psychotic to try to attempt to salve emotional pain through pregnancy. For people unconsciously motivated for parenthood as an attempt at psychological repair, self-cloning might be particularly appealing even in the absence of infertility. *But the presence of infertility as a conscious motivator for cloning does not preclude concurrent unconscious motivation(s)*, an example of the psychoanalytic principle of "multideterminism."[13]

ASSISTIVE REPRODUCTIVE TECHNOLOGIES

In the face of infertility, individuals and couples have a variety of options. The great majority of them, like Maria and Jack, first explore medical ones, whereas fewer initially decide to adopt or resign themselves to being childless.[14] Relatively few healthy white newborns are currently available for adoption, yet Caucasian couples may be reluctant to adopt a child from a different racial or ethnic background: "Many couples desiring a child want it to be both as much like themselves and as unlikely to attract attention as possible."[15] Even aside from the more obvious dissimilarities of appearance that might result from interracial adoption, many couples would prefer a child who is genetically "part of both of us."[16] A range of possible assisted reproductive technologies and arrangements gives these couples hope of either possibly or completely fulfilling that wish. Only after trying one or more without success do many consider adoption. The case of Maria illustrates these preferences and the commonly taken path.

In Vitro Fertilization Can Help to Produce a Genetic Child of Both Parents

When a woman has scarring of her fallopian tubes, in vitro fertilization (IVF) can be used to bypass them.[17] IVF is an uncomfortable and invasive process, in which the woman undergoes hormonal manipulation to "superovulate" mature eggs that are then "harvested" to be fertilized in a petri dish. Intracytoplasmic sperm injection (ICSI) involves the same procedure, but a single sperm is injected into an egg. In both procedures, the eggs are incubated through a number of cell divisions, and then one or more early embryos are inserted in the woman's uterus in the hope that one will implant and develop into a fetus.[18]

Many types of IVF are possible, the most common of which utilizes the aspiring mother's own eggs and sperm from her partner. A child conceived through IVF or through ICSI is the product of one of his mother's own eggs fertilized by her husband's sperm and so is genetically related to the parents no differently than any naturally conceived siblings the child might have.

Biologically, IVF is relevant by analogy to cloning in several respects. Reproductive cloning, like IVF, would require inducing superovulation in the egg donor. She may or may not be the woman who will carry the pregnancy. Like IVF, cloning would be conducted in a clinic and laboratory, and the resulting embryo would be inserted into the uterus of the woman who wishes to carry the pregnancy—most likely the woman who intends to rear the child.

Assisted reproductive procedures may be psychologically as well as physically demanding for the parents.[19] Citing several studies in this regard, Rosenthal and Goldfarb conclude that "grief following a cycle or pregnancy loss can be of considerable magnitude" and, more than some other types of mourning reactions, can be "a very personal loss for both the woman and her partner."[20] More than a few clinicians are so concerned about psychological risk for some men and women undergoing certain reproductive interventions that they believe that they should not be made available without psychological safeguards.[21] The psychological stress on parents going through the cloning process might be even greater given its high rate of embryonic loss and fetal abnormality, along with cloning's currently questionable legal status.

When it results in pregnancy, IVF does not appear to adversely affect infant attachment to the mother and interactions of mother and child.[22] However, we also need to consider how children fare beyond this early stage of psychological and interpersonal development.

Donor Insemination and Egg Donation Produce a Genetic Child of Only One Rearing Parent

IVF can also be done with surgical extraction of sperm, with donor sperm, and with donor eggs. Insemination with sperm of a donor has a better success rate than the 15–20 percent cited for IVF.[23] About 500,000 people in the United States today were conceived as a result of donor insemination. It was first done in 1884 in Philadelphia when a woman was inseminated with the sperm of the "best-looking medical student." She was never told that her husband was not the source of the sperm used.[24]

If successful, donor insemination or egg donation results in a child who is the genetic offspring of only one of his rearing parents, his mother. What might be the psychological and social consequences of this genetic fact? This question is very relevant to cloning. While stepchildren are also the genetic offspring of only one parent, they are not produced by an assisted reproductive technology. We will specifically examine the stepchild as the next model.

Donor insemination parallels cloning in that a child clone would be the nuclear genetic offspring of only one person, who quite probably would help to rear him—whether singly or with a spouse or partner who would become the child's other parent. If the child were a clone of the mother, then not only would the little girl's nuclear DNA be identical to her mother/progenitor's, so would her mitochondrial DNA: The child would be exclusively "hers." Going a step beyond so-called virgin births to single women through donor insemination, all births from cloning would be "virginal."[25] Conception would be achieved without sexual activity, and the mode of reproduction itself at the cellular level would be asexual.

If a mother who plans to rear a clone were not the source of the somatic nucleus inserted into an enucleated egg, then the child could still be "part" of her genetically if it was her egg that was the receptacle for nuclear transfer. It might still be biologically and psychologically meaningful to her that the child would carry her mitochondrial DNA. But carrying the pregnancy might prove to be the paramount biological contribution, even if the "mother" is not at all the "genetic" mother. Perhaps parallel to cloning in that regard is the new practice of embryo adoption. Although Maria is strongly against cloning and believes that her husband would be as well, it might allow her to literally realize her wish to have "his baby"—a baby almost completely genetically, essentially, and nearly exclusively "his."

Dr. Susan Golombok, a leading researcher investigating the psychological impact of assisted reproductive technologies, finds superior parenting in families created by donor insemination and IVF. Even the genetically unrelated "rearing" father tends to parent better. Golombok believes that her research demonstrates that "children conceived by IVF, donor insemination, and egg donation have good relationships with their parents and are not at risk for psychological problems," but the picture appears to be more complicated.[26]

Golombok and her colleagues compare families in which egg donation had been employed to those in which a child had been adopted or conceived by donor insemination or IVF. They find that couples in those families express more marital satisfaction and that the father is less stressed and anxious when there is no genetic link between the child and the mother, irrespective of whether there is a genetic link to him. When the child is not linked genetically to the mother but is linked genetically to the father, there is less parental distress and a greater sense of well-being. The parents are also more in accord in disciplining the child. In contrast, mothers and their genetically unrelated children have more severe disputes. The same research shows that a mother is less warm to her child when he or she is not genetically linked to the father. This is true whether or not the mother and child are genetically linked. The researchers view these findings as generally positive and believe they indicate that "a strong desire to have children may outweigh any negative effects arising from the missing genetic link."[27] That might be so for the parents' relationship with each other, but their findings belie that conclusion when it comes to maternal behavior.

However one interprets their research findings, it would be premature to conclude that these children will escape emotional harm related to their special status. That is because only now are these children beginning to enter adolescence, a key period in the formation of identity.[28]

DISCLOSURE AND SECRECY

What psychological consequences might there be for a clone to know or not know the truth of his origins? On the issue of disclosure, donor insemination might parallel cloning. Adoption also models this issue, and we will examine it further in that chapter.

Disclosing to a child that he or she is the product of donor insemination has not been the norm. This is true even in Sweden where since 1985 people conceived with donor sperm have had the legal right to identifying information about the donor.[29] Dr. Lindblad and colleagues did careful open-ended interviews of parents who had a child through donor insemination to try to elicit and understand their reasoning for disclosing or not disclosing the information to the child.[30] Parental attitudes toward disclosure versus secrecy and their reasoning about the issue could parallel how rearing parents of a child clone might approach the issue. In my view, the study could be a model for research on that question with people hoping to become parents through cloning.

Some patterns of reasoning regarding disclosure reflect an adult perspective; others, a child's perspective; and others, both. For parents who frame the issue of disclosure as one of honesty, disclosure is favored, whereas parents who frame the issue as one of confidentiality do not favor disclosure.[31] Other research has found that children who had not been told they were conceived through sperm or egg donation suffered no detrimental effects. However, because the children studied were eight or younger, we cannot exclude the possibility of future psychological harm.[32]

Some psychiatrists believe that children conceived through donor insemination may be at increased risk for mental health problems because of family secrecy about their parentage. The child may also suffer indirectly from a "sense of deficit" felt by his nongenetic father.[33] Sometimes the secret may be "blurted out" in a situation of divorce or a family crisis in a way that proves emotionally harmful to the child.[34]

Many clinicians recommend disclosure to the young adolescent, but in the main, clinical lore and experience guide this recommendation. However, a recent study looked for harmful psychological effects of nondisclosure of donor insemination on offspring.[35] The individuals were distressed by secrecy or by what felt to them like belated disclosure. A number blamed themselves for the rejection and distancing they had experienced from their rearing fathers. At the same time, many had an ill-defined suspicion that something had not been right in their families and believed that their having been conceived by donor insemination was probably at its core.

Because the participants in this research were recruited through support networks of individuals conceived through donor insemination, we may question how generally valid its conclusion is that nondisclosure is detrimental. Individuals belonging to such groups might not be representative of people conceived through donor insemination and, in fact, might have joined because of emotional difficulties. If that is so, then we also do not know whether the secrecy of their origin or their discovery of it is key to their emotional distress. It is conceivable that, at least for some, nondisclosure might be an incidental fact that became a focus as they tried to reach out for emotional support to others with whom they had something in common.

We might also suppose that many offspring of donor insemination never suspect that they were conceived this way, the secret having been successfully kept from them. Are there any ill effects to a successfully kept secret on this matter? This question will probably never be answered because it presents an insoluble paradox in human psychological research—obtaining informed consent from people who have been misled.

While there may be social, cultural, and national differences with regard to disclosure, how often and how do parents decide to truthfully tell a child of his origins in a modern Western society?[36] Research shows that 82 percent of parents of a child conceived by donor insemination plan never to tell the child, compared with 38 percent of parents of a child conceived by egg donation. While a wish to "protect the child" is commonly cited as the reason to maintain secrecy, it is most cited by egg donation parents, whereas a wish to "protect the mother" is not given even half as often. In a reversal of this pattern, parents through donor insemination indicate that they plan not to tell the child because of their wish to "protect the father" more than their wish to "protect the child." Consistent with these findings, fewer parents through donor insemination tell the child's paternal grandparents (the father's parents).[37]

This disclosure differential between parents who utilized donor insemination and those who made use of egg donation implies that a strong value is given to protecting the illusion of the rearing father as fertile. Perhaps this is for the sake of his self-esteem and how he may feel his parents might view him. In contrast, the parents of a child conceived through egg donation maintain the illusion that the rearing mother is the genetic mother for what they believe is the sake of the child. Perhaps this is out of a concern that knowing the truth might adversely affect a child's ability to feel close to her.

This research has implications for cloning. A man who tells his son that the boy is his clone implies that he, the father, may be infertile, a fact that couples parenting children conceived through donor insemination are loath to reveal. If, in order to protect his self-esteem, the father of a self-clone were to disclose that he is not infertile but, rather, chose cloning over natural parenthood, then his son might rightfully raise questions about his father's narcissism. As we shall learn in the Parent–Child Resemblance Model, a narcissistic parent likes to take credit for his child's successes, and so it might be difficult for a narcissistic progenitor not to disclose to his clone the nature of their special bond.

If a clone's parents decide not to disclose the truth about his origins, and if one of them is the clone's progenitor, then the clone might come to suspect this fact once he notices how strongly he resembles that parent—either in the present or from old images. More closely paralleling donor insemination and egg donation would be those situations in which the clone's progenitor is not a rearing parent. Unless the progenitor is well known, the clone might remain ignorant of his origin unless his parents disclose it.

SPERM AND EGG DONORS AND
THEIR SELECTION BY WOMEN DESIRING PARENTHOOD

By their choice of one particular sperm or egg donor and not another, the prospective parents express important preferences regarding attributes they hope their child will possess. Psychologically, perhaps sperm and egg donors and how they are selected might parallel people offering themselves as somatic cell donors or as egg donors for cloning and how parents who would rear a clone select them.

One may wonder how the seemingly unnatural process of selecting sperm and egg donors might be understandable as part of Darwinian natural selection. Interestingly, women select sperm donors by the same criteria by which they select mates.[38] They may choose donors by IQ, profession, appearance, or hobbies. If one accepts the idea that "these choices . . . [indicate] something of their wishes and fantasies for their offspring," then it becomes imperative to study the children conceived by donor insemination "to determine whether they have the desired traits or how these expectations affect them."[39] To my knowledge, this has not been done.

Such research would be quite relevant to cloning. Knowing that a child conceived through donor insemination shares only half the donor's genes, parents would help to temper their hopes that the child closely resemble the donor. In contrast, the rearing parents of a clone, knowing that child and progenitor are identical genetically, might expect and feel entitled to a child nearly identical to the progenitor, regardless of whether it is one of them or another person.

Women's choices of donor for insemination might also be relevant to understanding cloning choices to be made by individuals and couples wishing to parent a clone of someone else. If such research is any guide, then do not expect prospective rearing parents of a clone to select a progenitor rationally. Women value good health and good character, among other factors, when asked about selecting a sperm donor, and they choose character even though they do not believe it is inherited.[40]

When it comes to character, there is but a small literature on what may motivate men to donate sperm.[41] One researcher recommends that sperm donor applicants complete personality inventories to discern narcissistic traits. The implication is that because they believe they are superior, narcissistic men may believe their sperm is too.

In parallel to donor insemination, it might be important to psychologically screen individuals offering themselves as prospective somatic cell nuclear donors. Why might a person want to become a progenitor other than wanting to parent his self-clone?

There is no psychological research on women offering themselves as egg donors, including the more limited form of egg donation known as egg cytoplasm donation.[42] The latter technique makes the egg donor a "mitochondrial mother," a situation parallel to donating eggs to be enucleated for cloning.

As we have already seen, people seek the help of assisted reproductive technologies because they want a child that is genetically part of them. Perhaps cloning the husband with the wife being the egg donor and carrying the pregnancy might adequately fulfill the wish for a child that is "part of both of us."

SURROGATE MOTHERS

Biologically, there are two kinds of surrogate mothers. In the traditional type, the sperm of the husband of the infertile woman fertilizes a surrogate's egg, and when the baby is born, the surrogate surrenders the child to the couple. In gestational surrogacy, the surrogate agrees to carry an embryo unrelated to her genetically but produced through IVF with the intended mother's egg and her husband's sperm.[43]

Surrogacy is a complex matter, and clinical concerns need to be thought through for all involved—the fetus, the father and mother who will be receiving the baby into their family, and the surrogate. The surrogate mother's "incubating family," including her spouse or partner, and her children (siblings to the fetus) also need to be taken into account.[44] Because there may be instances in which the services of a surrogate might be enlisted in cloning, let us review this area to search for parallels.

From the parents' perspective, those who employ a surrogate are not on an "equal footing" in beginning their parenting roles.[45] A mother rearing a child clone of her husband might feel that way, even if she has carried the pregnancy. From the child's perspective, there may be psychological issues to face, though up to now it has not been shown that they suffer measurable psychological harm.[46] Once again, we do not know what problems might begin to emerge as these children reach adulthood.[47]

From the surrogate's perspective, she must be able to experience the baby she is carrying as "not mine."[48] This is not normal maternal–fetal attachment.[49] Women serving as surrogates for cloning would need to do the same as surrogates for sexually conceived pregnancy, and to a much lesser degree so would those donating eggs destined to become enucleated receptacles for another's nuclear material.

A woman may be motivated to become a surrogate for a variety of reasons. In addition to financial remuneration, she probably also values children, feels fortunate to be a mother herself, and wants to help infertile couples have children.[50] Not uncommonly, however, a woman offering herself as a surrogate also has additional and more complicated motives.[51]

Many women hope that becoming a surrogate will help them resolve or repair an earlier loss or psychological injury. The loss may include the passing of a family member or friend, but more commonly it is "reproductive related"—an earlier miscarriage, abortion, or having surrendered a child for adoption.[52] That some mothers want to have another child to replace the one they carried as surrogates is evidence that surrogacy failed them as an attempt at restitution.[53]

The sense of damage or loss may also be to the self, typically a consequence of having been abandoned or abused by a parent. By offering herself as a surrogate, a woman may attempt to repair herself by feeling worthy in doing something good and generous for others.[54] An emotionally injured and deprived surrogate may also identify with the baby, perceiving "the intended mother and father as idealized, loving, and providing parents not only for the unborn child, but also for herself."[55]

Beyond this, a woman may want to become a surrogate out of a need to feel "special, loved, or appreciated" or to have a moment of "glory."[56] An egg donor for cloning might also be motivated by one or more of the above. In addition, some might seek a sense of glory in being part of something new—to be one of the first to be a surrogate or "egg mother" of a clone. With surrogacy now less novel than it once was, the prospect of becoming one may have lost some of that sense of specialness.[57] We might expect the same to occur for women who might consider participating in cloning, should the practice become established.

A surrogate may tend to idealize the couple for whom she is carrying the baby. She may also feel especially attached to the man and competitive with his wife, the intended mother.[58] Not uncommonly, the prospective parents may welcome or encourage a relationship with the surrogate and her family that largely ends with the baby's birth. Some surrogates feel greater sadness over losing that relationship than over giving up the baby.[59]

Clearly, a woman offering herself as a surrogate may be suffering psychologically, and by becoming one she may risk further psychological harm.[60] A woman offering herself as a surrogate for a clonal pregnancy might well share those motivations and vulnerabilities, as might egg donors for cloning, though perhaps to a lesser degree. While an artificial womb would get around the above complexities, it would give birth to others, not least of which might be a deep sense of "motherlessness," perhaps most felt by a male clone.[61]

EMBRYO ADOPTION

Infertility treatments to obtain eggs for IVF typically produce more embryos than are ever used. The fate of these frozen embryos has been the subject of heated political and ethical debate. One possible option is embryo adoption, by implantation into a woman other than the genetic mother. Although an adopted embryo is not related genetically to either parent, it becomes profoundly tied biologically to the adoptive mother once it implants in her uterus and the pregnancy progresses. Because the mother gives birth to the child, she and the infant are also together from the very beginning of the infant's life. This might conceivably enhance the bonding of the pair. The mother's hormonal status immediately postpartum might further facilitate her attachment to an infant who, though not genetically related, would feel, biologically and psychologically, a "part" of her.

Given the low rate of success, multiple eggs are needed in cloning trials in animals. Likewise, multiple eggs would also be needed in cloning attempts in humans. If more than one egg began to develop normally as an embryonic clone, then the "surplus" embryos, like those in IVF, might be frozen and made available for subsequent implantation in the same mother or an adoptive mother.

CHILDREN CONCEIVED THROUGH
ASSISTED REPRODUCTIVE TECHNOLOGIES AS "SPECIAL"

Parents often regard offspring resulting from successful fertility treatment as "special" and treat them that way. It might feel good to be told one is special, but it may be a liability if a person feels especially different. Because "special" status may have mental health consequences, it would be essential to study children conceived through assisted reproductive technology throughout their development. This has been little researched, largely because most of the technologies are so new. The experience of these "special" children parallels that of clones in some respects. Richard Dawkins dismisses the concern that the first cloned child might feel like a "freak," by likening a baby clone to an IVF baby.[62] Although society has acclimated to IVF, IVF children are only beginning to reach adolescence, and it remains to be seen how they will feel about themselves as young adults. We will explore "specialness" further in the Adoption Model.

"VISUALIZATION" OF THE FETUS

Though not a technique that furthers reproduction, fetal ultrasonography does permit early visualization of its results and appears to affect the maternal–fetal relationship.[63] It has been argued that those early ultrasound pictures of the fetus may help parents bond earlier to the fetus and may make its loss more profound, a contention that makes intuitive sense.[64] Before our son was born, my wife and I had ultrasound image prints on which to gaze and try to imagine what he would be like. They begin his album of baby pictures.

If such images of a fetus do hasten attachment to the expected child and exacerbate the sense of loss following fetal death, then we might expect that being able to vividly visualize an expected clone would promote even more intense parental attachment to the unborn child. The images of memory, photographs, and video or DVD recordings of the clone's progenitor, along with contemporary ones, might make losing an embryonic or fetal clone absolutely devastating. We will expand on these themes in the Replacement Child Model.

USE OF ASSISTED REPRODUCTIVE TECHNOLOGIES AS HIGHLY INTENTIONAL AND DISSOCIATING SEX FROM REPRODUCTION

Aside from masturbation, homosexual sex, and heterosexual practices aside from coitus, sex has been dissociated from conception since the advent of attempts to control conception and birth, including methods both natural and artificial. Timing intercourse to fall during periods of the woman's menstrual cycle when she is apt to not be fertile (the "rhythm" method) has been supplemented by artificial methods of contraception. Barrier contraception, with a condom or diaphragm, the IUD, and chemical methods of contraception all produce this result. Assisted reproductive technologies take it a few steps further, with couples resorting to them because intercourse has *not* resulted in desired conception and pregnancy. Asexual reproduction would be the ultimate dissociation.

However one judges societal developments that may or may not be related to the dissociation of sex and reproduction, perhaps there might be some unmitigated positives to it. If conception is highly intentional, as it must be with assisted reproductive technologies, then parents must be heavily invested emotionally and otherwise in having a child. That, but not parental overinvestment, might be to the child's benefit. We will further explore these matters and their implications in the chapters "The Parent–Child Resemblance Model," "Intimacy, Sex, and Sexuality," and "Wider Social and Cultural Implications of Cloning."

SEX-SELECTIVE ABORTION AND PRESELECTION TECHNIQUES

Sex-selective abortion and sex preselection techniques give parents the power to choose what kind of baby they do and do not want. Gender is not a dreaded defect. Except in unusual circumstances, it is not a medical justification for abortion; yet sex preselection and sex-selective abortions are being done to express parental preference regarding one aspect of who their normal offspring will be.[65] As law professor Lori Andrews puts it, they are being used to enforce "admission standards for birth."[66]

Reproductive cloning promises to give the rearing parents the kind of preferences never before possible in a baby, a baby identical genetically to a progenitor chosen by them, either one member of the couple or someone of their choice. Genetic engineering might also be used to create a clone designed to be "like me but better" in some predetermined way. While parental choice is a feature of adoption, it is much more limited. When it comes to sexual reproduction, one may hope that the good choice one made in a spouse or partner will also prove to be a good choice as a parent to the child. It is then up to chance as to whether that child will be a boy or a girl or have some other trait one might have wished for it. It is not choice but simply blind faith, buttressed by prenatal testing and monitoring as to whether the child will be born normal and healthy.

Beyond hope and faith, with few exceptions, one never knows what combination of positive traits a child might manifest and what genetic risks she might carry. This would not be the case with a clone, whether of a particular progenitor or one intentionally modified genetically.[67] Preferences people exhibit in utilizing sex-selective abortion and sex preselection techniques might parallel parental preferences in cloning.

If the fetus is in the right position, doctors can frequently determine its gender by visualizing its genitals. In Western countries, parents may or may not want to know, but if they do, it is often a source of excitement. But parents can use that same technology in deciding to abort a fetus not the sex they prefer. There is now a significantly skewed ratio of births of boys to births of girls in certain regions of China and India.[68] The normally even girl–boy ratio in births in certain Indian states is now as low as 600:1,000. The strong desire for male offspring in their particular form of patriarchal family, coupled with the ability to determine fetal gender through ultrasonography, is responsible. Faced with these statistics, the Indian Medical Association has asked doctors to stop offering sex-determination services.[69] Although one may object that cloning is not abortion, preferences that parents exercise with sex-specific abortion might parallel parental preferences in cloning.

Just as a person's generation, culture, religion, region, or nationality influences his or her sexual attitudes and practices, we should expect that those same factors might also influence attitudes toward asexual reproduction.[70] The regional popularity of sex-selective abortion illustrates this well. For example, the traditional Chinese preference for male offspring combined with state restrictions on family size contributes not only to heavy use of sex-selective abortion but also to great interest in reproductive cloning.[71] Beyond the sex of the clone, fads and fashions in cloning selections might come and go, favoring progenitors with a particular image or even particular progenitors. We'll explore the wider social consequences of cloning further in a chapter specifically devoted to it.

It is possible to physically segregate sperm by what sex chromosome they carry, using the desired "enriched" pool for insemination. Except in rare genetic disorders, an egg always carries one X chromosome; and a sperm, either an X or a Y. If a Y-carrying sperm fertilizes the egg, the baby's sex chromosomes will be XY, those of a boy. If an X-carrying sperm fertilizes the egg, the baby's sex chromosomes will be XX, those of a girl. Early embryos created by IVF may also be selected for implantation in the womb based on their sex. Fertility specialists are actively debating the ethics of offering sex preselection of sperm and of embryos to prospective parents.[72] Parental choices in sex preselection might well parallel cloning choices.

We should also note possible genetic pressures underlying the psychology of the wish to alter sex ratios and the resulting social practices. These come from the related fields of sociobiology and evolutionary psychology, which attempt to explain social and psychological phenomena, respectively, as efforts to achieve evolutionary advantage. From their perspective, cultures where the middle and upper classes reduce the number of female babies tend to concentrate both wealth and women in the hands of a small middle and upper class. By taking women from the lower classes as brides, they also exclude the poorest males from the "breeding system."[73]

This seems to be more of a description than an explanation. Lower-class women have limited wealth to bring to a marriage. In addition, eliminating the female babies of upper- and middle-class parents by the practice of sex-selection techniques would cancel out whatever genetic advantage could possibly accrue to those parents by reducing the genetic contribution of lower-class males.

A FEMINIST PERSPECTIVE ON ASSISTED REPRODUCTIVE TECHNOLOGIES AND ARRANGEMENTS

Among other things, feminists see motherhood as a power base.[74] Because assisted reproductive technologies can enable previously infertile individuals to have children, these technologies could be a boon to the status of women. However, it is also argued that those technologies give "male-dominated culture" control over the procreative process.

Andrea Dworkin sees men as controlling women's sexuality and reproduction through a "brothel model" and a "farming model." The former is commodified unreproductive sexuality. The latter is illustrated by methods of assisted reproductive technology to enable reproduction of people much as they have helped farmers to reproduce animals. In her view, "cloning is the absolute power over reproduction that men have wanted."[75]

The data on sex-specific selection and abortion show how strongly female offspring can be disfavored and are not inconsistent with these feminist views. On the other hand, women have the potential through self-cloning to completely exclude men in the process of reproduction, a biological impossibility for men. We will explore feminism and cloning further in the chapter on intimacy, sex, and sexuality.

CONCLUSION

The Assisted Reproductive Technologies and Arrangements Model is built on the fact that cloning is itself a method of assisted reproduction. Just as many infertile people are motivated to utilize currently available assisted reproductive technologies and arrangements, some might be inclined to clone for the same reasons. Cloning is likely to present some of the same medical and psychological stresses as other assisted reproductive technologies and arrangements for its participants.

Having concluded this, *we must distinguish between the practices we have used to model certain aspects of cloning and the practices themselves*. It is important that we make this distinction with this and every other model in this book. With few exceptions, we should not interpret the psychological and social risks of any given practice relevant by analogy to cloning as reason enough to disapprove of the practice modeling it. For example, donor insemination incurs certain psychological and social risks for the offspring, but we need not conclude that the level of risk, and hence the practice, is unacceptable. However, by using these practices to model aspects of cloning, we could come to exactly that conclusion for cloning. That is because there are good reasons to believe that the risks of adverse psychological consequences the model suggests for cloning would be even greater than those for the practice used to model it.

The psychological and social consequences of several assisted reproductive technologies might model the same for human cloning because they make possible the conception of children related genetically to only one of their rearing parents. While the relationship of the rearing couple appears to benefit from becoming parents, some complications in the parent–child relationship may be to the child's detriment. A critical issue for these parents is whether and when to disclose this fact to the child. Learning more about surrogate mothers, sperm and egg donors, and their selection by prospective parents might help us understand people who would facilitate cloning and might help us predict whom, if not themselves, parents might seek as a somatic cell donor for cloning.

Like parents of a child born through currently available assisted reproductive technologies and arrangements, parents of a clone and society as a whole might view a cloned child as "special." For the parents, the child might be special simply because the parents overcame infertility to have a child. They might also feel that the child is special because of their feelings about the child's progenitor, whether it is one of them or someone else. At least when it is novel, society is most likely to view the cloned child as special simply because he is a clone. Regardless of what reason predominates, we should be aware that a person does not always experience being special as a good thing.

Psychological study of the impact of techniques permitting fetal visualization might allow us to anticipate the effects of prospective parents being able to "see" and "know" their child clone before birth through images and knowledge of his progenitor. Socially, cloning takes the intentionality of reproduction and its dissociation from sex to the limit. We will examine this in depth in the chapter "Intimacy, Sex, and Sexuality."

The practices of selectively aborting a fetus, sexing sperm for IVF, and selecting early embryos for implantation based on their sex might help us model the effects of parental choice in cloning. The findings suggest a bias toward male children, one consistent with feminist concerns. However, we might expect a full range of biases to emerge with cloning, based on the individual preferences of those seeking to employ it as a mode of reproduction.

A basic premise embraced by people seeking a child with the help of an assisted reproductive technology or arrangement is that they feel it would be "better" to have a child who is "part" of them or, failing that, as much "like" them as possible. The following three chapters, on the Stepchild, Adoption, and Parent–Child Resemblance Models look specifically at that issue and, as part of it, the question, is "genetic linkage . . . necessity or narcissism?"[76]

3

The Stepchild Model

A stepchild is related genetically to only one rearing parent, as would be the case with a clone.[1] For the same reason, we looked at children conceived through donor insemination and egg donation. Although those situations might better model cloning, the paucity of information on their psychological consequences makes it important for us to look at the stepchild, too. Perhaps aspects of the experience of stepchildren and stepfamilies might inform us by analogy of possible risks for clones and their rearing families.

Leon Kass, who has written extensively on human reproductive cloning, has asked, "What would kinship be without its clear natural grounding? And what would identity be without kinship?"[2] The Stepchild and Adoption Models shed light on these and related questions. This chapter deals with the Stepchild Model; the next, with the Adoption Model.

ROBERT

Robert was already a respected scholar at only thirty-five, but academic success had not brought personal happiness, and he came to me in the midst of a major depressive episode. His father had died in an accident when Robert was an early adolescent, and I expected that working with that trauma would emerge as key to helping him in psychotherapy. However, Robert was not ready to delve into that relationship and its tragic ending, and something else quickly presented itself as having been a powerful psychological influence—the fact that a stepfather came into his life about a year after his father left it.

Robert's mother's allegiance and loyalty to Jack appeared to override her attachment to Robert and his two sisters as she stood by while Jack berated them verbally and was, on occasion, also physically abusive. To be fair, his mother also proved incapable of protecting herself. Robert felt that there was no escape. He never ran away from home because he was afraid of what Jack might do to his mother and sisters.

The only place Robert felt he could go was further into his studies. That was the only place where he did not feel helpless and where he might receive approval, even accolades. Robert surmised that his mother thought he was probably OK despite what was happening at home because he never stopped getting straight A's.

Robert left home after college, but patterns established in relationships with family members repeated themselves with others, as they so often do. One such pattern had been established for Robert when Jack came into the family—feeling trapped and helpless in relationships that became unsatisfying and in which he was criticized and berated. Just as he had as a teenager, he continued to try to escape them by plunging obsessively into work. Of course, his workaholic style itself became a source of discontent for his partners, who complained of feeling neglected and essentially abandoned. These relational patterns clearly precipitated and maintained Robert's depressions, and his status as a stepchild appeared to be one relevant factor in understanding his plight.

THE STEPCHILD—A CHILD AT RISK

In their chapter "Cinderella Denied," Daly and Wilson, leading researchers on stepfamilies, point to societal reluctance to think that a stepchild and his family run certain risks.[3] Even some psychologists continue to deny the "myth of the wicked stepparent."[4] But myths, in disregard of political correctness, sometimes can be demonstrated to have a factual basis.

There is a wealth of data that demonstrates heightened risk of parental physical abuse for a stepchild, even to the extent of causing the child's death.[5] A review of that research shows that step-parenthood is the most important demographic predictor of child abuse, an effect barely diminished by statistically controlling for other variables.[6]

The magnitude of the effect is astounding. The risk of child abuse for children under three living with one genetic parent and one stepparent is *seven times* that of children living with both their genetic parents.[7] The risk of the most severe form of child abuse is even more powerful: Stepparents residing in the home were *seventy times* more likely than a genetic parent to kill a child less than two years old.[8] When stepparents murder their stepchildren, their methods are more brutal than those of other parents, and the murder is also less likely to be associated with mental illness.[9]

Stepchildren are more likely to be murdered by stepfathers than by stepmothers. There are some additional facts to keep in mind in interpreting these findings: Four to six times as many stepfathers reside with stepchildren and the children's genetic mother than stepmothers reside with stepchildren and the children's genetic father.[10] If one controls for this disparity in numbers of stepmothers and stepfathers and looks at less severe forms of abuse than homicide, then *stepmothers and stepfathers appear to be equally likely to mistreat stepchildren.*[11] Child mistreatment appears to be specifically directed at the stepchildren in the family, while the abuser's genetic children in the household are well treated, "just like Cinderella and her step-sisters."[12]

In addition to physical abuse, we should not overlook sexual abuse. Such abuse by stepfathers is being recognized increasingly.[13] Several studies document a greater incidence of sexual abuse by stepfathers than by genetic fathers.[14]

There are also other sorts of less striking ways a stepparent may tend to disfavor a stepchild compared with his or her genetic children in a phenomenon called "differential parenting." There are various ways "kin bias" may be seen in the provision of care to children.[15] For example, when a mother is rearing a nongenetically linked child (either a foster, step-, or adoptive child), less is spent on food in that household. The reduction in expenditure is thought to be at the expense of that particular child.[16] The concept of "kin selection" or bias is a key one in the fields of sociobiology and evolutionary psychology, scientific disciplines deriving from evolutionary theory that attempt to explain a wide range of social and individual behavior as evolutionary adaptations determined by "selfish genes."[17] We will explore these ideas further in later chapters.

The evidence is clear that stepparents do "not, on the average, feel the same child-specific love and commitment as genetic parents."[18] To some extent, a nonprogenitor parent jointly rearing the child clone of a progenitor parent might feel and act like a stepparent. As with stepparents and their stepchildren, the consequences might be an increased risk of child abuse.

The presence of a stepchild in a family appears to constitute a strain on the parents' relationship, increasing the risk that a marriage will end in divorce.[19] By analogy, a child clone of one of the parents might strain the marital relationship.

Even before a clone is conceived, trying to decide together whom to clone might create great conflict. Who is to be the progenitor: the wife, or someone she prefers, or the husband and someone he favors? Perhaps deciding between donor insemination or enlisting a surrogate mother might better model such a conflict, but couples never face a choice between those options. The infertility of the man or the woman presents them with one *or* the other.

Perhaps more of a real, though not ideal, model is represented by "blended families" in which one parent brings her own children to form a new family with the other parent and his children. Do parents in a blended family tend to favor their own and disfavor their partner's children, and if so, how do they manage the conflict?

FACTORS THAT MIGHT MITIGATE THE "STEPCHILD EFFECT" FOR A CLONE

The data on stepchildren and stepfamilies are certainly sobering, and by themselves should give pause to those couples planning to clone and rear a child of one of them. There is no disputing that, genetically, the non-progenitor parent would be as if a stepparent.

Some likely differences between stepfamilies and families of clones might lessen the potential for adverse stepparent-like effects for the clone and the nonprogenitor parent. Some of these differences might be the same as those that distinguish stepfamilies from adoptive families.

First, like an adoptive parent of an infant, the rearing parents of a clone are likely to have decided jointly to rear the infant clone. This would make the nonprogenitor parent very different from a stepparent who had no choice in the child coming into existence. Presumably, the decision to clone one member of a couple would be made jointly. A joint decision would clearly be reflected in a scenario in which the nonprogenitor parent is the gestational mother of her husband's clone. And second, being there from the beginning of the infant clone's life, especially when the nonprogenitor parent is also the birth mother, might likely facilitate the kind of bonding with the clone that stepparents cannot match with a stepchild.

Writing from an evolutionary perspective, the psychologist David Barash states that because adoption generally occurs when the child is an infant, the adoptive parents "can 'fool themselves' into responding as though the child is genetically their own." He contrasts this to the later age at which stepchildren enter the life of a stepparent and sees the greater age of the child as an active impediment to the stepparent developing feelings of parental love.[20] It would probably be very unusual for a child clone not to start her life with both rearing parents. Whether or not the child is her clone, the rearing mother might well carry her in a pregnancy.

The psychological notion of "attachment" is central to understanding much about human emotion and behavior.[21] Time together, while a factor in attachment, is not the only one, as Barash mistakenly implies in an attempt to refute attachment theory in favor of evolutionary psychology—a psychological theory that emphasizes genes and evolution.[22] Research showing stepfathers to be less benevolent and more negatively interactive with their stepchildren the more time they spend together only disproves attachment theory if the theory is misunderstood.[23] Attachment is not simply proportional to time together. There is more to it.

A key concept Barash neglects is that of the "critical period," though it is a concept congenial to both the psychological theory of attachment and evolutionary psychology. It is useful to both developmental biology and the psychology of behavior, as is the related behavioral concept of "imprinting."[24] Baby ducks follow the mother duck, having "imprinted" on her during a "critical period." How exactly the notions of critical periods and imprinting apply to the bonding of parents to their children, both genetic and nongenetic, is unclear and largely unexplored. There may well be hormonal influences on a mother during the process of giving birth that create a kind of critical period for her to attach optimally to her infant.[25]

Are there "critical periods" for others in attaching to the child? Only a mother is pregnant with the child, gives birth to it, and can be biologically postpartum, though fathers will attest to their own version of these experiences. It may be that stepparents are unlikely to be present during a "critical period" necessary to develop a "familial bond" with their stepchildren, a bond important to establishing boundaries to incest—or, for that matter, other forms of child abuse.[26]

How much of a delay between birth and living with the child can there be without adversely affecting attachment and its corollary, the observance of behavioral boundaries between parent and child? The relationship between siblings is obliquely relevant, where the incest boundary is more likely to be effective for siblings reared together before the age of two and a half than it is for those reared together after that age.[27] But that research relates to boundaries within the same generation, not between generations. The parental generation has the advantages, mentioned previously, of size, strength, and intelligence—all enormously helpful for rearing and protecting a child but also available for horrid abuse of power.

Better outcomes in adoptions of younger children are consistent with the idea that attachment is enhanced by the opportunity for parents and child to attach early with one another.[28] However, simply being present early on in a child's life does not facilitate attaching to it. Recall that the youngest stepchildren studied by Daly and Wilson, those under the age of two, were more likely to incur the most severe child abuse—homicide. Very young age is not an ameliorating factor but, rather, an exacerbating one in the abuse of stepchildren.[29] I am unaware of any research specifically on abuse of adopted children, though the findings on reduced household expenditures on food hold equally for step-, foster, and adoptive households.

In addition to the above-discussed concept of "critical period," it may also be that the sex of the biological parent is relevant to attachment. Among stepfamilies, including those with heterosexual stepmothers, heterosexual stepfathers, and lesbian stepmothers, the sex of the biological parent is more important than the sex of the stepparent in stepfamily adjustment. In general, the most beneficial situation for a child in a stepfamily is to reside with his or her biological mother.[30] This research implies that the mother–child bond may be more powerfully positive than any other attachment for a child. It has provocative implications, especially given the research on lesbian families with a child conceived through donor insemination showing that the co-mothers are more involved with their children than are fathers in two-parent heterosexual families.[31] However, it does not necessarily imply that it is better for a child to have two mothers than a mother and a father.

Concerns have been raised that a child related genetically to one and not the other rearing parent, the way a stepchild is to his or hers, might predict problems for a child resulting from certain assisted reproductive technologies. But, as we have seen in the chapter dealing with that model, research does not, so far, show the genetic parallel to stepfamilies to be nearly as strong as other factors in analyzing how such children are turning out.[32]

Some might argue that the relatively high socioeconomic status likely for people undertaking the expense of cloning would mitigate the risk that the nonprogenitor parent might be negatively inclined toward the child clone, much as a stepparent may be toward a stepchild. However, the data show that risk for harm from a stepparent holds even when controlling for socioeconomic status.[33]

FROM STEPCHILD TO ADOPTED CHILD

The stepparent data are not encouraging for the unrelated father, but the adoption data are not discouraging. In this regard, adoptions by stepparents are the most prevalent type of adoption in the United States, with stepfathers being the most frequently adopting parent. When stepfathers adopt, the child's genetic father is generally no longer in the picture.[34]

By adoption, the parent claims the child as "his," and the child commonly assumes his adoptive father's last name. We may suspect that claiming the child in this legal and symbolic way could beneficially impact the relationship. To my knowledge, no research has looked at whether adoption of stepchildren by stepfathers reduces the risk of harm to the children. Does the conversion of the stepfather into adoptive father represent a transformation of his relationship to the child that reduces the risk of the father abusing the child? If it does, then it might imply that the relationship of nonprogenitor parent to child clone might not have a "stepparent effect" cloud over it, or at least the cloud might not be so dark.

Half of all adopted children are adopted by relatives, and many stepchildren are adopted by their stepparent, but when one thinks of adoption, it is generally by couples with no genetic relationship to the child.[35] The next chapter deals with this form of adoption as a model of cloning.

CONCLUSION

It would appear that a deficient loving and benevolent attachment to the stepchild by a parent with whom he or she is not linked genetically underlies the higher risk of abuse or neglect. This may give one pause about the risks for child clones of one rearing parent, as well as children not linked to one rearing parent genetically.

However, there are many factors that might mitigate the "stepchild effect" for a clone. A child resulting from donor insemination or egg donation models cloning better than the Stepchild Model does. We will not know their full psychological and social consequences for some time, so we are forced to utilize this model in addition, as one in which the child is genetically linked to one parent and not the other. In other words, the Assisted Reproductive Technologies and Arrangements Model is more relevant by analogy to cloning, but there is scant evidential basis to judge its psychological and social consequences because the technologies are so new. In contrast, the Stepchild Model, while relevant along the same dimension of the child being genetically linked to only one parent, is less relevant given other mitigating factors. However, there is abundant evidence to judge its psychological and social consequences.

4

The Adoption Model

LANCE

Lance first walked into my office at forty years of age, but he was no stranger to psychiatry. In the midst of his third bout of major depression, he had been recalling painful aspects of his childhood with another psychiatrist, and they thought it might help him to contact his adoptive parents, from whom he had been estranged for a decade. His third episode of depression was the unintended consequence of his failed attempt to reconcile with them.

Lance's parents told him he was adopted when he was about ten. Knowing that helped him make sense of the fact that he had always felt as if he was not really part of his family. His parents' hugs felt as though they were at arm's length. Lance thought he detected quiet and cold anger and sad disappointment that he was not really "theirs," and when he got older he wondered if they resented him as a symbol of their reproductive inadequacy.

Lance felt lonely and anxious growing up, both in and outside the home. Contributing to this was his feeling that his parents "wouldn't let me be myself—whoever that was." His academic successes meant little to him in the absence of a meaningful connection to another person.

In junior high school he began to drink to alleviate anxiety, and in high school he began to experiment with cocaine and speed. Under their influence, he would transiently feel better about himself, more capable, powerful, and important. In disgust, his adoptive parents put him out of the house when he graduated and never again initiated contact with their only child.

After quickly failing out of a junior college, Lance resolved to quit drugs and alcohol. He worked as a store clerk and, after a few years, began taking college courses in the evening. He finally earned his bachelor's in computer science at the age of thirty. Feeling more successful and secure, he tried to reestablish contact with his adoptive parents. When they rebuffed him, he briefly relapsed into addiction.

After a few years of sobriety, he was fortunate to meet and marry a loving and supportive woman with a young son. He has been an ideal stepfather and thinks that being adopted gives him a perspective that helps him to make sure his stepson feels his love and support.

As mentioned earlier, shortly before coming to me, he again sought out his adoptive parents. When he called their home, his adoptive mother angrily told him that she had "no use" for him and that his father had died the previous year, and then she abruptly hung up.

About a year later, and while in treatment with me, Lance decided to try to locate and contact his birth mother. Although he could no longer hope for a relationship with his adoptive mother, he hoped that his birth mother would be receptive. He managed to find her through court records and discovered that she did not live far from his childhood home. When he phoned to try to arrange a meeting, she was courteous but declined. Clearly, she had never wanted to hear from the child she had put up for adoption so long ago. Following this rejection, more frequent therapy sessions and increasing his antidepressant dose helped to reverse a slide into depression.

Throughout all this, Lance continued to avoid substances he had abused in the past. He was also working with some sense of satisfaction as a computer programmer, though he often felt frustrated with his boss. In therapy, it became clear to Lance that getting more control over his work life was important and that there were limits to the possibility of doing so in the job he held. So, with some reluctance, he began to explore other opportunities, landing a position as a senior programmer in another state.

Lance was apprehensive about leaving the area, but his wife's support and encouragement proved instrumental. The last I heard from him, he was in treatment with a psychiatrist in his new city, was continuing on his antidepressant, and was worried that he would relapse without it.

At a deep level, Lance still felt unwanted and insecure. Despite success and recognition at work, he did not feel comfortable with his new title of senior programmer. *The only thing about his identity that he knew for certain was that he was adopted.*

INTRODUCTION

About 1 percent of adults in the United States have adopted a child, and 1–2 percent of all children are adopted.[1] In the total U.S. population, more than five million children and adults were adopted.[2] Family members and stepparents adopt about half of the children. Genetically unrelated individuals adopt the rest.[3] We will be focusing on adoption by people unrelated to the child.

There is a wealth of psychological and social theory and data on adoption that is relevant by analogy to reproductive cloning.[4] We shall see that the Adoption Model can help us anticipate some important psychological and social consequences that might result from cloning.

At first glance, it might seem unlikely that adoption could bear any relationship to cloning. After all, adoption is not parallel to cloning biologically. In fact, it would appear to be opposite to it, with an adoptee being unrelated to his rearing parents while a clone, unless himself adopted, would be related to one and only one parent. In unexpected ways, however, sometimes oppositions can help to enlighten us as much as parallels.[5] In this chapter we will see that we can utilize both parallels and oppositions between cloning and adoption to try to anticipate the psychological and social consequences of cloning.

We will address the psychological meanings of adoption from multiple perspectives. In the process, we'll refer most often to the work of Betty Jean Lifton, an authority on the psychology of adoption and herself an adoptee. We will see that a psychological understanding of adoption is relevant by analogy to understanding the possible psychological and social consequences of cloning—for the clone, his rearing parents, the progenitor, and society.

Betty Jean Lifton wonders if there may be an "adoption syndrome"—"a series of traits peculiar to adopted people."[6] Analogously, even if the range of biological abnormalities and "syndromes" described in cloned animals can be prevented in human clones, psychological and social factors alone might be able to cause a psychological "cloning syndrome" in humans.

There are good reasons to wonder about a possible "adoption syndrome." Adopted children are two to five times more likely to be referred for psychological treatment than their nonadopted peers, most typically for impulsive, provocative, aggressive, and antisocial behavior.[7] Adopted individuals or adoptees are substantially overrepresented in the psychiatric population.[8] Children placed at an earlier age are less at risk.[9] It is not only among "clinical" populations that we find more adoptees.[10] While many adopted individuals are well adjusted, a large national representative school survey clearly confirms that more adopted adolescents have psychological and behavioral problems compared with their non-adopted peers.[11]

Biological factors, such as poor prenatal care and possibly even a genetically based increased incidence of psychiatric disorder, might be relevant factors.[12] Without a doubt, the psychological consequences of having felt cast off must also play a role.

To the extent that adoptees are at risk for behavioral and psychiatric disorders because they were unwanted and relinquished by their genetic parents, we might expect that clones would not share the same legacy and its consequences. As discussed previously, cloning as an assisted reproductive technology would be highly intentional, and so we might expect that a cloned child would be very much wanted by its parents. On this dimension, adoption would not model cloning.

But there are other areas to consider in adoption that might parallel psychological aspects of cloning. These include possible complications in parent–child attachment and in the child's ability to attach to others, as well. There might also be additional complications for the child in resolving what psychoanalysts call the "Oedipal complex" and in developing his own identity.

ADOPTEES AND CLONES—BOTH "CHOSEN" CHILDREN

The Japanese and ancient Romans considered adoption to be an acceptable substitute for reproduction and, in ancient Rome, sometimes even a superior one. There, the adopted child or "alumnus" was the object of "special affection" "by virtue of having been chosen."[13] Such is not generally the case in modern Western societies.

To help the child feel special and wanted, adoptive parents typically promote the myth of having chosen the child. I am not using the term *myth* as meaning an invented story with no factual basis. The power of myth lies in its presentation of an idea in an emotionally resonant story. The story may be completely or only partly fictional.

On the issue of having been "chosen," one adoptee told Lifton, "It's a burden to be chosen. Its very specialness isolates one."[14] However true the myth, unspoken aspects of the adoptee's "chosen" status include the likelihood that adoption was not the parents' first choice and that they probably did not have many babies from which to choose.

Clones, adopted or not, would truly be chosen, with specific expectations that the child clone be like or at least resemble his progenitor. Some parents might even come to prefer cloning as a route to parenthood over having a child whom they jointly reproduced sexually. In any case, at least genetically, the degree to which a clone could be "chosen" by his parents would be much greater than the degree to which an adopted child may be chosen by his. Compared with the stark reality of parental choice in cloning, the notion that an adopted child is chosen seems flimsy. Nonetheless, the way adoptees feel about the "chosen" myth is relevant for clones. Lifton quotes one adoptee as saying: "Where do you connect with the human condition when you are chosen and everyone else is born?"[15] One might imagine a clone asking a similar question—just substitute the words *cloned* for *chosen* and *conceived* for *born*.

Adoptive parents must be able to put aside their longing for the genetic child they hoped to have and invest emotionally in the child they adopt. They must see that child for who he really is, and not who they want him to be, and provide him with empathy and support.[16] While this is good advice for any parent, it might prove more difficult for parents of a clone to follow because they might well have stronger convictions about who that child is supposed to be based on his genetic identity.

The force of parental expectation is greatest when parents feel entitled to the kind of child they want. The issue of "entitlement" has been raised as an important one in adoption, if only because without genetic ties to the child coming under their care, the adoptive parents must have their entitlement to control and regulate the child legally conferred. Unfortunately, little research has been done on it.

If genetic linkage to an offspring naturally confers parental entitlement, then adoption and cloning would appear to be conceptually opposite. Parental entitlement needs to be responsive to the child's needs, sensitive to his developmental stage, and respectful of his individual self and identity as they develop. In chapter 5—"The Parent–Child Resemblance Model"—we will see how narcissistic parenting takes normal parental entitlement to pathological extremes. With the strongest of genetic ties possible between progenitor/parent and clone, and with specific expectations that the clone be like the progenitor, we might expect that the clone's parents might feel a heightened sense of entitlement.[17]

While adoption and cloning would be opposite to one another on the dimension of genetic linkage, they parallel one another in that both adopted children and clones would have been "chosen." Because they have chosen their child, some adoptive parents may feel especially entitled to regulate that child's behavior.[18] Although a variety of biological factors, including genetics and prenatal exposure to drugs and alcohol, might underlie the antisocial tendency of some adoptees, from a psychological perspective, it might also be in opposition to pressure to comply with their parents' expectations of them as "chosen" children.[19] If this is correct, then we might also expect an increased incidence of antisocial behavior from clones.

Even if an adopted child is not his parent's first choice, his parents must have wanted a child enough to adopt him. However, whether the adoptee comports with what the parents wanted in a child is another matter. On the issue of being wanted, a nonadopted friend of Betty Jean Lifton told her: "I think the obsession to find out who you are is universal. . . . It is an obsession to re-create yourself. To give birth to yourself under another set of circumstances. . . . I always thought adoptive parents must be better than natural ones, because at least these people wanted a child."[20] It sounds like her friend wishes she could have adopted and mothered herself and believes that she might have been a better parent than her own father or mother. Perhaps some people who feel similarly might wish to clone themselves.

Oral contraceptives have helped to decrease the number of unwanted babies, at least the kind of babies most couples prefer to adopt.[21] The pool of potential adoptive parents is also smaller than it might be without the successes afforded by assisted reproductive technologies, helping couples have babies, fully or partially their own genetically. The number of adoptions of unrelated children sharply decreased between 1970 and 1975, and in the United States it has leveled off at the rate of about 50,000 per year.[22] Cloning might increase the number of children not "chosen," except for a new kind of adoptee, clones created on demand, for the purpose of adoption. Abandoned and unwanted children would still be with us.

ATTACHMENT, BONDING, AND TRANSFERENCE

As mentioned earlier, parents, especially mothers, go through a process of bonding to their infants.[23] It is the foundation for parental attachment to the child and the child's attachment to the mother.[24] I say "mother" and not "parent" because there has been more research on attachment and bonding in mothers than in fathers, where it is now getting some overdue attention.[25]

One manifestation of bonding is "mirroring."[26] On a concrete level, when a mother mirrors her infant, she often literally mimics the infant's facial expressions and echoes his verbalizations. But the psychological concept of mirroring means more than being a mirror reflection. Mirroring includes being emotionally responsive, too. Mirroring is the mother's way of helping to define for the child what he is experiencing and who he is as a person. By responding empathically to her infant, she helps define his experiences and affirms his very existence. The baby discovers himself in his mother's eyes: If they are full of love, then he feels loveable; if they are empty, inattentive, or disapproving, then he learns something quite different about himself. When a mother regards a young child's efforts approvingly and praises them, she affirms his value and developing sense of self. Through the apparently innate process of mirroring, the mother shows that she knows how the baby feels, and this helps him begin to know himself through her.

Adoptive parents generally prefer to adopt an infant rather than an older child. There may be a number of reasons for this, but one probably has to do with the belief that bonding and attachment of mother to child and child to mother would be easier with an infant than with an older child. In fact, a child adopted at a late age may have difficulty attaching to his or her adoptive parents and is more likely to have behavioral problems.[27] We do not know how much of this might be caused by the parents finding it difficult to bond and attach to their child.

Is the bonding of a mother to her adopted infant any different than that of a mother to her own genetic infant? Golombok cites studies demonstrating "the lack of difference between adopted and nonadopted infants in the proportion that may be classified as securely attached to their mother."[28] While this is reassuring, to my knowledge there is no research on attachment of children adopted as infants beyond their infancy.

Although genetic ties or the lack thereof may, to a large extent, underlie the findings demonstrating that parents relate differently to their own children as compared with their stepchildren, it is presumably mediated by differences produced in the inclination of parents to attach to the child. Remarkably, there appears to be no published research with adoptive families on a phenomenon so clearly demonstrated in stepfamilies—the "differential parenting" of a child genetically linked versus one not genetically linked to a parent.[29] We just do not know whether parents bond and attach differently or differentially with a "chosen" adopted child compared with a child who is genetically "theirs."

Lance felt that his parents were not much attached to him. While it is clear that many parents do attach strongly to their adopted children, perhaps Lance's parents had feelings about him not being "theirs" genetically that interfered with their attaching to him. On the other hand, it may be that they might not have been able to attach and relate empathetically to any child.

What might we suppose for a child clone of a parent? If the mother gives birth to the child, then we might expect that she would bond to it as well or as poorly as to a child she might have conceived sexually. If the child was a self-clone, then perhaps the mother/progenitor would have a heightened ability to mirror and be empathic to her child, but this might not be as good as it sounds. One could imagine a kind of reverberating feedback circuit as the mother sees herself re-reflected in her mirroring. This could intensify the normal mother–infant symbiosis and interfere with the separation-individuation process discussed in the Identical Twin Model.[30]

If the mother gives birth to a clone of her husband or in the case of a lesbian couple, her partner, we might expect that her bonding and attachment to the child would be influenced by her feelings about her partner, the child's progenitor. This would surely be true regardless of the progenitor's identity and relationship to the mother.

Those feelings would be important because, in some respects, the mother might well relate to the child as if he were the progenitor in a manifestation of a universal phenomenon called "transference." Patients have experienced and related to me as a psychotherapist in ways indicating that they were perceiving and experiencing me as if I were a father or a mother, a sibling, or even one of their children. I do not think that any possible transference projections would be impossible even if the patient and I were clones of the same progenitor. The fact that the adopted child is not a genetic relative does not make it immune from transference projections from the parents. Transference always exists, to some degree. The adopted infant could become a kind of "blank slate" for myriad transference projections by the parents. When it comes to transference, adoption and cloning would not be opposites of each other. Rather, cloning would tend to specify and direct parental transference, though it would be unlikely to limit it.

Might adoption shed light on a clone's ability to attach? Once an adopted child knows he is not his genetic parents' offspring, he might have difficulty committing himself to belonging to his adoptive family.[31] Once he learns his progenitor's identity, it might even be more difficult for a clone to be reared by parents, neither of whom is his progenitor. The fact that his rearing parents selected that progenitor over either of them might imply to the child clone that he should hold the progenitor in higher esteem. While the clone might feel good about what he knows or fantasizes about his nonrearing progenitor, his attachment to the people really in his life, including his rearing parents, might suffer as a result.

THE CONCEPT OF "GENEALOGICAL BEWILDERMENT"

Knowing that one has two sets of parents produces conflicts in identification for an adoptee.[32] By analogy, knowing that he has only one nuclear genetic "parent," his progenitor—who may or may not be a rearing parent—might create conflicts and confusion in identification for a clone. To this, add the complexity of possible dissociation of the roles of mother. There would be the egg mother, gestational mother, and rearing mother. And let us not forget that the clone's progenitor's parents are also the genetic parents of the clone. Taken together, one has all the ingredients of what has been referred to as "genealogical bewilderment" in adoptees, and then some.[33]

Coining this term in 1964, Sants reasoned that if we accept Freud's hypothesis that there is a universal deep-seated fear of incest, then one reason underlying intense desire to identify the natural family is to avoid incest. As a result, people who do not know who their parents are would have a fear of unknowingly committing incest.[34] A clone's genealogical bewilderment might also lead to problems in this area, especially when he learns, as all children do, that his parents have a relationship from which he is forbidden. Many children have difficulty with this, but when the child clone knows that he and one parent are genetically the same and that the opposite-sex parent is not a genetic parent, we might expect further complications.

INCEST AND THE "OEDIPAL COMPLEX"

Oedipal conflict tends to be intense in adopted children. The oedipal period corresponds with the stage "initiative versus guilt," stage three of Erikson's eight life cycle stages. It occurs from age three to five years. During this time, awareness of gender differences emerges, and gender identity consolidates. During the oedipal period, the child wishes to exclusively possess the parent of the opposite sex and feels jealousy and rage against the parent of the same sex. Freud conceived this period from the legend of Oedipus, an adoptee who unknowingly killed his father and married his mother. During this period, parents must control the child's incestuous desires without completely humbling him. Play and fantasy are vital to the child, and he develops a conscience during this period. For those who have not negotiated this stage well, Erikson's theory predicts defects in initiative, guilt, or conscience.[35]

Psychoanalytic experience also demonstrates later difficulties for people with unresolved oedipal issues in relationships. They are especially likely to have problems in relationships with the opposite sex, as well as with those of the same sex whom they experience as authorities. It is important to note that some people dispute the whole concept of the Oedipal complex, but it has been of enormous utility in psychoanalytic psychology, and some objections to it can be refuted.[36]

Fully resolving the Oedipal complex is important to developing a stable sexual identity for the adopted adolescent.[37] If the child knows that he shares no genetic ties to his parents, then he may feel less restrained by the incest taboo. Of course, it is the parents' responsibility to set the boundary and help the child deal with those feelings. Unfortunately, a child does not have the reciprocal ability, given her huge disadvantage in power and dominance with respect to the parent and her dependence on, implicit trust of, and wish to please, or at least not displease, the parent. It is nearly impossible for her to effectively resist the incestuous actions of an adult.

Though some parents may also feel less restrained with an adopted child, and with varying degrees of awareness may be seductive, we do not know if adopted children fall prey to incest by their adoptive parents more than nonadopted children are similarly victimized by their genetic parents.[38] This apparently taboo question has not been asked, much less answered, by research. In fact, a computer search of the psychological literature on the intersection of "incest" and "adoptees" yielded articles only on incest suffered by children *prior to adoption.*

Incestuous wishes sometimes emerge for adoptees as adults in another context—after reuniting with their biological families.[39] By analogy to adoption, the risk of incestuous longings—in both clone and opposite-sex parent—might also arise from their lack of genetic linkage. In addition, if a child were a clone of a parent, then the parent of the opposite sex might be at risk to feel sexually attracted to the clone as a younger version of the spouse.

A parent might also experience incestuous feelings toward a child clone of his or her own opposite-sex parent, as the child might, through no fault of its own, reactivate the parent's original oedipal wishes. Perhaps, the risk of incest might be greatest if the child were the clone of a nonrelative selected because a parent admired or found the progenitor especially attractive. If part of the pull of sex psychologically is seeking self-completion through uniting with the opposite sex, were it ever to become possible to create a self-clone genetically engineered to be of the opposite sex, that child might be at great risk of incest.

THE "FAMILY ROMANCE FANTASY"

At some point, all children wonder whether the adults they have always known as "mom" and "dad" are their "real" parents. The fantasy of having parents, usually of higher standing, is called the "family romance fantasy."[40] In it, the child typically imagines that he or she was orphaned or simply adopted by the people they know as their parents, whereas others imagine themselves to be stepchildren. The fantasy begins as early as age two but always prior to the fifth year, at the same time that the child begins to identify with the parent of the same sex.[41] The fantasy is maximal during the period called "latency," years that correspond to middle childhood.

Psychologically, the family romance fantasy serves a number of functions for the child. By giving him ideal parents, it defends him against feelings of disillusionment with his real parents.[42] Those fantasized ideal parents never disappoint the child or arouse sexual feelings.[43] The child also bolsters his self-esteem with the fantasy because "idealized parents would most certainly have no less than an ideal child!"[44]

Reality and fantasy "merge" for the adopted child when it comes to the family romance fantasy.[45] He truly does have other parents, and that reality interferes with his fantasy possibilities.[46] Knowing one is adopted may interfere with the defense that the family romance fantasy offers against incestuous and aggressive urges elicited in the parent–child relationship, as well as the anxiety and guilt that go with them.[47] For similar reasons, we might expect that a clone would also be deprived of the beneficial effects of family romance fantasies.

In oedipal fantasy, a child may imagine the mother in a secret love affair with a man who is the boy grown up.[48] For a father's self-clone, such fantasies might be too easy to entertain and also may be likely to evoke more anxiety and guilt because, in a sense, the boy's father would be the child grown up.

A father's self-clone might be hampered in trying to move beyond oedipal fantasies like this. Here is why: The young boy denies his oedipal wishes by identifying with his father as the kind of man he may become. This identification, essential for a boy in resolving his Oedipal complex, might become especially problematic because his father would also be the boy's progenitor. As such, the father, in a sense that would only be possible in cloning, would be the man the boy will become. As a result, the boy might identify too strongly with him, including the fact that his father is in an intimate relationship with the boy's mother, a sexual relationship forbidden to the boy.

Consequently, identifying with his father might create intolerable conflict for a father's self-clone. In order to defend against his incestuous wishes and resolve his Oedipal complex, the boy needs to identify with his father, but doing so might make those very wishes harder to deny and more confusing. Thus, a man who clones himself might bequeath his son an insoluble oedipal paradox.

The family romance fantasy might also play a role for some progenitors and rearing parents of a clone. The progenitor parent has an opportunity through cloning to try to create an ideal version of herself and simultaneously try to make herself into her own ideal parent.

Reality and fantasy might converge for a cloned child of a famous person or person of allegedly superior ability or achievement. Some parents might choose to adopt such a child, his rearing mother perhaps giving birth to him. In addition to the child perhaps preferring his renowned or "ideal" but real genetic progenitor over his rearing parents, his parents would also have signaled to him that they consider his progenitor to be a superior "real" parent to either of them, separately. We will explore this area further in this chapter, under the section "'Adopted' Clones," and in a later chapter, "The Child of the Famous Model."

A clone with a progenitor other than either of his rearing parents might be deprived of the salutary effects of the family romance fantasy in dealing with the real-life disappointments and conflicts with his parents. He might not feel safe enough to fantasize about something he knows is reality—that he is not only *not* of these parents but, in fact, of some*one* else, his progenitor. In a world with clones in it, noncloned children might also entertain new variations on the family romance fantasy.

IDENTITY

In its impact on identity formation, cloning might be parallel to adoption in some ways and opposite in others.

Identity Defined

In Erik Erikson's life cycle schema, a person does critical psychological work in forming his or her identity in adolescence during stage five, "identity versus role confusion." Elaborating on his understandings, Ruth Ellen Josselson defines identity as "the stable, consistent, and reliable sense of who one is and what one stands for in the world. It integrates one's meaning to oneself and one's meaning to others; it provides a match between what one regards as central to oneself and how one is viewed by significant others in one's life."[49] "The experience of identity is one of meaningful continuity over time and place" within a particular social and historical context.[50]

Identity formation, though most prominently a feature of adolescence and young adulthood, does not begin and end there. An adoptee often revisits identity issues regarding adoption during life transitions.[51] We might expect the same type of identity issues related to cloning for cloned individuals.

Adoptive Identity and Clonal Identity

"Adoptive identity" is said to include "the adoptee's sense of continuity of generations . . . integration of influence from parents of rearing and parents of birth, and a sense of how" they fit into "a family in which some may be related by birth and others not."[52] By analogy, all clones would have a "clonal identity." This would fall into the category of what has been called "assigned" components of identity, that is, components over which individuals have no choice and cannot change, such as gender, race, and adoptive status. These are things around which the self must construct meaning. Within the category of assigned components, one can differentiate between those shared by many and those that "may set the individual apart from most people."[53] Being adopted is one such component; being a clone would be another.

One aspect of clonal identity would be generic—the simple fact that the person is a clone. Another would be that of being a clone of a specific progenitor. A third aspect of clonal identity relates to how the clone fits into his family, in which one of his rearing parents may or may not be his progenitor and any siblings may or may not be related to his rearing parents as he is. This last feature might parallel how adoptees think of themselves in a family with nonadopted siblings.[54]

Adoptive parents want to think of their adopted child as their own and of their family as an "ordinary one." This may lead them to underestimate the importance of being adopted for their child.[55] A clone of a rearing parent would be thoroughly the couple's "own" genetically—but of only one of the rearing parents and not the other. If the adoptive family is not an ordinary family, then the family with a clone would certainly not be ordinary. Hence, clonal status might well prove to be at least as important as adoptive status.

In developing his own identity, an adolescent requires "concrete facts and information," a special challenge for an adopted individual.[56] Certain facts help us formulate an identity by providing material we can use to construct a narrative about who we are.[57] Lacking facts about his origins, the narrative of the adoptee has a confusing start.[58] Lance knew he was adopted but knew next to nothing about his genetic mother and nothing about his genetic father. His parents' emotional distance made him feel adopted. Lance's adoptive identity was devoid of anything positive. It was a large, empty, negative space—in which he alone resided.

A clone might also face difficulties constructing a personal narrative from the very beginning. Unlike the adoptee who must grapple with narrative uncertainty regarding his or her parents, the clone would have to deal with *narrative certainty* regarding his progenitor. If that progenitor also happened to be famous, then aspects of his narrative would also be public knowledge. A clone's self-narrative would also be based on subjective perceptions of the progenitor and most certainly on his parents' expectations of him as a clone, those expectations being one more "assigned" component of identity.[59]

It is normal in adolescence to reevaluate one's assigned identity and make major strides in developing one's own. We might expect that at this time in his life, a clone would want to go beyond being a clone of his particular progenitor and try to steer his own life to make it uniquely his. All people, including clones, might be happiest if they could fully agree with Dr. Seuss that "there is no one alive who is youer than you!"[60]

THE COMPLIANT ADOPTEE

A so-called good adoptee is always grateful to his adoptive parents for the life they made possible for him.[61] A "good clone" might also be grateful, but his indebtedness would be singular—to his progenitor—and so possibly greater. Accompanying it might also be a sense that the gift of his life is contingent on him being like the progenitor.

In an effort to be "good," an adoptee may construct an "as if" or artificial self: grateful, compliant, eager to please, and afraid to express feelings, especially anger and sadness.[62] While there is a dearth of research in this area, one researcher finds that many adoptees manifest a "tendency toward a false self or exaggerated persona."[63] I should hasten to add that adoptees are not the only people who may develop a "false self" or even an "as if" personality.[64] Perhaps more than noncloned children adhering to parental expectations, a clone might feel compelled to be "as if" the progenitor, as expected by the rearing parent.

A clone might be inclined to feel like there is no room for him to be anything but a replica or simulacrum of his progenitor. Parental expectation can so mold individuals' identities that they do not try to explore values and attitudes at variance to those of their parents and do little to define themselves as their own person. These people may be described as having a "foreclosed identity."[65] A clone might be at great risk to develop one.[66]

THE DEFIANT ADOPTEE

Adopted individuals, often cut off as they are from key pieces of themselves, may reject positive role models in their adoptive family and in their "identity hunger" may choose a "negative identity" as a way to regain some mastery.[67] One psychological root of the antisocial tendency of some adoptees may be anger about feeling that his parents did not allow him to express grief and anger regarding being cut off from his origins.[68] A clone might be inclined to act out angrily for analogous reasons, feeling deprived of a natural origin whereby he would have had two genetic parents and not just one progenitor.

An adopted child may also react angrily to a sense of not being connected to his or her adoptive parents.[69] Reacting against too much connection to the progenitor parent, a clone might do the same to assert his separate identity.

THE ADOPTEE'S SEXUAL SELF

Adolescence is a challenging time for many people. Sexual maturation brings with it intense and often confusing emotions and acute feelings of self-consciousness. For an adopted individual, the process also often stirs up questions about his or her sexual origin.[70] Perhaps it is in this regard that the fact of being adopted seems to be more important for girls than for boys. Analogously, we might predict similar gender parallels for how important being a clone is to identity. In addition, any clone, and perhaps especially a female clone, might have a hard time when she confronts the asexual nature of her origin. The result might be an alienation from herself as a sexual being.

THE MOTHERLESS AND FATHERLESS SELF

For an adoptee, searching for biological parents, particularly a mother, may be critical for identity formation. One female adoptee put it this way: "Trying to find my mother is connected to trying to find my sense of self."[71] This poignant statement takes on added meaning when we understand that for the infant, baby, or very young child who has not differentiated herself from her mother, the mother "not only plays the role of the child's self, but naturally is that self."[72] Self-cloning might particularly appeal to an adopted individual who has not worked through her feelings at having been relinquished and who has not been able to find or establish a satisfactory relationship with her birth parents.

Many adoptees experience what Lifton calls a "motherless self." She asserts that adoptees have two mothers, both, in a sense, unreal. While identifying with the many mythic heroes who lost their mothers at birth may comfort an adoptee, having lost her mother may sometimes interfere with her being able to imagine herself as a mother.[73] As one woman told Lifton, "As I see it, a child who has no mother can mother no child."[74]

Unlike an adoptee, a clone would probably know her genetic mother as her rearing mother. If the roles of egg mother and gestational mother had been dissociated from genetic and rearing mother, then it might conceivably also be important for a clone to know her other biological mothers to reduce the risk that she might share the adoptee's experience of a "motherless self."

Although both male and female adoptees know that they have a mother somewhere, a male clone might be left with a void that not even fantasy can fill (although genetically speaking, his progenitor's mother would also be the clone's mother). The mournful song "Sometimes I Feel like a Motherless Child" might come to serve as his anthem, and he might be at greater risk for clinical depression. In his natural desire to imagine a mother, a male clone would have to bend and twist the rules of reality much more than do conventional adoptees in their "family romance" fantasies. This might put him at greater risk for thought disorder and psychotic illness.

A man dealing with the effects of fatherlessness as a result of being an adoptee might find self-cloning especially appealing. By cloning himself, an adoptee might be trying, in a sense, to create his father, a man he may have never known, as he makes himself a father of himself as a clone. In this regard, it is interesting to note that Erik Erikson, a half adoptee, was thwarted in ever learning the identity of his genetic father. The father of identity theory and research said he changed his last name from that of his adoptive father, Homburger, to Erikson in order to be his "own originator," father of himself—Erik's son.[75] Were cloning possible at the time, perhaps he might have cloned himself in reality as well as in name. We will specifically explore the implications of "namesaking" for cloning in the Namesake Model. As for an adopted woman who never found her biological mother or who may have been disappointed in whom she found, self-cloning might be an attempt to salve and solve her motherlessness by becoming the mother of herself as a clone.[76]

The search for the birth father can be particularly difficult for an adoptee, as the birth mother may not wish for him to be known or found. The mother of a self-clone would also be preventing her child from knowing a father because her clone would be truly fatherless. The clone might share some of the same feelings a female adoptee may harbor about having a biological mother who refuses to let her know anything about her father.

DISCLOSURE

All adoptive parents must decide whether and how to disclose to a child that he is adopted. Some answer the question both by telling him that he is adopted and by including the child's birth mother, and sometimes father, in his life in what is called an "open" adoption.[77] Such an arrangement neither enhances an adopted child's self-esteem nor diminishes it, and it does not appear to confuse the child about who his real "rearing" parents are.[78] Those researching the issue of disclosure hasten to point out that because the children studied have been in middle childhood, follow-up in adolescence is necessary to determine long-term effects, especially on the development of identity.[79]

In her chapter "Telling the Child," Lifton's subheadings give us the possibilities regarding disclosure and knowledge regarding cloning: "knowing too much," "knowing too little," "not telling," and "what to tell—the true story."[80] I will not go into the specifics but, rather, refer the reader to her book, as the arguments under each would be relevant to informing the clone of his or her status. In any case, there would no doubt be a "burden of knowing," but knowledge is always a burden, though usually one that lessens others.[81] In her chapter "The Right to Know," one may draw the analogy that a clone ought to have the right to know that one is a clone, who one's progenitor is, and if one has clonal siblings.[82]

In "the ancient and traditional concept . . . at the moment of adoption, the child is literally born again. It has *by law*, been given a second birth. Its past life is cancelled out. The child is a new person."[83] By analogy, a progenitor might perceive his self-clone as himself "born again." However, the progenitor would still be living his own prior life alongside the child's—a fact that he might or might not disclose. So, *while the adoptee may know too little about his parentage, the clone may know too much.*

The literature on open adoption is relevant to the knowledge dimension of cloning. Parents in fully disclosed adoptions appear to benefit from the arrangement, and the children may as well.[84] By analogy, were a clone to be adopted, it might be beneficial to all concerned to fully disclose that information to the child. That would include informing the child that he is an adopted clone and telling him about his progenitor. A progenitor of an adopted clone should expect that the rearing parents and their child, his clone, might want to contact him.

When should a rearing parent tell a child that he or she is a clone of one of them or of someone else? Adoption provides a partial parallel but, as yet, no clear answer. Our earlier examination of donor insemination also gets at it partially and indirectly. However, it is one thing to know that one's father is not one's father and another to know that one's father is oneself, genetically at least. In my view, prematurely knowing that one is a clone of a rearing parent could be detrimental to the development of the child's sense of identity. Similarly, knowing too soon that one is a clone of someone other than one's rearing parents might interfere with the normal process of attaching to and identifying with the latter.

CAPACITY FOR INTIMACY

Regarding attachments later on in life, suboptimal trust may interfere with an adopted person's ability to develop satisfying intimate relationships as an adult.[85] For an adoptee, the prospect of intimacy may stir up deeply distressing feelings regarding having been relinquished or abandoned by their biological mother and the fear that this will be repeated.

Although a clone would not share with the adoptee such facts regarding his origins and fears of relinquishment or abandonment, one circumstance regarding his existence might lead him to feel insecure about attaching to and being intimate with others: Both an adoptee and a clone might be a "replacement." Just as adoption is "an institution built on loss" at least in part, the same might be said of cloning.[86] A clone might be sought to salve the loss of a loved one or the childless state—if not actually to replace the lost or hoped-for loved one.[87] We visited the issue of loss in the chapter on assisted reproductive technologies and will delve deeply into it in the Replacement Child Model.

Paradoxically, an adopted child may come to view the adoptive mother as the abandoning mother and see adoption as synonymous with abandonment.[88] Similarly, a child clone might come to view cloning as a method for generating replacements and view himself as *replaceable*. Thus, just as adoptees may avoid intimacy out of fear of abandonment, a sense of being replaceable might lead clones to feel insecure in their relationships.

Another reason adoptees may avoid intimate relationships as adults relates to having felt controlled by parents who expected gratitude and could not or would not tell them what they needed to know about their origins. Such an adoptee may be determined to never feel emotionally indebted again or risk getting close enough to anyone who might want to control him.[89] I can find no research comparing parents of adopted and nonadopted children on this issue, but to the extent that a clone's parents might assume an attitude similar to that of the adoptive parents described, a clone would be at risk to feel like those adoptees. Those feelings might interfere with the kind of mutuality needed for successful personal and work relationships.

A person needs to have a secure sense of self to be able to become close to another "without being unmasked as a fraud." Adoptees may feel like impostors, as they earnestly try to pretend to be natural members of their adoptive families. A clone who adopts the identity he feels is expected of him, to be a "junior" version of his progenitor, would be prone to imposture. This would necessarily interfere with his ability to establish true intimacy as an adult.

As to whether adoptees as a group are less capable of intimacy, there has been no research to my knowledge on their rates of marriage and divorce compared with the rates of nonadopted individuals. With the question unasked, the absence of hard data is not a reassuring answer.

"ADOPTED" CLONES

In the United States adopted children are commonly viewed as marred by the "taint of bad blood," poverty, and illegitimacy.[90] Clones would have whatever "bad blood" a progenitor possesses, although eventually genetic engineering might permit the correction of genetic defects at the earliest embryonic stage so that a parent could become a progenitor of a "new and improved" self-clone. Until such a time, parents could "adopt" a clone to try to avoid passing on their own faults, genetic and otherwise. Some people might also wish to rear an allegedly superior baby, perhaps the clone of a famous or accomplished person.

It is unlikely that a clone would be adopted in the same way a child is now. As with the new practice of embryo adoption, a clone would probably be born into his "adoptive" family by his rearing mother. If a surrogate mother were employed, then the intended rearing mother might provide the egg to be enucleated for insertion of a donor nucleus.[91] However she might be viewed socially, an adopted clone would be without the taints associated with a conventional adoptee. Clones would not be the "superior babies" one adoption agency once advertised its children to be but, in fact, even better, made to order for the rearing parents, as clones of famous individuals or those with allegedly exceptional health, talents, and achievements.[92]

Knowing about and having experiences with one's parents helps an adolescent establish a firm sense of identity, even if he mainly uses that knowledge and experience to repudiate the parents.[93] This makes it important for adopted children to seek out their biological parents.[94]

When an adopted clone seeks out his progenitor, how might he be received? If the experience of sexually reproduced adopted children contacting their birth parents is any predictor, then unqualified receptivity might not be the norm.[95] Lance is a case in point. The progenitor of a clone might feel even less of a sense of connection and desire to meet his offspring than a sperm donor might feel about meeting one of his. Also analogous to sperm donors, some donors of cells for nuclear transfer for the purpose of producing a clone for adoption might not want to imagine their clone(s) as anything more than exuviae, to which they felt no attachment—perhaps one of several tissue samples contributed or sold for the purpose of cloning.

An adoptee may also experience a sense of let down after reunion with his or her birth mother. Finding this missing part of themselves may lead to a loss of fantasies about a mother who is a "vital part" of the self.[96] A clone might be similarly disquieted confronting a woman who was not his rearing mother but was his egg mother or surrogate mother.

Beyond this, finding one's progenitor might be an uncanny if not disturbing experience, as the clone discovers that she is not biologically unique and gets a preview of how she may age. We will discuss the implications of the age difference between clone and progenitor in this regard in the parent–child resemblance chapter.

An adoptive father, like any father, to some extent represents authority, power, and control to a son. However, because his father was not the impregnator, the adopted child may also see him as deficient in masculinity, "asexual" and "henpecked." On the other hand, in his fantasies, he may perceive his genetic father as "virile" and "macho."[97] Analogously, the adopted clone's progenitor would be *virile in his asexuality* and in fact might be the kind of man about whom the child would have chosen to fantasize—powerful, accomplished, and famous.

Just as some birth mothers who put their children up for adoption later search for them, the same may prove to be the case for adopted clones.[98] Gestational surrogates and possibly even some egg donors might also want to find their "child."

Adoptee, adoptive parents, and birth parents all confront many emotional and practical issues when the child meets his birth parents. Adoptive parents must face the fear of losing their child, and the child must grapple with his fear of being cast off by them.[99] The same issues would be relevant for an "adopted" clone and his family.

SOCIAL ASPECTS OF ADOPTION

This chapter follows soon after the one on assisted reproductive technologies for many reasons. Several have to do with similar social significance of the practices. The key link is that couples desiring parenthood are much more likely to adopt if they have already tried to have a child through one of those technologies.[100]

If cloning becomes one more assisted reproductive technology, those same couples might choose it instead of adoption. Others might prefer to be the adoptive parents of a clone of known genetics, from an accomplished, talented, or famous progenitor. This would be a real test of Richard Dawkins's "selfish gene" theory. The basic tenet of this theory is that the sole goal of genes is to propagate themselves into the future and as widely as possible and that all behavior of organisms containing those genes, in one way or another, serves that "selfish" goal. Dawkins views adoption as an "extreme mistake" from the perspective of that theory or as evidence against it.[101] David Barash attempts to explain the paradox that adoption presents for this theory when he speculates that evolution has not caught up with the fact that we now no longer live in small bands of highly related individuals. He points out that for our ancestors, adopting a child meant adopting a child who likely shared many of the adoptive parents' genes.[102]

Some people might choose cloning out of the wish to defeat the genetic uncertainty intrinsic to sexual reproduction. Others might choose to rear a clone of someone else out of the wish to avoid the near certainty of passing on their own genetic defects, overriding the Darwinian pressure to propagate one's own genes. For others, the narcissistic aggrandizement they might experience in rearing the clone of a famous person might win out over passing on their own genes, either in part (through sexual reproduction) or in full (through self-cloning). There will be more on this in the "Child of the Famous Model."

Social support is an important dynamic in adoption, and some adoptees feel socially marginal, a factor that adversely impacts self-esteem and identity.[103] Perhaps being teased and ridiculed because they are adopted is a major cause.[104] Analogously, we might expect that a lack of social support—or downright social opprobrium regarding reproductive cloning—would adversely affect both the clone and his or her rearing parents. Cloning is insular genetically, but in reaction to perceived societal disapproval, we might expect that the family of a clone would become even more so socially. Child clones might also run the risk of ridicule and, like many adoptees, feel that they have to hide a terrible secret about themselves. For this reason, perhaps like those of any marginalized minority, clones might want to affiliate with one another.[105]

There might also be social implications for the ability of the clones to form their own families. One adoptee told Betty Jean Lifton, "I never had the sense that I was born. I was adopted."[106] By analogy, a clone might say, "I never had the sense that I was sexually conceived. I was cloned." Cloned individuals might have a hard time becoming parents themselves, especially of a sexually reproduced child. Research on this question with children born through assisted reproductive technologies might help to answer it.

It is conceivable that a clonal origin, sanitized as it would be of sexual intercourse, might come to be a preferred story of origin over the sexually conceived one. Cloning shields the child from the uncomfortable linkage of sex and reproduction, but let us not forget that the primary shielding would be of the clone's parents, spared from the awkwardness of explaining sexual reproduction. For the same reason, some parents of sexually conceived children might prefer the scientific myth that the children are clones to the myth that a stork brought the child. We will look at this possibility further in the chapter "Intimacy, Sex, and Sexuality."

CONCLUSION

Though being a stepchild more closely parallels cloning than being an adoptee, adoption and cloning as genetic conceptual opposites have much in common, psychologically and socially. It is worth repeating that it is important to distinguish between the practices we use to model aspects of cloning and the practices themselves. Perhaps this is nowhere more important than with adoption. Although theory, research, and clinical experience indicate problems inherent to adoption, being cared for by an adoptive family is usually better for the child than the alternatives. In general, children who are put up for adoption are not deliberately conceived, and for their genetic parents the social consensus that it is best to avoid unwanted pregnancy is of little value after the fact. If the analysis by analogy indicates a heightened risk of psychological and social problems for clones, then it is easier to raise a flag of caution regarding their conception because any clone would have to be deliberately conceived. There could never be an "accidental" clonal pregnancy.

The Adoption Model suggests the possibility of problems for clones in having been specifically chosen. It also suggests that there might be a deficiency in the quality and strength of attachment to the clone by his rearing parents. In addition, there are many indicators of potential difficulties for clones establishing a true identity of their own. Bewildering family connections might further interfere with identity formation and, prior to that, might complicate how clones and their rearing parents resolve incestuous feelings and oedipal issues. The issue of disclosure of clonal status is a thorny one, but the myth of cloning as a sexually sanitized story of origin might come to have general social appeal. The experience with adoption and cloning as its opposite also suggests that clones might have difficulty in developing intimate relationships with others and in their ability to be successful in parenting sexually reproduced children. Combining actual aspects of adoption with adoption as a model of cloning, we might anticipate the risk of some additional difficulties for "adopted" clones.

What we know about the psychological treatment of adopted individuals and adoptive families might prove to be directly relevant and clinically useful when it comes to offering psychological treatment to clones and their families. Clinical conceptual thinking about adoption to frame clinical issues and treatment strategies might be quite relevant for the rearing parents of a child clone and for the child clone him- or herself.[107]

5

The Parent–Child
Resemblance Model

Because a clone would be identical genetically to his progenitor, we might expect him to greatly resemble the progenitor in appearance and even in temperament and other partially heritable behavioral traits and cognitive capacities.[1] The Identical Twin Model addresses the psychological and social consequences of genetic identity but fails to reflect generational differences and other important aspects of the parent–child relationship. This is a major deficiency because those missing elements are responsible for huge disparities in power and dependency inherent in that relationship. While not as reflective of cloning biologically as the Identical Twin and Assisted Reproductive Technologies and Arrangements Models, the Parent–Child Resemblance Model will better illuminate psychological and social issues relevant to cloning.

In this chapter we will explore the important role of resemblance in adoption and in kin identification as part of evolution and then venture beneath the surface of resemblance for a psychological examination of resemblance and identification. By exploring pathological narcissism, a condition in which there is a strong preference for resemblance to the self, we will discover what might unconsciously motivate some individuals to want to clone. By looking at the consequences of narcissistic parenting, we will be able to anticipate specific sets of problems for clones and their parents.

Chapter 5

JERRY, AGAIN

Recall Jerry, the identical twin whom we met in chapter 1. His case is also relevant to a number of other models of cloning, including this one. Recall that Jerry physically resembled his father, Sam, as did George, Jerry's identical twin. However, because Jerry acted more like Sam than George did, he gave a greater overall impression of resembling him.

Some difficulties Jerry encountered establishing his own identity were the result of so greatly resembling his father. As a young boy, he would relish it when people commented on how much he looked like his father, the mayor. But when Jerry was a teenager, he wished he was not so readily recognizable as "one of Sam's boys." Moving from his home state as a young man proved essential in his strivings to see himself and be seen by others as his own person.

Sam died when Jerry was in his thirties. More than one thousand people turned out for his funeral, as the city went into collective mourning. Jerry and his brothers and sister mourned too, but it seemed that their grief for the father they knew could not match the public's sorrow for the loss of their beloved mayor. Some people at the funeral gasped when they saw Jerry, who by then seemed to resemble his father even more.

Jerry took up smoking cigars not long after his father died. Most people could not think of Sam without envisioning him with a cigar in hand or clenched between his teeth. In psychotherapy, Jerry came to realize that he was holding onto the memory of his father by trying to look and act like him. Once again, Jerry liked hearing that he reminded people of his father. They were particularly likely to remark on their resemblance when Jerry was smoking a cigar. He became uncomfortably self-conscious of the fact that he was using cigars in an imitative way when an old friend of Sam's asked Jerry if he would light up a cigar so that he could take Jerry's picture smoking it.

Years after Sam's death, Jerry's youngest brother, Harry, received a community service award. Jerry was there to honor his brother with the rest of the family and many townspeople. Jerry was taken aback when several people seemed to pay more attention to him than to Harry. It was just that he looked so much like his father, the people told him. Jerry replied, "I haven't done anything. This is Harry's day, not mine."

☾

Parent–child resemblance is an important issue in Jerry's psychological makeup. Like most boys, when Jerry was young he was happy to be told that he was "just like" his father, but once he became an adolescent, like most teenagers, he became ambivalent about being told that he strongly resembled his "old man." In Erikson's schema of psychological development through the life cycle, adolescence is a time when people begin to try to define their own identities and being a son of one's father is no longer good enough. Moving far from home as a young man helped Jerry in this process: people could get to know him without comparing him to his locally well known father.

Smoking cigars was a way for Jerry to hold onto the memory of his father by becoming like him in a very concrete way. The recollections of others evoked by seeing him, the son who most resembled Sam, reinforced his own memory of his father. Smoking a cigar, Sam's trademark habit, brought back memories of him for both Jerry and others. When Sam's friend wanted to pose Jerry with a cigar for a picture, Jerry had the uncomfortable insight that he was not only identifying with his father but also imitating him. Years later, in psychotherapy he came to realize that he could hold onto the memory of his father in other, healthier ways than through the superficial resemblance that cigar smoking afforded him, and he was able to quit.

When asked about the psychological meaning of his own cigar smoking, Sigmund Freud replied, "Sometimes a cigar is only a cigar." For Jerry, that was not the case, and it probably was not for Freud either. We could conjecture about other possible psychological meanings of cigars for Jerry, though without the support of convincing clinical material. Perhaps a more intensive therapy than the one in which Jerry was engaged might have allowed such material to emerge.

Jerry's comment, "This is Harry's day, not mine," is interesting for what it does not say as well as what it does: Instead, he might have said, "This is Harry's day, not Sam's." Jerry had been approached not for anything he had done as an individual but because he resembled his father.

We already know that Jerry is a clone by embryo splitting, but the effects of his close resemblance to his locally famous father have been a more potent factors in his psychological life than his even closer resemblance to his identical twin. The difficulties Jerry has faced in developing his own identity might be like those likely to be faced by a clone created by somatic cell nuclear transfer.

FAMILY RESEMBLANCE IN ADOPTION

At least on a genetic basis, we would expect the least parent–child resemblance between a child and its adoptive parents. Behaviorally as well as physically, adopted children are more likely to resemble their biological relatives than their adoptive family members.[2]

As an important criterion used in placement, adoption agencies used to try to decide whether a child needing adoption would come to physically resemble a child whom the prospective parents might have sexually conceived. Perhaps this is based partially on millennia-old traditions that promote the fiction that adoption creates an actual blood tie when one never actually existed.[3] The more similar the child is to the adoptive parents, the easier it is to hold that fiction. Based on the child's resemblance to them, the adoptive parents sometimes pretend that their child is a genetic offspring.[4] Physical resemblance is still a factor in adoption, reflected in attempts to "match" the child and adoptive family where one criteria of "likeness" is race.[5]

In developing one's identity, a person compares himself with others, especially his parents and siblings, and asks how he is the same and how he is different. For the adopted child there is an added dimension of difference.[6] Transracial and transnational adoptions may further add to dissimilarity between parents and adopted child.

As we have seen in chapter 4, adoption and cloning, though conceptually opposite in the biological dimension, highlight the same psychological and social issues. The greater degree of difference between adoptee and adoptive family appears conceptually opposite to the degree of "sameness" between clone and progenitor, yet as extremes on the dimensions of resemblance they highlight the same psychological issues. In addition, a clone might have to contend with dimensions of difference with his nonprogenitor parent parallel to those between an adopted child and his or her parents.

RESEMBLANCE AND IDENTIFICATION OF KIN

The core of the theory of evolution is a process known as "natural selection," and parent–child resemblance is critical to the theory. Here is how: Parents possessing traits most likely to enhance their survival and reproductive capacity contribute more offspring than do other parents to the next generation. To the extent that offspring resemble their parents, more individuals in the offspring generation will possess those same traits that furthered their parents' adaptive advantage.[7]

Being able to identify kin has evolutionary importance.[8] That is because "copies of genes are much more likely to be carried by kin than non-kin." Being able to favor kin requires being able to recognize them. Therefore, favoring others who resemble one provides a great "selective advantage" to those like oneself.[9]

Richard Dawkins elaborates on this idea in his influential book, *The Selfish Gene*. He offers that *animals, humans included, tend to behave al-*

truistically toward others who physically resemble them. He believes that there may be a "kin-selected tendency to identify with individuals *resembling oneself.*"[10] From this perspective, one's genes ensure their survival into the future by biologically programming the organism they are within to be able to identify closely related kin and favor them. If physical resemblance is a likely marker of a high degree of genetic identity, then hardwired neurobiological mechanisms are likely to have evolved for individuals to favor others bearing them a strong physical resemblance.

That is the theory. Although many facts are consistent with it, much more research needs to be done in this area.[11] In the realm of parent–child resemblance studies, the data would seem to support the view that there may be biological or social psychological mechanisms designed to try to persuade the alleged father of a baby that the baby is his. Research exploring the issue of assuring domestic fathers of their paternity is contradictory on whether or not infants and babies objectively resemble their fathers more than their mothers or vice versa.[12] However, a mother and relatives on her side of the family tend to emphasize to the father that the child looks like him and, by doing so, make implicit that it is his.[13] While this principle may possibly be evolutionarily derived at a deep level, the explanation need only be psychosocial.

Whatever the mechanism, assurance of paternity might increase the chances that the father would invest emotionally and materially in the child. Beyond hoping that the father will behave beneficently toward the child, let us not forget the "dark side" of resemblance. When a child is singled out for abuse in a family, the child is most likely the one who resembles the father least.[14] This is best demonstrated for stepchildren, where a lack of resemblance to the stepfather would be expected. In families of convicted spouse abusers the men were apt to have a better relationship with their own children when they believed the children resembled them.[15]

All of this is quite relevant to cloning. Self-cloning would be the only way a man could produce offspring with nuclear DNA identical to his. From the "selfish gene" perspective, it might be preferred, but what could be in it for the nonprogenitor mother? Except for the mitochondrial DNA she might have contributed in an enucleated ovum, not much would genetically tie her to the child. But psychologically and socially, what better way could there be for her to try to guarantee her child, and hopefully herself, continued emotional and material support from her partner than to absolutely assure her partner of his paternity by having him cloned? But, of course, cloning could never prevent cuckoldry and guarantee fidelity.

THE ROLE OF RESEMBLANCE IN IDENTIFICATION
AS A PSYCHOLOGICAL PROCESS

Do biology and evolution make a psychological examination of identification unnecessary and superfluous? Do they so completely underpin psychology as to make the mind and interpersonal interaction mere epiphenomena? From the selfish gene perspective there are certainly evolutionary benefits of altruism directed toward others resembling us, but as an all-encompassing explanation of altruism and identification, biology and evolution do not by themselves appear to be up to the task.[16] We will see that a psychological examination of identification can further add to our understanding of the significance of parent–child resemblance.[17]

It is a normal part of a mother's experience with her new baby to perceive or imagine ways the child resembles her, the baby's father, or other family members. In making these identifications, she claims the child as hers and a member of her family.[18]

Physical resemblance as a factor in the quality of interpersonal relationships, and in identification, is a relatively new area of psychological research. However, over forty years ago, Dr. R. W. Parnell found that where physiques of father and son are similar, 60–70 percent of the sons assess themselves as getting along well with their fathers. In contrast, where the physiques of father and son are dissimilar, only 20–30 percent of sons rate themselves as getting on well with their fathers.[19]

Discussing this finding, the eminent psychoanalyst Joseph Sandler has said, "It is much easier to identify with someone who resembles oneself." He links this to the resolution of the Oedipus complex, where the child identifies with the parent of the same sex. He goes on to suppose that the great degree of mutual identification shown by identical twins is based on their mutual resemblance.[20]

Evolutionary psychologists exploring parental investment as a function of presumed genetic relatedness are doing more contemporary research on parent–child resemblance. In the previous section, we reviewed examples of such research on the attribution of paternal resemblance by mothers and their family members. A recent study looked at self-resemblance. It presented men with an array of children's faces morphed by computer to resemble their own to varying degrees. The men reacted more favorably to those children's faces that contained 25 percent or more of their own characteristics.[21]

It is but a short step beyond the earlier thoughts and findings of psychoanalysts and this more recent research to presume that a progenitor rearing his own clone would strongly identify with his "son." Given the natural propensity of young sons to identify with their fathers, it is also likely that it would be mutual, at least for a time.

The dimension of same gender is implicit in father–son resemblance and identification. It might also be key in the predominant preferences of parents (perhaps especially fathers) for male children in some patriarchal societies. We took a look at this in the Assisted Reproductive Technologies and Arrangements Model, when we focused on the practices of sexing sperm for artificial insemination and of determining sex through fetal ultrasound for the purpose of gender-specific abortion. Same gender is also a crucial aspect of the mother–daughter connection, about which Ruthellen Josselson says: "The problem of separating is the problem of not only becoming different but of becoming different and maintaining connection at the same time. To the little girl, *being liked means being like.* Attachment implies sameness. . . . With becoming like mother and therefore pleasing her comes the assurance of remaining attached forever."[22]

The heightened resemblance of a clone to the progenitor parent might increase that parent's attachment to it. This may not be an unqualified positive. Intensity is but one dimension of attachment. To understand a parent–child attachment psychologically, we must consider its dynamic complexity, as the following case illustrates.

Julie

Julie, a woman of thirty, was mildly depressed and struggling to find her own way. She loved her new husband but was still largely trapped in orbit around her beautiful and successful mother. Julie relied heavily on her mother, whose idea of providing emotional support seemed to be to tell Julie what to do, think, and feel—but especially how to look. Her mother had turned to Julie for emotional support since Julie was quite young, confiding more to her than she should have. To avoid her drug-abusing husband, Julie's father, her mother frequently slept in little Julie's room.

Everyone comments on how much Julie highly resembles her mother, but to Julie, her mother is beautiful and she is merely pretty. To make sure that her mother always comes out on top despite their resemblance, Julie underplays her looks, wearing baggy clothes and no makeup. Talented in the same area as her mother, she also holds herself back from fully pursuing professional success in it.

Parent–child resemblance appears to be a double-edged sword. While it helps guarantee a connection, it also raises the specter that the daughter could be as good as her mother, maybe even better. But the idea of competition brings with it the dread of separation or emotional abandonment.

The first person the child attaches to and loves is generally the mother. The daughter identifies with her same-gendered parent. A son must pull away from his mother and identify with his father so as to identify himself as male. The daughter does not need to separate for those reasons. The identification of a mother and daughter is mutual and probably is why a mother maintains a more enduring relationship and intense attachment to a daughter than to a son. Mothers and daughters tend to only partially separate from each other.[23]

Though delayed in coming to it, Julie is confronting the dilemma that adolescents face regarding dependence and conformity. Simply put, it can be expressed: "If I am not the same as someone else, what will I be like?"[24] As a corollary to this, a clone at adolescence might ask, "If I am the same as someone else, how can I be different? How can I be myself?" Compounding this dilemma is Julie's concern that her mother would not gracefully accept Julie doing as well as or even surpassing her in certain ways. The few times Julie allowed herself to surpass her, her mother responded with what seemed to her like jealousy disguised as emotional withdrawal.

For Julie, the possibility of truly being herself lies somewhere between resembling her mother and not resembling her too much. Julie has been caught between being dependently attached to her mother and striving to be herself. Straying too far from what she believes her mother feels is best for her causes Julie to fear her mother's disapproval, emotional disengagement, or abandonment. These fears lead to feelings of nearly unspeakable loneliness and almost intolerable dread.

Given that daughter and mother only partially separate normally, we might anticipate that a self-clone might separate even less from her "mother"/progenitor. In a family like Julie's, emotional problems for a child related to separating and individuating are likely. On top of this, the high degree of parent–child resemblance between Julie and her mother can only approximate what would be an even greater degree of parent–child resemblance of a progenitor and her clone. This might contribute to even less complete separation of the two than occurs between a mother and noncloned daughter. Julie is a case in point for one possible psychological fate of a self-clone. By the same token of heightened parent–child resemblance, we might also predict that a boy clone and his father/progenitor would separate less from each other, paralleling what is more typical of a mother–daughter relationship.

We might wonder whether the very strong resemblance of Julie to her mother has anything to do with her mother thinking that she knows what is best for Julie. Consistent with that sort of view is Dawkins's argument for cloning: "Young Einstein could be given an education to match his genes and an inside track to make the best use of his talents from the start." As a result, "he might turn out to be better than the paragon."[25]

Although special empathy of a progenitor for his self-clone has been claimed as possibly advantageous, Julie's case illustrates how special empathy can go awry.[26] Her mother apparently felt she had special empathy for Julie because of their strong mutual resemblance. A parent of a clone, even if not pathologically narcissistic, might blur the line between true empathy and narcissistic identification in responding to a self-clone.

Rachel

Rachel, a highly intelligent and attractive woman in her early forties, suffered from Body Dismorphic Disorder. She was convinced that she was ugly and believed that various parts of her body were not the right size or shape. She had plastic surgery done many times to try to accentuate or reduce various bodily or facial features and sometimes had them redone because she was dissatisfied.

Rachel described her father as warm, loving, and supportive and her mother as harshly critical and cold. Her mother seemed to delight in criticizing her appearance and undermined any hint of self-confidence. Oh, and one more thing about Rachel and her mother: They highly resembled each other physically.

In one session, Rachel criticized my choice of artwork, complaining that there was too much light brown in several prints on the walls of my office. I asked if there was something about that color that troubled her. Indeed, there was. It was the color of her mother's eyes! Knowing how her mother had found fault in the patient's appearance, I asked Rachel what she had thought of Rachel's blue eyes. Her mother had been disappointed, Rachel told me, but she was glad to be different from her mother in this way. Otherwise, she might have worn tinted contact lenses to break that resemblance.

Just as she had learned to hate the sight of her mother, resembling her physically she had come to hate her own appearance and could not look at herself without seeing her mother. What she hated the most were her mother's light brown eyes, the eyes her mother had used to dissect her body, soul, and self. But behind all the alterations in appearance, Rachel was still just Rachel.

RESEMBLANCE IN PARENT–CHILD SEXUAL ORIENTATION

We cannot examine the question of same gender and resemblance without venturing into the area of sexual orientation and homosexuality, in particular. It appears that children raised by gay and lesbian parents turn out pretty much no differently than children raised by heterosexual parents.[27] Hence, we might expect that a child clone reared by a homosexual couple might turn out much the same as one raised by a heterosexual couple. However, we should not find this assessment terribly reassuring.[28]

Familial factors, partly genetic and partly environmental, influence sexual orientation.[29] Fully 52 percent of male monozygotic or identical twins share the same sexual orientation, compared with 22 percent of dizygotic twins and 11 percent of adoptive brothers.[30] This illustrates that the more genes a person shares with another person, the more likely it is that the two will share the same sexual orientation. It does not mean that an identical twin is more likely than a non-twin to be homosexual. About 2 percent of males in the general population are homosexual.[31]

The about even chance that identical twins will share the same sexual orientation might lead us to conclude that the clone of a homosexual progenitor would have about a 50 percent chance of sharing his sexual orientation. However, this research demonstrates not only a strong genetic contribution to determining sexual orientation but also a presumably powerful social influence of having a homosexual brother.[32] A boy with an older brother who is homosexual is himself more likely to be homosexual.[33] Because a progenitor as a rearing father would also be in a sense an older identical brother, the incidence might turn out to be greater for a self-clone of a gay man.

A heterosexual progenitor/father might be disturbed to learn that his clone is homosexual. Many fathers of homosexual sons implicitly ask, "Did I somehow contribute to this?" This question often gets elaborated into ones like, "Am I not manly enough either in how I act or in the genes I passed on?" Some fathers may panic, fearing that they, too, might also be homosexual. Others may disavow what they fear about themselves by responding homophobically to their gay sons.

By analogy, a heterosexual progenitor of a homosexual clone might be even more likely to panic with the thought that because he and his clone are genetically identical, deep down he shares his clone's homosexual tendencies. Many fathers have found comfort in assigning responsibility to the genetic contribution of their wives for the homosexuality of a son, an option not open to male progenitors of homosexual clones.

SELF-RESEMBLANCE AND NARCISSISM

"Narcissistic" attachment is one way a parent may be connected to his child and is directly relevant to the issue of parent–child resemblance. Before we explore it we first need to examine the term *narcissism*.

In the Greek myth, Narcissus falls in love with his own reflection in the water and pines for it to the exclusion of all else until he dies. Narcissus not only could not love others but also, in loving his image, a false rather than true self, failed even to love himself.[34] Narcissism has been the subject of interest and study by many psychiatrists. Sigmund Freud wrote about it in his classic paper, "On Narcissism."[35] Otto Kernberg and Heinz Kohut have led in further developing the concept and devising treatment approaches for individuals with pathological narcissism.[36] It is a difficult area, but it is vital to a psychological examination of cloning and not just to the current model.

Narcissism has become a derogatory term in common parlance, but having high self-esteem, that is, a good opinion of oneself, would not be pathological if it is realistic and not a defensive attitude. This kind of narcissism can be contrasted with pathological narcissism, which, simply put, is defensive pride.[37] People can be justifiably proud of themselves without being boastful or dismissive and condescending to others.

The unifying principle of the concept of narcissism is the focus on the self. This focus may be found in various ways: First, it may describe a stage of development in which the boundaries between self and other are as yet indistinct. As babies, all people experience themselves, the world, and others in this way. Second, narcissism may also be a kind of choice of whom one wishes to associate with, that is, wanting to be with others resembling oneself. Third, narcissistic withdrawal describes the withdrawal of interest and attention to the world and others in it onto oneself. Fourth, a narcissistic relationship describes a kind of immature relationship that distorts real aspects of others and in which the boundaries between self and other are weakened. And fifth and finally, narcissism may be the linking of positive feelings with thoughts of oneself.[38]

The second form of narcissism, preferring to be with others resembling oneself, is obviously relevant to this chapter. The fourth type, that is, a kind of relationship, is important in understanding potential pitfalls in the rearing of child clones. The fourth type, when it is an expression of defensive pride in a rearing parent, may have a range of adverse consequences for the child.

We will focus primarily on this fourth form of narcissism as it is found in the pathologically narcissistic person. In the process, it will be easy to illustrate the other types of narcissism of interest to us. That is because people who are pathologically narcissistic also manifest it in its other forms. Specifically, they tend to relate to others narcissistically and to prefer self-resemblance in others.

Narcissistic Personality Disorder is the clinical diagnostic term for a pathologically narcissistic personality. To compassionately understand these people, we must not give pejorative meaning to the word *narcissism*, and above all we must try not to hold people suffering from it in contempt. This does not mean that we should condone how people with this disorder treat others, but disapproval alone does not help a person to change and by itself serves no useful purpose.

A person need not have Narcissistic Personality Disorder to possess a lesser degree of narcissistic personality traits. Also, people with other personality disorders may utilize narcissistic defenses. Although we will focus on the unalloyed narcissism of Narcissistic Personality Disorder, lesser degrees of narcissism are also relevant when it comes to parenting.

People with narcissistic personality traits up to and including Narcissistic Personality Disorder tend to make narcissistic choices of whom to be with and how to relate to them. This might be quite relevant to cloning choices made by parents and how those individuals relate to the child clone. Finally, we will try to anticipate the mental health consequences for the clone. Let us look, then, at Narcissistic Personality Disorder.

People with Narcissistic Personality Disorder have a sense of grandiose self-importance and are preoccupied with grandiose fantasies. They believe that they are special and unique, only want to associate with other special high-status people or institutions, and may show arrogant haughtiness in the way they relate to others. Requiring excessive admiration, they have an unreasonable sense of entitlement, lack empathy, and are interpersonally exploitative. In addition, they often envy others or believe that others envy them.[39] Self-righteous rage, exhilaration, and shame admixed with anxiety are states of mind characteristic of people with Narcissistic Personality Disorder.[40]

It is felt that a person begins to develop pathological narcissism sometime between the ages of three and five. Difficulties in maintaining constant consistent conceptions of caretakers and problems in negotiating Mahler's separation-individuation subphase (explained in chapter 1) are critical in the development of pathological narcissism. To cope with these difficulties, the child develops a highly unrealistic and idealized self-concept, what has been called a "grandiose self." "He dissociates, represses, or projects onto others all the negative aspects of himself and others" with whom he identifies. Cold, rejecting, yet at the same time admiring parents foster these developments.[41]

People with Narcissistic Personality Disorder might come to prefer cloning. For them it would be *the most natural mode of reproduction psychologically*, serving them in their quest for perfection and also providing a mode of reproduction without the demand for intimacy and reciprocity.

Self-consciousness is exaggerated in narcissism, and "the perception of the self includes the body self as a kind of double."[42] Thus, the body of a self-clone would represent a double of the progenitor's self, and any independent action of the clone a potential threat to the self-esteem of the progenitor. Her mother apparently viewed Rachel's body and self in this way, and Rachel's hatred of the physical traits she had in common with her mother illustrates a response we might find in clones of narcissistic progenitor/"parents."

CONSEQUENCES OF NARCISSISTIC PARENTING

The narcissistic parent relates to her child as if she were a part of herself and as such feels absolutely entitled to control and regulate all manner of the child's behavior. The lack of or inaccurate parental empathy for the child is the common denominator for a variety of narcissistic scenarios of parenthood. Such parents undermine their children's development as autonomous persons, and the children themselves are at risk for what has been called "pathologies of self," including Narcissistic Personality Disorder.[43]

It is normal for parents to project their own ideal of what they would like to be (their "ego ideal") on the child, and it is normal for the child to identify with this ideal. This and other narcissistic scenarios of parenthood coexist in the background with the normal parent loving and relating to the child as a unique separate person. When narcissistic pressures from the parents predominate, the child is bound to have difficulties, though perhaps not until much later, when he ceases to play his assigned role.[44]

Parents with Narcissistic Personality Disorder are not the only parents whose narcissism intrudes on the child, behaving toward their offspring in ways that risk fostering narcissistic pathology. Parents with less pronounced narcissistic personality traits do so as well, though in less prominent ways. Parents without pronounced narcissistic personality traits may also parent narcissistically when in the throws of depression or alcohol or other substance abuse, a time when they may be so focused on themselves and their needs as to not be attuned emotionally to their children.

At the most fundamental level, people with narcissistic disorders are preoccupied with themselves and self-centered. Because their main purpose in relating to others is to try to repair a sense of defect in themselves, they are unable to feel concern or care for others.[45] It seems almost inconceivable that they could successfully parent an emotionally healthy child.[46]

Pathologies of parenting that give rise to narcissistic disorders in children exhibit the common denominator of what has been called "depersonification"—not being treated as a separate person with a separate self. Some types of depersonification put a child at risk for developing narcissistic disturbance. These include being treated as if one were a spouse, a parent, a possession, or an endlessly dependent and infantile baby.[47] The child may also be viewed as an extension of the parent's self.

The failure of the mother to establish a truly empathic connection with the child leads to the development of what D. W. Winnicott calls a "false self." The false self is a false self-image developed as a complementary response to the mother's misperception and mishandling of the child. Eventually, what emerges as a pseudo identity for the child is the developmental result of accommodating the mother's own false identity.[48]

An empathic connection of mother to child requires what Winnicott calls "mirroring."[49] In mirroring, "the mother gazes at the baby in her arms, and the baby gazes at his mother's face and finds himself therein" but only if "the mother is really looking at the unique, small, helpless being and not projecting her own expectations, fears, and plans for the child." Deprived of an attentive, generally accurate, and empathic "mirror" in the person of a caretaker, the child tends to futilely seek one for the rest of his life.[50] But only their mothers could have mirrored them they way they needed it as infants and young children. As long as the individuals are unaware of their repressed need, they continue to try to gratify it through other means. Too often the most available and malleable person they eventually find for this role is their own child.[51] Their child cannot see the unconscious manipulation, and as a result a mother can bring up a child so that "it becomes what she wants it to be."[52]

When a mother is a mirror for her baby, she is allowing herself to be experienced as a "selfobject" for him. *Selfobject*, a term coined by Kohut, refers to the fact that the baby does not experience the mother as a separate person but, rather, as part of himself.[53] Being a selfobject for the baby requires the parent to put the needs of the child first. The unavoidable and developmentally normal narcissism of the infant reciprocally requires selfless, empathic attentiveness from the mother, the antithesis of parental narcissism.

When a mother unconsciously tries to meet her own needs through her child, she may love him "but not in the way that he needs to be loved. The reliability, continuity, and consistency that are so important for the child are missing from this exploitative relationship."[54] Furthermore, "the inability to be a needed selfobject to the developing child . . . involves a failure to match the child's needs with the caretaker's capacities. . . . The category of parental shortcomings range from the parent who is narcissistically preoccupied because of his or her own pathology to one who is unavailable to the child for whatever reason."[55] Typically such parents were themselves parented in this way. To the narcissistic mother, the child is "someone at her disposal" who can "be used as an echo" and controlled. He is "full of attention and admiration" for her, and she is at the center of his world.[56] This is what these parents should have been able to expect from their own mothers when they were babies, but it is too late. The baby cannot be his mother's mother.

Because narcissistic parents use their children to try to compensate for their sense of having themselves been inadequately parented, even a child's earliest signs of developing independence and a desire to separate threaten these parents. The implicit message from the narcissistic parent to child is: "You may go through the motions of separating from me and appear accomplished and successful, but only if everything you achieve is ultimately in relation to me." Thus, the two processes of separation and individuation are dissociated. Children subjected to these pressures come across as "pseudo mature." They tend to be obsessional, depressed, and emotionally aloof, all concealed beneath the false affect of a "false self."[57]

If what the mother is looking for is the kind of mother she wished she had, then the child will be treated as if she or he were that mother in what are called "transference enactments."[58] When the child is passive and compliant, she will be viewed in a positive way. When the child begins to exercise her own autonomy, she is no longer behaving like the perfect parent her mother unconsciously wishes her to be. Then the mother may experience the child as if she were the parent she had herself experienced: detached, unempathic, or selfish. She may treat the child with icy contempt or worse, having never been able to express such feelings toward her own parent.

Julie had experienced her mother in this way whenever she deviated from what her mother wished of her. It would seem that her mother did not consider Julie as a separate self but, rather, as a kind of junior version of herself. Why was her mother this way?

Julie's mother had been raised by a mother (Julie's grandmother) who was unprepared to be attentively and empathically attuned to her young children. That is probably because Julie's grandmother suffered the loss of both parents (Julie's great-grandparents) in an accident when she was only five years old. It was a remarkable achievement that Julie's grandmother, raised in the sterile, regimented, and emotionally impoverished environment of an orphanage, did as well as she did in raising her own family. But the emotional consequences of what she was not able to do have cascaded down several generations, with Julie its latest victim. The vehicle for the multigenerational transmission of narcissistic trauma is the tremendous intergenerational asymmetry in power and dependency between parent and child.

People with narcissistic personalities may be quite distrustful of others because they seem to only be able to perceive others as "duplicates" of themselves, that is, as entitled, selfish, and exploitative as they themselves are.[59] This tendency would probably be even greater for a narcissistic progenitor relating to a junior physical duplicate of self in the form of a clone. Apparently unintentionally foreshadowing this possibility, Dr. Paulina Kernberg has written that the narcissist appreciates other people either as fulfilling their needs or as "echoes or *clones* of themselves."[60]

NARCISSISTIC TRANSFERENCES
AND PROCESSES IN RELATIONSHIPS

The narcissist is prone to particular kinds of transferences in his relationships to others, including his children. We first encountered the concept of transference in chapter 4. Let's explore it further here. Transference is the perceiving, experiencing, and behaving toward another as if he or she is an important person from one's past. There is some degree of transference in every relationship. If it is powerful enough, then the person is hampered in seeing other people for who they really are and so cannot relate to them with accurate empathy and full rationality.

"Mirror transferences" ask others to reflect or respond to us. The young child who says, "Look at me!" for approval is an example. As he matures, he can feel proud in his accomplishments himself and employs this kind of transference less often. A narcissistic person is not likely to have achieved this and requires excessive praise from others to assure his self-esteem. He needs others to be good mother surrogates to him, all positively reflecting mirrors.[61]

The notion of "idealizing transference" is also very relevant here. When a narcissist knows people whom he idealizes as selfobjects or as part of himself, he sees them as all-powerful and all-knowing and wishes to merge with them.[62] As a psychotherapist, one must be aware that the dark side of this kind of transference emerges when the therapist inevitably falls short of expectations. The resulting de-idealization may be rageful, devaluing, and even paranoid.[63] This state is a therapeutic challenge for a trained therapist. Imagine what the child of such a person goes through when faced with it. Because the raison d'être for a narcissist to self-clone might well be to try to create an ideal self, we should expect him to have an idealizing transference toward his clone, frequently punctuated by the kind of states just described.

Psychoanalysts sometimes speak of a "twinning transference" or "twinship transference," not as a phenomenon confined to identical twins but, rather, as a form of transference in which another person is experienced as identical to oneself.[64] In its healthy form it may result in a feeling of belonging, but in its pathologic form the person perceives and relates to another person as if that person were literally a copy of himself and not someone with his own ideas, feelings, and attitudes.[65] Even if a progenitor/parent were normal psychologically, a clone might be at risk for becoming the object of a "twinning transference" from that parent. Such risk might be much greater were the progenitor/parent to be narcissistically disturbed.

Everyone's psychological growth requires mirror, idealizing, and twinship transferences, and it is normal for them to persist to a degree into adulthood. However, when these types of transferences are used persistently and urgently with others in a demand to make a person feel complete, they reach a pathologic level.[66]

To the extent that any of these transference propensities are present for a progenitor, rearing a self-clone would be likely to elicit them. It is developmentally normal for a child to become disillusioned with her parents and imagine that she might really be the child of other, more ideal parents. I examine this phenomenon as part of the "family romance fantasy" in the Adoption Model. One variation on it is not to invent other parents but, rather, another self, a phenomenon called a "twin fantasy."[67]

It is possible that for some people, cloning oneself, whatever other motivations there might be, could also be an unconscious attempt to realize a "twin fantasy" with the hope for an ideal self in the person of the clone. The delayed double would be desired as an extension of the self but also welcomed as a "not me" through which the progenitor could live vicariously as the clone does what the progenitor had forbidden himself.

Eventually, the clone might come to be dreaded as an alien self over which the progenitor has no control. The experience might be akin to the "doppelgänger" phenomena, like those described in literature where characters contend with their "doubles."[68] It also might approximate the experience of alternate personalities in Multiple Personality Disorder.

As we have seen in chapter 2, parents begin to develop expectations of their children before birth. When parents of a sexually reproduced child find him to be incongruent with their representations of him, if they refuse to alter their expectations, they are unable to mirror him adequately. Even in the absence of narcissistic pathology, parental expectations about a clone would likely be more specific and firmly held because they would map onto a clear mental "template" of the clone's progenitor. Although there can be no exact, materially existing original template for a sexually reproduced child, such would not be the case for a cloned one. If a clone were of the self or of another person in whom a rearing parent were invested narcissistically, then the rearing parent would probably keenly wish that the child match that template of expectations.

The rearing parents of a clone might be even more likely to perceive a mismatch between their expectations and the reality of their child clone. We should expect such parents to have problems mirroring the child. And as explained already, children who have not been adequately mirrored are likely to develop narcissistic or borderline psychopathology.

Problems in mirroring a child may arise from a phenomenon known as "projective identification." Commonly employed as a defense mechanism by people with Narcissistic Personality Disorder or narcissistic personality traits, it involves projecting onto others desires, wishes, intentions, and thoughts one finds unacceptable in oneself. The narcissist can then disavow them more readily, as he takes others to task for allegedly holding them. But it is not just a matter of how the person doing the projecting perceives the other person. Projective identification is not only an intrapsychic process; it is also an active interpersonal one: The recipient feels pressured "to experience himself and behave in a way congruent with the projective fantasy" of the person doing the projecting.[69]

As a narcissistic wish, the desire to self-clone would entail the wish to create a projective vessel. Sexually reproduced children have too often been in this position, but a progenitor parent might be even more likely to narcissistically identify with a self-clone, making projective identification more likely. As a vivid external representative of the progenitor, a self-clone might even more easily become the repository for the progenitor's disavowed wishes and desires. The child might become a ready scapegoat, *a developmentally delayed, projectively identified double of the self.* Because children respond to projective identification by reprojecting back onto the parent, a vicious cycle of escalating intensity might arise with each demonizing the other and inducing the other to act the part.[70]

The relationship of rearing parent to child is full of possibilities, for good and for ill. As a psychiatrist, I have become all too aware of the lasting harmful effects of abuse of parental power. At its worst, the confrontation of the powerful parent and the helpless and dependent child, a situation inherent to the parent–child relationship, goes beyond simple lack of empathy or more complicated depersonification to what has been called "soul murder."[71] This non–capital offense is defined as "the deliberate attempt to eradicate or compromise the separate identity of another person."[72]

Cloning, by itself, might not be soul murder, but it could facilitate it if parents were even less inclined than they might have been with a non-cloned child to permit a child clone to have his own separate identity. Like other victims of soul murder, a child clone might suffer the typical consequences—a life without a sense of individual identity and hence authenticity, true passion, and vitality.

The soul murder of horrendous child abuse can lead to a dissociative splintering of the self whereby alter selves whom the child creates experience and remember what the rest of the self does not. I am referring to dissociative disorders, the most severe form of which is called Multiple Personality or Dissociative Identity Disorder. We might think of the alter selves as "psychic clones." A woman I once treated whose father had repeatedly raped her actually had some dissociated parts of her mind she called "Dead Souls." What they had experienced and knew was so unbearable that it felt as if they had died. Shattered even more than other parts of her self by terrible pain and unspeakable knowledge, these parts of the mind were buried, as it were, in a psychic graveyard.

A number of specific features may be described for a family likely to foster the development of serious psychopathology in children. Several of them may be directly relevant to families that would choose to reproduce through cloning. These include "various degrees of disarticulation from the extrafamilial world" and "confusion or obfuscation of . . . age and intergenerational roles and functions."[73] A "symbiotic container for soul murder" occurs within this kind of "closed system."[74] Clones would likely be born into a system with a greater degree of closure of various types and degrees than could be found elsewhere. In addition, a narcissistic parent is typically exclusively and possessively close to his or her child, whereas the other parent is excluded or remote.[75] Narcissistic parents may so actively interfere with a child identifying with any other person that we could view their attempts to exclusively possess their children as a variant of "soul murder."[76] This sort of relationship can only compound the harm the child suffers as a "need-fulfilling extension" of that parent.[77]

The physical reality that a clone is genetically of only one parent might foster this sort of exclusive relationship and even where parents are not narcissistic, could at least shade parental relationships in that direction.[78] A parent who clones herself might well want the child to identify predominantly, if not exclusively, with her.

When a parent fails to respond empathically to the evolving self of the child within certain reasonable limits, the child experiences a narcissistic injury. A child develops self-esteem from his parents relating genuinely to him as a unique and valuable person. Because deep down the narcissistic person's self-esteem is poor, rather than relating authentically to others, he tries to present "idealized self-images."[79] Such parents often try to use their children as self-esteem-enhancing devices. Parents raising a self-clone or a clone of another with whom they are narcissistically invested might be even more likely to try to use the child to try to bolster their shaky self-esteem and cover their deep narcissistic wounds.

People with pathologic narcissism feel entitled to control, possess, and exploit others without guilt, including their own children.[80] A narcissist might feel an even greater sense of being entitled to control and exploit a child clone.[81]

A main characteristic of narcissistic personalities is intense envy of others, including their own children.[82] This may be seen when the parent perceives the child as being as successful as or more successful than the parent academically or socially and, later on, professionally. A narcissistic parent may be especially envious of a child's growing independence. Indeed, such a person in middle age envies his own past.[83] We would expect a narcissistic parent to respond with heightened envy to a child clone that he may have created to literally represent and embody his successful past.

Barry

Barry, a man in his late thirties, came to me for the treatment of a major depression. In our first meeting, he also begrudgingly admitted to abusing alcohol. He was the first born of a high-ranking executive. Key to understanding him psychologically was the fact that he had broken both legs in an accident when he was five years old. As if this was not difficult enough, he subsequently developed an infection in the broken bones. He was in a cast for many months and was pretty much bed ridden for most of a year. After this, the bone infection recurred several times, and he had to miss school for long periods of time. The growth of his legs somewhat lagged behind that of the rest of his body. As a child, his physical activity was restricted for several years, and later, his damaged legs did not allow him to keep up with other children.

Barry told me that his father had been terribly disappointed to have a sickly son. At first, I was not sure how much of the disappointment was really Barry's, but over subsequent visits it became clear that his father had said and done things that had given Barry good reason to feel that his father viewed him as "not good enough." For example, his father, in apparent disregard of Barry's painful and deformed legs, encouraged him to run, timing him with a stopwatch. His father was harshly critical of Barry's performance, not just in this example but in many others. He could not accept that his son had any imperfection, and if he were to admit that Barry had any, if they were limiting him in any way, it was only because Barry was not trying hard enough.

When Barry, the firstborn son, entered adolescence and young adulthood, his father was quick with advice and could never understand why Barry often did not choose to follow it. His father was sure he knew what was right for him and could not understand why any child of his would want to make an independent choice. He seemed threatened by the fact that Barry, more than his other children, had ideas and opinions of his own. Maybe Barry's independent-mindedness had something to do with the fact that he had so often tried and failed to please his father.

Even so, Barry still had not fully abandoned hope that he might someday win his father's approval, yet his father was displeased when Barry elected a premedical course of study in college, rather than following his father's choice, business. When Barry got into medical school, his father offered no financial help, which he could have easily provided.

Barry abstained from alcohol, and his major depressive episode fully remitted with an antidepressant medication; but he still felt "empty" and unsatisfied with his professional accomplishments and family life. We began to see that this was directly related to an "inner voice" he heard when alone or when doing routine well-practice tasks that no longer required his full attention. It was his father's voice, criticizing him, his choices, and the most important people in his life, his wife and daughter. Neither they nor Barry measured up. Whenever Barry felt any hint of justifiable pride, the "voice" minimized his achievements or derogated them entirely. Barry had been able to make the "voice" go away only by full attentive engagement in a task, such as in his work as a physician, or by the excessive ingestion of alcohol.

This sort of voice in a well-functioning person without signs of serious depression or schizophrenia is typical of a dissociative disorder. There was a part of Barry's mind that was autonomous of his conscious will. It felt like an alien, though familiar, presence.[84] He had heard it since childhood and felt it as the "presence" of his father. Though Barry had not spoken to his father in many years, inside his mind a psychic clone of his father continued to relate to him in the same way he had earlier experienced his father in reality.[85]

In psychotherapy, I tried to reassure Barry's psychic clone of his father that it was not our intent to get rid of it but, rather, to enlist it in treatment, recognizing that it, too, must have to be suffering. First, I tried to make clear that it was not actually his father but part of Barry's mind and that it was there for a purpose. As this "alter" or psychic clone put it, that purpose was to prevent Barry from "fucking up," exactly the attitude Barry's father would have had about him. Eventually, however, this part realized that it had formed to help young Barry anticipate his father's behavior so that Barry could try to avoid being abused. Paradoxically, that frightened boy had tried to protect himself by incorporating the abuser into his mind, represented as this psychic clone.

We began a long process of helping this part reconsider its attitudes by developing empathy for Barry and at the same time helping Barry develop empathy for it. As therapy progressed, more "psychic clones" of the father emerged. Recalcitrant about engaging in the therapy process, they needed to be made part of it. Barry had a right to a life and a self that was his own and not his father's. Fostering an internally empathetic connection helped us to begin to integrate Barry's self to make that possible.

Barry's case exemplifies how, as children enter adolescence and begin their quest for independence, a narcissistic parent may experience a sense of loss, the child having served as an important source of self-esteem. The adolescents may come to experience their striving for independence as a threat to the parents' self-esteem and may fear parental withdrawal or abandonment.[86]

Though it is impossible for a psychiatrist to diagnose relatives of patients based only on how they describe them in psychotherapy, it is possible to arrive at some working hypotheses. It sounded to me as though Barry's father was pathologically narcissistic but that he came by it honestly, having been subjected to narcissistic parenting by his father before him. As reviewed earlier, pathological narcissism begets pathological narcissism. Such parents are apt to idolize their children, consider them their possessions, and discourage strong peer relations.[87]

Barry's father was thwarted in being able to idolize his firstborn son because of Barry's injuries and illness. It was impossible for Barry to provide his father with the outstanding physical performance he craved from his son. His father refused to recognize this fact and also discouraged Barry's friendships, viewing his friends as "losers" or bad influences who interfered with his own attempt to mold Barry as he desired—for Barry's own good, of course.

Barry frustrated his father's attempts to use him to bolster his own shaky self-esteem. His injuries and illness played a role in saving him from being able to successfully gratify his father's narcissistic needs. Without those physical limitations, he might have been more successful in this regard, and although that would have spared him both considerable physical pain and emotional distress, he might have ended up narcissistic, like his father. When a child comes to feel that he is important for the parent's well-being and self-esteem, feeling important in that sense can foster pathologic narcissism in the child.[88] Some progenitor parents of clones might view their clone as a kind of ego ideal or ideal self, probably engendering narcissism in the clone.[89]

Having repeatedly failed to gratify his father's narcissism, Barry did not become a narcissist himself, but he did develop a dissociative disorder. His father came to exist as a kind of psychic clone in his mind.

For the prospective progenitor, the prospect of self-cloning offers the fantasy of fulfilling the narcissistic desire not only to create an ideal self but also to live not just one life. Self-cloning might be the hoped-for fulfillment of a narcissistic desire to create a kind of "best of all possible selves." In this case, the psychic clone of his father was attempting to do this very thing with Barry.

NARCISSISTIC PARENTING AND PARENT–CHILD RESEMBLANCE

By integrating resemblance and identification as psychological processes with what we know about parental narcissism and its influence on children, some clear mental health implications emerge for the relationship between a rearing narcissistic "parent" and clone and for the clone himself. Although research is lacking on the subject, my clinical experience suggests that narcissistic parents, more than others, tend to notice how much their children resemble them. The parent may engage in a self-referential kind of depersonification of the child, whereby the narcissistic parent has categorized the child as either "like me" and good or "not like me" and bad. But falling into the first category is not necessarily an enviable position, as it commonly sets the child up to be "not good enough." Such parents are quick to claim a child as their own when the child is good, obedient, or successful and just as quick to disown the child as not theirs or like them when he or she presents a problem. In such cases, a parent may view the child as being like a despised parent of his or her own or a hated ex-spouse. Often, however, the parent is unconsciously reacting against despised aspects of herself that resonate with the child's behavior.

Researchers have explained the appalling data on abuse of stepchildren as a consequence of the genetic unrelatedness of stepparent to stepchild. Perhaps pathological narcissism in a stepparent is the psychological variable mediating the sometimes murderous effects of genetic unrelatedness.

With the genetic resemblance of a self-clone to progenitor as absolute as one can get, it might seem hard for a person parenting her own clone to disown the child as some stepparents do their stepchildren. However, that would not stop a progenitor/parent from blaming others involved in the clone's upbringing for behavior or accomplishments deemed less than ideal.

Preoccupation with one's image, in the context of interpersonal insecurity, is the hallmark of narcissism.[90] When the progenitor beholds his clone, the image of the clone may serve as a mirror by which he hopes to counteract deep feelings of personal insignificance. We would expect a progenitor parent to want a self-clone, as a kind of self-reflection or double, to behave in a way the progenitor feels reflects positively on him.

Because the primary concern of narcissistic parents is with self-reflection, when their children manifest troubled behavior, the parents do not attempt to understand what the behavior means for the child but, rather, experience it as a reflection on themselves.[91] When that child happens to be a self-clone of a rearing parent, we might suppose that the parent would be especially likely to feel this way.

But couldn't parenting a replica of a younger self turn out to be easier and more satisfying for a parent and more beneficial to the child? The philosopher Michael Tooley argues: "Knowledge . . . of the way in which [a parent] was raised can be used to bring up the child in a way that fits better with the individual psychology of the child." He assumes that there would be "greater psychological similarity" between self-clone and progenitor parent and that this would help the parent "be better able . . . to appreciate the child's point of view"—and so better understand him. As a consequence, Tooley believes that the clone would have a happier childhood.[92]

This chapter contradicts that line of reasoning. While the psychological similarity Tooley presumes for progenitor and clone would presumably arise largely from their genetic identity, psychological similarity does not necessarily enhance empathy. When their narcissism is engaged, highly similar people are least inclined to be tolerant and understanding of one another. In addition, parent–child conflict arises not only from the fact that half of the genes between them are different but also from the fact that *parent and child are different selves with different wishes and needs.*

Whatever conscious "good" reasons they might give for cloning, individuals choosing to self-clone as their path to parenthood might also be enacting narcissistic transferences by doing so. If so, the clone would become the living embodiment of those transferences and the object for behavior bound to undermine his chances for a normal upbringing and happy life.

While one category of resemblance is of the child to a parent, another is of the child to a person who is not a parent. We all probably know a child who suffered the consequences of being perceived as resembling someone other than the parent in question. We might anticipate a tragic outcome if the resemblance were to a relative with whom that parent had a poor or clearly conflictual relationship. But why would anyone want to clone someone other than himself if he did not love and admire that other person? Perhaps no one would do so, but a rearing "parent's" strong positive feelings about a clone's progenitor would not obviate the mental health risks to the clone. Indeed, resembling one of these well-loved others would carry its own significant risks for the child, and even the most positive and loving relationship has some degree of negativity and ambivalence.

CONCLUSION

We would expect a greater degree of resemblance between a self-clone and progenitor than is possible between any parent and child. In this chapter, we have revisited the limitations of the Identical Twin Model to assess the effects of resemblance. The Parent–Child Resemblance Model captures the most salient and powerful features of the parent–child relationship—the intrinsic asymmetries in power and dependence that go with generational difference and the roles of rearing versus being the one reared.

We have seen that the ability to identify kin is a cornerstone of evolutionary theory. Presumably, it is underlain by neurobiological mechanisms favoring resemblance to self, but a psychological analysis is essential to understanding the role of resemblance in identification.

Parent–child resemblance appears to enhance identification as a psychological process. Gender is the variable most studied, and sexual orientation is another important dimension. While biology plays a strong role in sexual orientation, based on the findings with identical twins, it appears to account for only half of what determines it. Parental attachment to a child is essential to the child's welfare, but not all attachment is beneficial. We have examined the form of attachment based on resemblance—narcissistic attachment. When present to a pathological extent, it is a means to bolster shaky self-esteem in the parent.

Narcissistic parenting is incompatible with being truly empathic to a child and produces real psychological harm. At the least, the child may have trouble separating from the parents and developing a separate autonomous self and identity. Too often, a multigenerational pattern ensues.

Even if parents of a clone were not pathologically narcissistic, they might be more inclined to engage in narcissistic scenarios of parenting with a clone because of heightened expectations that the clone resemble its genetically identical progenitor. This might lead these parents to be more narcissistically attached to the child clone than to any sexually reproduced child they might have had. Consequently, a clone might be more likely than a sexually reproduced child to be subjected to narcissistic parenting and to have fewer means to escape it.

For the sexually reproduced child, perceived dissimilarity appears to diminish parent–child identification, whereas resemblance enhances it.[93] When the parent is not narcissistic, we may reason that dissimilarity to the child might be a modest disadvantage and resemblance, a modest advantage. But a child too dissimilar to a narcissistic parent may find his dissimilarity quite detrimental. However, resemblance may carry its own problems, enhancing parental expectations that the child be like the parent.[94] Such is apt to be the plight of a self-clone, the ultimate exemplification of parent–child resemblance.

The "selfish gene" would collide headlong with the self in self-cloning. There is a good chance that the progenitor/parent of a self-clone would not be capable psychologically to rear the child in a way that acknowledges that the clone has a separate autonomous self. This would more than negate the enhanced altruism that the selfish gene theory would predict from a parent who knows that his offspring has identical genes.

That said, we must acknowledge that it would not be impossible to parent a self-clone as well as a child reproduced sexually. However, from what we have seen in this chapter, it would appear that a parent would need to clear very high psychological hurdles in rearing a clone, and as a result, in many instances, the clone would run substantial risk of psychological harm.

In the next three chapters we will look at other dimensions of resemblance. In the Child of the Famous Model we will see the psychological and social consequences of similarity and difference to a famous parent. In the Replacement Child Model we will examine resemblance to someone other than oneself, in addition to dealing with the issue of loss and grief. Then, in the Namesake Model, we will look at an assigned symbolic dimension of resemblance.

6

The Child
of the Famous Model

JERRY, YET AGAIN

Recall from the previous chapter that Jerry's father was locally famous and that Jerry's close resemblance to him has been more important psychologically for Jerry than being an identical twin. For him, strongly resembling his father and being the son of this famous person have been psychologically and socially synergistic factors and the most defining facts of his life.

INTRODUCTION

In this chapter, we see how being a child of a famous person might model important psychological and social aspects of cloning. Fame is primarily a social phenomenon but can powerfully affect the famous person psychologically, as well as those close to him, including his children. Societal expectations that a clone would resemble his progenitor might parallel those that a child of a famous person would resemble his famous parent.

Society yearns for some of its members to make outstanding contributions, and this hope produces special interest in children of famous people. In this chapter, we will see how the psychology and sociology of fame and the children of the famous may not only inform how we think about clones of famous people but also contribute more generally to our consideration of clones and cloning.

Society has a spectrum of expectations of its children. It has relatively few generic expectations, and these mainly concern certain minimum standards of appropriate behavior. The parents, relatives, and acquaintances of any child may have certain specific expectations of that child, some based on perceived parent–child resemblance. However, when a child's parent is famous, societal expectations tend to be greater and more specific. Children of famous individuals may experience intense social scrutiny and expectations based on parent–child resemblance.

What is fame? A person probably has achieved a degree of fame, or at least notoriety, if many people know him whom he does not know. Beyond this, one may describe types and levels of fame, as well as a variety of trajectories it may take over time.[1]

If they should choose to disclose it, then the first successful instances of human cloning would make the child, progenitor, rearing parents, and those involved in her creation famous. Besides general public curiosity, we might anticipate that there would be careful scientific scrutiny to see how much the cloned child resembles her progenitor in appearance, physiology, personality, and aptitudes.[2]

The Child of the Famous Model is also socially relevant to cloning because part of society's interest in cloning relates to the possibility of cloning famous people or ones of special ability. The hope is that such a clone would have the same positive traits as his famous progenitor and the potential to achieve as much or more of social or scientific value, presumably in the same or a similar area. Some philosophers speculate that parental expectations of a clone of a brilliant scientist or a leading statesman would put the clone at risk of great psychological harm.[3] In this model, we will address that concern with theory and data relevant by analogy.

BIOLOGY AND FAME

From an evolutionary framework, a famous person may be preferentially sought after as a mate for reproduction of offspring. Additionally, nonmonogamous pairings may enhance a famous person's genetic contribution to the next generation. In some instances, their sexual exploits have been prodigious, if not actually productive of progeny. Through cloning, famous people might get to disproportionately contribute genetically to the next generation without sex or infidelity.

A child of a famous person would have no greater chance of inheriting genes for their parent's special abilities than any other child would have of inheriting genes for any particular ability their parents might possess. Socially, however, there are hopes and expectations that the child of a famous person would be just like him. With exactly that intent, cloning would literally take a nuclear "chip off the old block."

While society often hopes that the child of a famous person will strongly resemble the famous parent in abilities, aptitudes, and achievement, it may not appreciate that the child might also inherit a greater risk of mental illness. Society learns about the important achievements of famous people but seldom hears about their private misery, except to misinterpret or take glee or morbid fascination in such signs as the tragic failings of a hero or a deserved comeuppance for moral failings. However, research indicates that artists and writers have a higher prevalence than the general population for mood disorders, especially Bipolar Affective Disorder.[4] There might also be a higher incidence among famous people of milder forms of bipolarity that may come less often and less dramatically to psychiatric attention. The kind of drive, enthusiasm, energy, and willingness to take risks that such people possess could be an asset in the achievement of fame. A clone of a famous person would share all his progenitor's genes, possibly including those for mental illness.

THE DESIRE FOR FAME AND ITS RELEVANCE TO CLONING

The desire for fame might be one motive among others for individuals wishing to clone themselves or wanting to facilitate cloning endeavors. It might also predominate among those individuals wishing to clone a famous person or be involved in parenting such a clone. To understand the wish to clone a famous person, we need to examine not only the desire for a famous offspring but also the desire to achieve fame oneself, as well as the wish to have famous parents. Let us assume that most prospective parents of clones of famous people would be adoptive.

THE DESIRE TO ACHIEVE FAME ONESELF

People who feel estranged within their culture may have a particular desire for greatness.[5] David Giles, an authority on the psychology of fame, is unconvinced that the selfish gene theory, reviewed previously, adequately explains individual desires and efforts to attain fame.[6] He argues that fame is another way of preserving one's identity for future generations. It may be "a way of defying death" and symbolically realizing "the basic human desire for immortality."[7] In addition, perhaps narcissism, and not necessarily pathologic narcissism, is an important motivation. In other words, fame may be sought to enhance one's self-esteem in the case of an emotionally healthy person, whereas others may seek it to compensate for deep feelings of low self-esteem. If a person feels unappreciated, alienated, or insufficiently loved, then he may strive to be appreciated, included, and loved on a grander scale. For such a person "fame is not a successful defense against feelings of inadequacy. It only appears to be."[8]

Of course, many people become famous incidental to doing what they want to do and not because they were motivated to do it by a wish for fame. Even so, when fame greets such people, the way they handle it is largely a function of the above issues.

THE WISH TO HAVE FAMOUS PARENTS

As mentioned in the chapter on adoption, the "family romance fantasy" is universal. All children question whether the people who claim to be their parents truly are. A child prefers to think that he is really an offspring of people who are somehow "better" than the adults who claim to be his parents. In the case of an adopted child, reality may correspond to an extent with fantasy, and family romance themes can persist into adult life. Erik Erikson's daughter says that her father, who never knew his biologic father, "found much comfort in the thought that his father might have been of noble birth: thus an abandoning parent was transformed into a source of pride."[9] In people with a more tenuous hold on reality, the parents might be presumed to be specific people. One adolescent boy had the fantasy that he might be the son of Marlon Brando because many of his classmates thought he looked like the actor.[10]

The fantasy of being the clone of a famous person might also be a reality in a world with clones. That possibility would be relevant not only for clones but, given the universality of the family romance fantasy, for any and all children. If a child were, in fact, the adopted clone of a famous person, then he might seek out his progenitor and any other clones of the same to claim his rightful place in the "royal" clan. A severely emotionally disturbed individual might falsely believe himself to be a clone of a famous person, with unpredictable results for the alleged progenitor. The sort of individual who would stalk a famous person might also be inclined to falsely believe he is a clone of one.[11]

One may liken the adoptee's quest for his origins to those of the hero as described by Joseph Campbell.[12] But knowing his origins and the unique identity of his progenitor too well, an adopted clone of a famous person would be deprived of a true quest, critical to the formation of his own identity. The same might be said for any clone who knows he is a clone. Those raised not knowing they are clones might still be affected by the reality of their clonal origin if it influences how their parents interact with them. This could occur despite the parents' best efforts to the contrary, as the literature on disclosure and secrecy in adoption illustrates.

THE DESIRE TO HAVE FAMOUS OFFSPRING

The desire to have famous offspring combines elements of wanting to be famous oneself and wanting to have famous parents. Elements that might be related to wanting fame for oneself could be (1) trying to attain some measure of immortality as the parent of a famous person and (2) trying to narcissistically enhance one's self-esteem though a sense of ownership of one's famous child. These two elements are not completely separable.

Those who have failed to achieve fame for themselves may try to achieve it by proxy, parenting children whom they hope will become famous.[13] A parent raising a celebrity clone might feel herself to be a celebrity. Parenting a clone of a famous person might ultimately be a disappointment for the parents and a disaster for the clone.

In addition, an adult's desire to parent a famous offspring might be a disguised reprise of a childhood wish to have famous parents. The family romance fantasy is universal because every child becomes disappointed in his or her parents, both in those early years in which that fantasy occurs and again during adolescence. However, the child destined to become pathologically narcissistic may be profoundly disappointed for very good reasons.

As explained in the last chapter, the narcissistic parent relates to her child as if that child were a parent, hoping to get from the child what she was unable to get from her own parents. A narcissistic parent might want to adopt the clone of a famous progenitor she unconsciously wishes had been her own parent. However, the clone would not be this idealized parent of fantasy but, instead, a child who not only would be unable to meet the parent's needs but would be in need of care herself. As with all narcissistic parents, she would be emotionally unequipped to empathically parent her child.

Narcissistic parents often feel frustrated and angry with their children for failing to fulfill their assigned role of helping the parents feel better about themselves. This tendency might be exaggerated for people who would choose to adopt a clone of a famous person. That is because for them this child should have been the fulfillment of the ideal fantasy—a marriage of the family romance fantasy and the romantic fantasy of fame.

THE PSYCHOLOGICAL IMPACT OF
A PARENT'S FAME ON A CHILD

How do famous parents parent their children? What some children of famous parents have written about their experiences suggests that many famous people narcissistically parent their children with the kind of unfortunate results described in the previous chapter.

Narcissistic Parenting

Anecdotes are not strong evidence on which to base generalizations. Are famous people really more likely than others to narcissistically parent their children? They might be if pathological narcissism was more common among the famous than among the rest of the population, but there has been no psychological research to confirm or deny this.[14]

A person may have been driven by his narcissism to achieve fame. Alternatively, the heady experience of fame might have encouraged a latent potential for narcissism. Regardless of whether our presumptions about fame, narcissism, and narcissistic parenting are true, we might expect that people wishing to adopt a clone of a famous person would be narcissistically invested in the child. As a result, they would presumably parent that child narcissistically.

While some celebrities neglect or abandon their children, others are overly involved. Parents of a clone might be more likely to be overly attentive. In celebrity families, such attention may have nothing to do with the child's own needs but, rather, with those of the child's parent who views the child as a narcissistic self-extension.[15]

"Absent but Present" and Missing Parents

Children often experience their famous parents as having been "absent but present."[16] Jerry described this quality in his father's relationship to him and his siblings. If a couple were to select a progenitor for cloning based on his fame or presumed genetic superiority, then the progenitor would also be absent for the child clone.

The rearing parents of a sexually reproduced adopted child may indistinctly sense the "presence" of the child's genetic parents in the child, and the child may as well as he compares himself with his adoptive parents and nonadopted siblings. The adoptive parents of a clone of a famous progenitor would also sense the progenitor's "presence" in the child but with vivid images of that progenitor in mind. For a child clone, with possible exceptions, the famous progenitor would not be a real, interactive part of his upbringing. However, he would be present in the life of his clone in the way famous people are present, presented, and represented in and to society in the media and books. In addition, the famous progenitor would be ineffably and uncannily present, sought after or detected by the parents as inhabiting or possessing the child. At the very least, he would hover imperceptibly over the child's view of himself.

Children of famous parents frequently feel parentless.[17] This might be partially true for clones. Unless we think of the progenitor's parents as also the clone's biological parents, a clone would in fact be either motherless or fatherless, an issue looked at in the Adoption Model. Just as an adopted child feels the loss of his original parents, a clone might feel the absence of what never was, a mother or a father, as the case may be.

An adopted clone of a famous person or carefully selected progenitor might feel about both his parents and himself much as an adopted child may feel about his biological parents and himself. As a genetically enhanced clone would not be a perfect genetic copy of a progenitor, he might also have feelings and fantasies about the genetic engineers, reproductive scientists, and clinic responsible for his creation and abilities, perhaps seeing them as parents of a sort.[18]

Psychological Responses

Developing and expressing a self that is "unique, or distinct from others" is a basic component of the sense of personal identity.[19] This can be a particularly difficult struggle for a child of a famous person. Psychoanalytic treatment of children of famous parents reveals a variety of ways in which a child of a famous person may attempt to cope with feelings about being a "son of" or "daughter of" such a person. Sometimes they merge their own identities with those of their parents and so have difficulty thinking of themselves as separate human beings. Another child may try to share the spotlight or attempt to follow in the parent's footsteps. Often he ambivalently identifies with the parent, idolizing them, on the one hand, and envying and hating them, on the other. Alternatively, a child may compete with his famous parent or try to succeed in a different area or endeavor.[20]

Erik Erikson's daughter writes that while "it has sometimes been a source of great pride" to be his daughter, "more often it has overwhelmed my sense of myself. . . . My father's fame is always there to be reckoned with, a powerful force in my life."[21] She "felt deflated by my father's fame—not enhanced" by it.[22] Sigmund Freud's son, Martin, writes similarly that "the son of a genius remains the son of a genius, and his chances of winning approval of anything he may do hardly exist if he attempts to make any claim to fame detached from that of his father."[23]

If a child of a famous person fails to be either like his famous parent or successfully different from him, he may fall into alcohol or other substance abuse, severe mental illness, and even suicide.[24] Such a struggle for personal identity might be great for all clones and further magnified if the progenitor were famous.

The parents of any clone might think of his progenitor as a prototype or "ideal form" for the child they are rearing.[25] The narcissistic parents of a clone of a famous person might take this one step further. They might view the child's famous progenitor not only as a prototype but also as an ideal form, rather than only the first expression of a particular genetic and human potential.

Paralleling the child of a famous person's psychological alternatives, a clone might either attempt to emulate or even outdo the progenitor or reject the progenitor completely and engage in antisocial behavior. A clone might construct a "false self" in order to conform himself to the outward identity demanded of him as a clone of the particular progenitor.[26]

Children of celebrities acquire celebrity status through no effort of their own, "by birthright" taking on "the accoutrements of the celebrity in the absence of any personal achievement." They may be reluctant to strive to achieve, lest they risk not reaching the same level of success as their famous parent.[27]

An "all or nothing quality" in the personality of children of celebrities may "pervade their functioning" not just in the area in which their parents are famous. That quality develops as they adopt a "why try" attitude if they are unable to achieve the status of their celebrity parent.[28] Whereas some people are "wrecked" by their own success, clones could be "wrecked" by the progenitor's fame.[29]

Some children of celebrities may imitate rather than truly identify with their famous parent, magically thinking that they can be the parent without having to become like him through efforts of their own. If this magical identification persists, it interferes with their psychological maturation. Such children find being forced by reality to recognize their limitations jarring to their grandiose self-image.[30] A clone of a famous person would run the same risks.

"Exceptions"

People endowed with exceptional beauty, talent, or a physical deformity may not develop a realistic conscience or superego. They may feel that they are "exceptions" and, as such, entitled to rebel against the rules that apply to everyone else.[31] A clone of a famous person or one selected from a progenitor of exceptional beauty or ability might feel the same way. However, as an exception to the rules of nature, any clone might feel like an exception. Even if a person did not know he was in fact a clone, he might still feel like an exception if his parents treated him as if he were one.

Loss of Fame

The loss of fame, once achieved, can be a "ghostly prospect." Some individuals may "self-handicap" to excuse themselves, whereas others may commit suicide.[32] A clone famous for nothing but being a clone of a famous person might be haunted by the specter of his alleged potential—to be as famous and accomplished as his progenitor. Having started out famous but not for anything he did, he might feel any recognition for anything he himself might achieve to be false and undeserved.

Siblings

The way siblings in families of celebrities relate to one another might have implications for the sibling relationships of clones.[33] Some children of famous parents find that siblings are important nurturing companions, whereas others develop a stronger rivalry than normal. We might suppose a clone would encounter more sibling rivalry and envy than nurturing support from non-clone siblings.[34] If there were more than one clone of the same progenitor in a family, we might expect such dynamics to be even more intense.

A clone would not need to be genetically enhanced or a clone of a famous person to evoke this response. Any couple choosing to add a clone to children they may already have would be sending the sexually reproduced children an implicit message: "You weren't exactly what we wanted . . . but he is." If the clone were of a parent, then his siblings might be inclined to displace onto him the negative charge of their ambivalent feelings toward that parent.

THE SOCIAL CONSEQUENCES OF FAME
FOR FAMOUS PEOPLE AND THEIR CHILDREN

Very Public Expectations May Encourage the Development of a "False Self"

The child of a celebrity must cope not only with what he expects of himself compared with his famous parent but also with what an adoring or envying public expects of him. Treated by the public as the "son of" the famous person, rather than who he is, he may feel like an impostor and construct a "false self" to deal with the world.[35] The painful split in identity between a public and private self may cause him to feel that others do not know him. It may lead him to suspect their motives and discount their flattery. He may feel deeply insecure and inadequate and develop a fearful, even paranoid stance toward the world.[36] These problems might only be magnified for a clone of a famous person.

Public Expectations May Be a Mixed Blessing

Whereas a child of a famous person may be less likely to get realistic feedback from those who know his famous parent, a clone of a famous person might meet with something worse. He would risk being viewed as an empty caricature of the original, expected to be without substance and abilities of his own.

While some social advantages may accrue to a child of a famous person, the blessing can be quite mixed.[37] Expectations that a child share a parent's talent may bring the child special notice and give him special opportunities, but those same expectations constitute a specific standard by which they are judged, sometimes unfairly.

For example, some publishing house editors may pay more attention to manuscripts by children of well-known authors, in the belief that "talent is handed down." One literary agent, Suzanne Gluck, acknowledges that the literary "marketplace . . . is very interested in new voices" but at the same time "acknowledges pedigree."[38] That said, it is often difficult for a child of a famous author to establish his or her own literary identity: "Reviewers compare their work to that of their parents, usually unfavorably." For example, David Updike, son of John Updike, struggles to be successful in his own right and says that it can be distracting "to be related to someone who's so successful" while trying to do the very thing at which he has been so successful. Understandably, many such children fight against their urge to write.[39]

There are a number of prominent political examples of the power of societal expectation for children of famous politicians. In the 2000 presidential election, both George W. Bush and Al Gore faced difficulties related to this issue. That status undoubtedly gave them an initial advantage because "American voters seem to like political dynasties" but only "up to the point where the romance of father–son continuity" seemed "to verge on something resembling royal entitlement."[40] Perhaps even more important were the largely unspoken oedipal implications, depicted in the less edited, more emotionally primitive world of political cartoons: Is this son a potent man in his own right or a boy in his father's house, obedient to his father's rules and direction? Both Bush and Gore needed to try to convince the voters that they were autonomous and substantial on their own.

In an apparent sign of how important that issue was for him, George W. Bush commissioned a secret report by one of his father's aides regarding what happens to the children of presidents. That forty-four-page report "All the Presidents' Children" showed some children of American presidents to have flourished with the help of the special opportunities that status provided them. Others were "crushed by heightened expectations and scrutiny." Bush was said to have been "chilled by the implications of the report and to have ordered all extra copies destroyed."[41]

Children of famous people in almost every area of accomplishment encounter powerful social expectations. If they go into the same field, children of famous athletes, actors, and actresses are likely to face expectations both within that field and from the general public.[42] Those children of parents who achieved renown within a smaller sphere might face it only there.

PARASOCIAL INTERACTIONS AND RELATIONSHIPS

The term *parasocial interaction* describes how media consumers form relationships with media characters.[43] A parasocial interaction results from blurring the distinction between the media character and the real person, resulting in the real person "being treated by the general public as though 'in character.'" The famous person experiences the insistent and sometimes bizarre aspects of parasocial interaction as a loss of privacy.[44] Adoring fans impinge on his personal life and, in the worst cases, sometimes become threatening stalkers.[45] Leonard Nimoy, of *Star Trek* fame, describes his firsthand experiences of parasocial interactions in his aptly titled book *I Am Not Spock*.[46] Even those close to a celebrity may some of the time interact not with them but with their mass media representation or replication, leading the celebrity to feel lonely, isolated, and lacking a confidant.[47] A clone of a celebrity would be a real living replication.

As a person becomes famous, he may develop not only a public persona but also, as he becomes more self-detached and self-aware, a "gloried self."[48] It arises from the "reflected appraisals" of others, some of whom seek a "pseudo intimacy" with the famous person in order to bask in his "reflected glory." These same risks would exist for clones of celebrities and other famous people. The same sort of person who has always wanted to have a close relationship with a celebrity might wish to adopt a clone of one. Some might want to rear a clone of a celebrity because they admire the progenitor's public persona or, in the case of an actor, the characters she played. Their preexisting parasocial relationship with the celebrity would certainly influence how they would relate to her clone.

Once the general public knows a person is a clone of a famous individual, and especially once they note a resemblance, the clone might experience parasocial interactions based on the public's parasocial relationship to his famous progenitor. A clone of a famous actor might be treated not only as if he were that actor but also as a fictional character portrayed by that actor. Such a clone would be doubly estranged from his true self. But simply being a clone, when cloning is still novel, would also create parasocial possibilities.

A clone of a famous person might seek to alter her physiognomy through plastic surgery in a belated and doomed attempt to claim autonomy. A clone of a celebrity might be more likely to do this, as such a person could barely go out in public without being "recognized," though not for her own separate and unique self.

SELF AS COMMODITY

An alienating aspect of the celebrity "star system" is the "commodification of the self" as a product.[49] Not only might a clone of a famous person be "commodified" in her creation, she might be at risk of further commodification as a clone of an already commodified progenitor.[50] A progenitor would not have to be a celebrity to be commodified. Images of famous people have long been appropriated to sell products. For instance, George Washington's is used to sell automobiles on Presidents' Day, and Albert Einstein's is employed to market products to aspiring highbrows.

CHILDREN OF THE INFAMOUS

The dark side of fame is infamy. Because not everyone is in agreement about who is evil and infamous, it is possible that infamous people might also be cloned. Children of infamous parents may feel that they are serving a "life sentence" for their infamous parents' misdeeds.[51] The six children of Shoko Ashahara, founder of the murderous Aum Shinrikyo cult, have faced tremendous social difficulties.[52] Difficulties might be even more extreme for a clone of an infamous person. Even a clone of a famous person might feel that he is serving a "life sentence"—but for his adoptive parents' narcissism. In addition, if an infamous person were ever cloned, in an inversion of the family romance fantasy, any child might struggle with the fantasy that he could be such a clone, and with the attendant feelings of shame, guilt, and anger.

DEATH AND FAME

For a child of a famous person, that parent's death may be an especially powerful experience, traumatic for some and liberating for others.[53] To the extent that a clone might have had difficulty establishing his own identity, his progenitor's death might be both a traumatic crisis and an opportunity for truer self-definition and maturation. One woman, upon learning of the death of her famous father, exclaimed: "Thank God, I don't have to be his daughter anymore."[54] A clone might say, "Thank God, I don't have to be *him* anymore."

Because a clone would very closely resemble his progenitor, the progenitor's death might end some direct comparisons and expectations, but it might bring others with it. One we might call the "Elvis is alive phenomena": being sighted by the public, photographed by the press, and pressured to reenact aspects of the deceased celebrity progenitor's life.

Recall how Jerry was asked to smoke a cigar so that he would look more like his famous father for a picture. The death of a clone's progenitor would not put an end to many of the psychological and social forces bearing down on the clone to maintain a "false self."

In Christian ideology, the son is the "soul bearer as well as the perpetuator of his father's life on earth."[55] In an era of asexual reproduction, a clone of a famous man might get to share in his father's immortality but be denied his own.

By the same token, the parents of a clone of a famous person might feel that they own a share of the progenitor's social and cultural immortality because the clone is theirs. I am doubtful that such blatantly narcissistic exploitation could ever replace heroic struggle or honest faith in the quest for immortality. The sacrificial "soul murder" of the child would bring neither redemption nor everlasting life to his parents.[56]

CONCLUSION

The Child of the Famous Model takes the hopes of parents and families for parent–child resemblance into the social sphere. We have seen how the prism of the parent's fame may accentuate and distort wider social hopes in ways that can be psychologically and socially problematic for the child. Although this model is most relevant by analogy to understanding a clone of a famous person, it is also relevant to cloning in general and to any clone, parent of a clone, or progenitor. In some respects, it might also be relevant for understanding the psychology of any child in a world with clones.

In order to try to fathom what might lead individuals to want to clone a famous person, we have explored what leads people to wish to be famous themselves, to wish for a famous parent, and to wish for a famous offspring. In the process, we have found the concept of the "family romance fantasy," so useful in the Adoption Model, once again helpful.

The psychological impact of a parent's fame illustrates dilemmas likely for a clone of a famous person in establishing autonomy and identity. But it also models the plight of any and maybe every clone, even if only within her family. Psychologically and socially, "everyclone" would be "everyman" in facing central issues of being human—but with some very specific handicaps.

The social and psychological dimensions of fame intertwine, and so do their consequences for the child of a famous person. Although the model is most relevant for the clone of a famous person, it is more generally applicable to any clone in familial and smaller social spheres.

Fame is one way to live on after death, a social and cultural road to immortality. People of means who have felt thwarted in their own search for personal fame or efforts to attach themselves closely to a famous person might be tempted to rear a clone of a progenitor who has already achieved a measure of social immortality.

The next chapter, "The Replacement Child Model," models many of the same psychological and social themes as this one but with a different emphasis. Like a child of a famous person, the child meant to replace another person would have to meet some very specific expectations. For a child of the famous person, the expectations would be personal, familial, and, on a grand scale, social. For a replacement child, social expectations may be less, but for him and his parents, death and loss figure as larger issues.

7

∞

The Replacement
Child Model

He wasn't there when we went to bed, and he isn't in his bed this morning. It can't be. . . . Our hearts are left with an empty room that will never be filled again. . . . The urge to beat on the door and cry out "Please come back" is almost too much to bear.

—Klass (1997), p. 151, quoting a bereaved parent

How can we ever hear such anguished words, spoken by a bereaved parent, and know how to respond? Psychotherapists must open themselves to this kind of despair and meet the parents in their lonely desolation with compassion and empathy. Clinical psychological theory and research help the therapist grasp the ramifications of the parents' awful loss, and that knowledge also helps the therapist steady himself so that he can try to help. Otherwise, he might scream with the bereaved or, worse, turn away from them and their nearly unbearable pain.

INTRODUCTION

The idea of cloning a dying or deceased child may seem strange or repugnant to those unfamiliar with the overwhelming grief caused by such a loss. However, bereaved parents are among those most interested in making cloning humans a reality, and it is important that we try to understand their motivations.[1]

Presumably, they hope that a clone of their beloved child, if he cannot literally *be him*, will be very much *like him*. Psychological theory and clinical and research data on parental bereavement may help us to make sense of this wish. One way parents may try to cope with a child's death is by attempting to "replace" him. Perhaps psychological insights into the "replacement child" and his parents may inform our thinking about the psychological and social issues inherent in cloning for the participants and the replacement clone.

In this chapter, we will see that the concept of "replacement child" is relevant to more than this specific and very likely cloning scenario. That is because, almost by definition, any clone would be a replacement. Therefore, our understanding of the "replacement child" might shed considerable light on the likely psychology of cloning—for anyone wishing to clone and for the clones themselves.

PARENTAL LOSS AND GRIEF

Why Do Mothers Need Babies?

It goes without saying that the death of a child is a profound loss for a parent. But before addressing the grief caused by such an event, we ought to consider the sources of parental attachment. The varieties of ways parents may be attached to their child can help us to understand both their awful grief at losing one and how they cope with their loss. One coping strategy, the topic of this chapter, is to try to replace the child.

The more experience a parent has with a child, the more there is to grieve should that child die. The generally held view that parental grief is greatest when the child is old enough to have had a clearly defined personality, existing, as it were, as a distinct person, leads to the misconception that miscarriage, stillbirth, and the death of a new baby are not profound losses. The grief parents feel in response to perinatal loss, especially stillbirth and miscarriage, implies that a strong attachment to the child exists even before the child is born.[2] The literature on parental grief to the loss of a child before birth may help us to understand why parents are attached to their children before they even have a chance to get to know them *ex utero*.

These parents' grief cannot come from an attachment to the child as a real child whom they have known and loved for his unique self, a child about whom they have memories. In such circumstances, the parents' grief says more about what they were hoping for or expecting in the child, and that probably reflects why they wanted or what they needed from a child in the first place. We have already explored some aspects of what leads people to become parents in the Assisted Reproductive Technologies and Arrangements Model, but we will go further here.

Much more has been done with women than with men to explore the question of what may motivate a person to become a parent of a new life, so we will focus on the motivations of women. Some but not all of these motives may also apply to men. Aside from the likelihood that evolutionary forces dictate an innate biologically based drive for motherhood, a mother may also need a baby psychologically.[3] A woman may seek social acceptance and status in reproducing and rearing the next generation, and her sense of personal maturation as an adult and as a woman may also require it. Motherhood offers a second chance to resolve internal psychological conflicts (though, too often, it provides an arena in which the mother acts out those conflicts with her children). It can also provide a woman with a sense of reunion with her own mother and, at the same time, help her to more fully separate and individuate from her.

Motherhood may also be a means by which a woman tries to enhance herself. Creating a new life may bring with it a sense of omnipotence. It can also be a way for her to deny death by projecting herself, if not into perpetuity, at least into the next generation. Additionally, pregnancy and motherhood may be a means of narcissistic self-extension as the woman expands her boundaries to include the unborn child and infant. It may also be a vehicle for self-aggrandizement, an expression of exhibitionistic and grandiose wishes by the mother for her child, imagining that he will accomplish great things.[4] Our exploration of pathological narcissism in the Parent–Child Resemblance Model covered a variety of ways a parent may attempt to enhance herself through her child. Clearly, "a child, living or dead, plays many roles within the family and psychic system."[5] Mothers (and fathers, too) may unconsciously wish for a baby for the wrong reasons—for example, to live life as the parent's ideal self or to take care of the parent.

All clones would start life as babies, and so the above should apply to any woman wishing to mother a baby clone. However, cloning would make it possible for a woman to choose to mother a baby clone of herself, her husband, or someone else, even if she were capable of having a sexually reproduced child. There may be additional motives for clonal motherhood, but they would be in addition to those just listed.

Loss of a Baby and Parental Expectations of Babies

The premise of the prior section, that parents have no real experience of their baby prior to his birth, is useful in considering why parents want or need a baby, but it is not completely true. Aside from experiencing fetal movements and even temperament, parents, especially mothers, have distinct images of and expectations for their offspring prior to birth.[6] In fact, the mother's gestational experience and prenatal fantasy of the future infant more strongly influence her conception of her infant's personality than actual interaction with the child, even when the child is as much as six months of age.[7]

Such expectations might be based partially on the mother's experience with the fetus during pregnancy, but they must derive largely from other factors. For example, how could there have been any cues detectable prenatally to lead one father to comment that his thirty-six-week-old fetus would be "sensitive to other people, to the surroundings. Not selfish. Obedient"? And how could a prospective mother know that her unborn child might be "mischievous"?[8]

Prospective parents, especially mothers, psychologically prepare for their new baby during pregnancy. The preparation revolves around constructing a mental representation of the new family member. This is done on the most generic level from prior experiences with babies. More specifically it draws from the parents' own experiences in their families of origin, along with indirect experiences of the baby-to-be, such as gross fetal movements. Prospective parents may daydream about the new baby. Such daydreams express specific hopes and expectations for the child. The couple also imagine themselves in a new social role, especially if the baby is a first baby. If the baby is not their first, the stories they tell the baby-to-be's siblings help create a new psychological reality for everyone in the family.[9]

Parental Grief

Dealing with the reality of one's own impending death is a difficult and lonely task.[10] However, mourning for one's own self always has an ultimate and inevitable ending. Grieving for the loss or impending loss of another requires a different kind of resolution, one that allows the survivors to continue to live without that person. Sigmund Freud described the essential psychological features of mourning in his classic paper "Mourning and Melancholia," and demonstrated psychological parallels between grieving and clinical depression.[11] Mourning is not simply an adaptation to loss; it is a process of adapting to it.[12]

The following are the critical ingredients of the classic psychoanalytic conception of adequate mourning: First, fully realize and accept that the other is lost, experience the painful feelings associated with this, and ultimately give up trying to regain the lost other. Second, resolve anger and any irrational guilt feelings associated with the loss. Third, loosen emotional attachment to the lost other and withdraw emotional investment from them. And fourth, adjust to living without the other and redirect interest to new others.[13]

The process of mourning is highly individual and need not proceed in neat stages. It is doubtful that grief can ever be completely resolved, most especially grief for a dead child. Perhaps this is because such a loss turns the normal order of things on its head—parents are supposed to not only keep their children from harm but die first.

When a baby dies before or soon after birth, the parents' mourning process may abruptly halt, but halting is not the same as ending—or, rather, concluding with a sense of resolution. These parents are handicapped in their grieving by having fewer mental representations of their dead baby. With few memories to mourn, they instead mourn the hopes and expectations they had of the child. This aspect of the parents' experience of perinatal loss informs current recommendations by professionals caring for bereaved parents. In trying to help the parents grieve more effectively, these clinical experts urge parents to see and touch the dead baby, name and memorialize him. It is hoped that such actions will help the parents distinguish the real baby from whatever fantasy images they hold, freeing them to better parent a subsequent child.[14]

When a child dies, especially perinatally, parents may feel as if they "lost their futures."[15] The passage of time between perinatal loss and a subsequent child's birth does not appear to diminish overprotective and replacement feelings of the mother.[16] It is not enough to follow the frequently given recommendation that parents postpone a subsequent pregnancy to allow adequate time to mourn the child who died. Time alone does not heal all wounds, especially the death of a child. Indeed, grief over the death of a child does not end and can only be "resolved as much as it can be."[17]

Bereaved parents establish an inner representation of their dead child that survives him. As a result, they "often have a sense of double connection; they are bonded both to the child as the child was when alive and to the child who is immortal."[18] Parents resolve their grief as much as they can by transforming how they represent their dead child in their minds and in their social world. They do not break their bond with him but, rather, integrate him into their private and social lives differently than when he was alive.[19]

In the process, parents struggle with letting go versus holding on. One bereaved mother is quoted as saying: "I was afraid to let go. Afraid I would forget the details of him, the peculiar color of his eyes, the shape of his nose, the sound of his voice."[20] Parents well along in their grief have let go but not completely. They hold on in a different way, one that is less painful and that celebrates the child's life rather than in a way that focuses on the loss. One mother said of her children: "One will always be four and a half. I heard him laughing the other day in the giggles of some preschoolers."[21]

Some parents find joining a community of other bereaved parents helpful. Sharing their memories and grief with others who are doing the same gives them a sense that the dead child also has a place in the community as a valued and beloved member.[22]

The grief of parents who believe that they were less than adequate is likely to be quite difficult. It may also be complicated for those still troubled by negative or ambivalent attachments to important people in their own pasts. Perhaps such bereaved parents might be inclined to clone a child. Some, in a kind of repetition compulsion, might even want to clone someone from the past with whom they had a negative or ambivalent attachment, hoping to redo and master that relationship with the person's clone.

A COPING STRATEGY FOR PARENTAL GRIEF: A REPLACEMENT CHILD

The Original Concept of "Replacement Child"

The concept of the "replacement child" is the organizing principle of this chapter. Cain and Cain proposed it in 1962, after studying six emotionally disturbed children. All of them had been born subsequent to the death of a sibling. The parents, especially the mother, had been unable to adequately resolve their intense grief and had consciously conceived a child with the intent to replace a dead one. Their unresolved grief over the death of the previous child appeared to determine how the parents related to the replacement child.[23]

Although the replacement infant's arrival may "interrupt, distort, and delay the mourning process," it cannot resolve it. The parents transfer the expectations they once held for the dead child onto the living replacement.[24] The parents' image of the lost child dooms their relationship with the substitute, who is "virtually smothered by that image and the expectations that go with it."[25]

Characteristics of Parents of Replacement Children and the Children They Lost

In their original description of the "replacement child," Cain and Cain suggest several characteristics of the parents and the children they lost that might be considered risk factors for children born subsequent to the death of a sibling.

The Parents of a Replacement Child

Prior to their tragic loss, most parents of a replacement child had suffered from phobias, obsessions, and depressions. Understandably, having lost a child exacerbated the mothers' tendency to worry, but they were unable to contain it. They fretted over the replacement child's health, overreacted to minor illnesses or accidents, and worried that this child, too, would die. These fears led the mother to impose overprotective and severe restrictions on him and warn that if he did not comply, he could die. To underscore the message, she would recount to him how the previous child had done just that.

In a few cases, mothers gave evidence for unconsciously viewing the substitute child as responsible for the death of the other. Their irrational unconscious reasoning process went something like this: "This child is alive *instead* of our dead child. He has *taken his place*. This child is not our dead child. . . . It is his fault he is not. It isn't fair that he should live and our other child die. He is responsible for *all* this, it is *all his fault*."[26]

These parents also had highly idealized the dead child, suggesting that they had been quite narcissistically invested in him. The new child had the futile task of competing against his perfect predecessor. The parents tended to compare the replacement child with the child he was meant to replace, and the replacement's achievements produced wistful expressions of how much he was like the dead child.[27]

As a group, the parents were older. As older parents, they may not have been equipped to have another child, especially an infant. But in my view, diminished stamina for child rearing would not, by itself, have been enough to create replacement dynamics. The parents might have been older simply because they had to have been old enough to have had an older child, the first of the child-related factors below.

The Dead Children

Several factors relating to the child who died are thought to increase the risk that the subsequent child would be viewed as a replacement. One of these is the child's age at death.[28] Older children have more developed and distinctive traits and a history of specific accomplishments. These parents were unprepared for a new infant probably because what they wanted and at an unconscious irrational level even expected from the next pregnancy was not really a new infant—*it was the child they lost, at the age he was when he died*. The persisting vivid image of that child could not easily be projected onto an already living sibling of the child. A new life appeared to be needed for that purpose—a biological offspring, though in one case, the parents hoped to adopt a child of the same age and similar appearance as their deceased eight-year-old son.

Sudden and unexpected death of a child is another factor that might contribute to the risk of replacement status for a sibling born subsequently. Research has shown that unexpected death makes grieving more difficult.[29]

Implications and Limitations of Research on the Replacement Child

The seminal study of Cain and Cain is limited by a small and perhaps nonrepresentative sample. Indeed, the parents and children came to the authors' attention because of psychological disturbance. Nevertheless, as we have seen in other models, clinical case studies are useful in identifying psychological phenomena. It then becomes important to systematically study larger groups of people from nonclinical populations. Surprisingly, in-depth psychological research on a more representative sample of children born subsequent to the death of a sibling has yet to be done.

In the absence of such research, especially one that follows the children over time, we must rely heavily on clinical case studies, like that of Cain and Cain, while trying to get what we can from the limited research that has been done. None of that research has directly assessed children born subsequent to the death of a sibling, that is, children we might think would be at risk to be replacement children.

Nevertheless, that research raises questions as to whether a child born subsequent to the death of a sibling is, in fact, at serious risk to become a replacement. However, we should note that the populations studied are quite different from those described by Cain and Cain.[30] First, Cain and Cain's case parents were older, in their late thirties or early forties, and may have previously decided to have no more children. Second, the decision to have a subsequent child in Cain and Cain's study was the mother's, whereas in the research samples of parents, it was almost always mutual. Finally, the dead predecessors of the children in Cain and Cain's case series were latency age to early adolescent, while those in the other studies for the most part died perinatally, and the oldest was three.

We should also note that any conclusions from this research are limited by their reliance on parental report and lack of assessment of the babies and young children. Of course, interviewing these children into childhood and beyond would be the best way to determine whether a child became a replacement child.

Even research that downplays the risk that a child born subsequently to the death of a sibling might become a replacement gives some important evidence to the contrary. For example, in one study, most mothers who lost a baby reported difficulty separating the live baby from the dead one, and several reported that they held an idealized image of their dead baby in their minds and compared their subsequent child to it with a sense of disappointment.[31]

We might be reassured by other research in which couples who had lost an infant or very young child one to two years earlier and wanted to have another understood that they could not replace the dead one. The parents studied were not adjudged to be apathetic or withdrawn and were excited about the prospect of having a new child, and yet they still grieved for their dead son or daughter. Furthermore, some parents said that they felt guilty that their child had died and hoped to expiate their guilt with a new child, saying things like: "We will be better parents this time."[32]

With regard to the risk that the next child would be perceived and treated as a replacement, it is not reassuring that these parents "wanted another child of the same sex and as quickly as possible."[33] In a world where sexual reproduction (unassisted or assisted) or adoption are the only means of trying to replace a dead child, the greatest degree of resemblance for which parents can rationally hope is that the subsequent child be of the same sex as the one lost. Recall the parents in Cain and Cain's sample who sought to adopt a child resembling the child they had lost. As cloning could assure a greater degree of resemblance, such couples might be inclined to pursue it.

One study of parental adjustment to perinatal loss finds that parents either try to maintain a connection to the dead baby or try to replace the loss created by its death. Two different patterns are found for maintaining the connection: The first pattern is to maintain the space the child filled as an empty void, though the centrality and importance of that space may diminish. The second is to try to continue the relationship with the dead child through rituals, memorials, and storytelling about the dead child, including other children in the family.

The researchers assume that the first pattern of parental adaptation to loss would not result in a subsequent child being a replacement. However, the void this pattern maintains might envelop a new child from birth. The *dead child's absence* in such a family *may turn out to be a suffocating presence for the new one.*

The researchers find that parents following the second pattern do not report anything to indicate that they treat the subsequent child as a replacement. They offer that this might be because those mothers make a distinction between "replacing a loss and replacing a person." The mothers may have been able to do so because their children died shortly after or before birth and not in childhood, when parents are more likely to have a clear view of their children as distinct individuals.[34]

Given the many limitations of this research, we may not want to join the researchers in their belief that parents who fall into the second pattern of trying to replace their loss would be unlikely to view the next child as a replacement. A child born to parents attempting to maintain a connection to a dead sibling might well feel compared with the memorialized child and would also be inclined to compare himself negatively with him. *The replacement child is not only a replacement; he is also a living memorial to the dead. A clone of a dead child would be an even more representative memorial.*

One way in which parents maintain a connection to the dead child is through a "linking object" such as a possession closely identified with the dead child.[35] *The replacement child is both an animate link to and supposed to be the lost child. Clones, as replacements, would also presumably serve both functions.*

Characteristics of Replacement Children

The prolonged unresolved grief of parents having a baby subsequent to the death of a child may impact the new child in three major ways. First, it may impair mother–infant attachment and maternal responsiveness.[36] Second, it may result in what has been called the "vulnerable child syndrome," whereby the mother is overprotective of the subsequent child. Third, it may result in the replacement child syndrome, whereby the mother perceives the subsequent child as a replacement for the dead one.[37] Although these outcomes are theoretically separable, in my view, the first two would always be part of a replacement child's experience.

Let us start by looking at the original replacement children, those six described by Cain and Cain. They were all aged seven to twelve. Four were boys, and two were girls. Four were moderately neurotic, and two were psychotic. They all had been "born into a world of apathetic, withdrawn parents, a world focused on the past and literally worshipping of the dead child."[38]

The children were phobic and generally fearful. For these children, overprotective restrictions and fearful warnings from their parents had made them infantile, immature, and practically homebound. They were passive and dependent and had difficulty being able to think for themselves.[39] They also felt inadequate and vulnerable, living in a world they perceived as full of "constant unpredictable dangers."[40]

In addition to being generally preoccupied with death, the children strongly identified with the physical symptoms the dead child had, and several expected to die at the same age as the children they had replaced.[41] The age of the child who died is an important factor in the replacement child's sense of foreboding. Substituting for a child who reached five or more years of age involves the replacement child nearing the age of death of his predecessor with some conception of the reality of death. The child's fear intensifies with the approach of that age, abetted by the developmentally normal increasing awareness and growing comprehension of death. Until that fateful anniversary passes, the child must contend with his identification with a real predecessor who died at that age, along with parents influenced by their idealized memories of that child. But the passing of the anniversary age does not free the replacement, especially because his parents are apt to still wish that the replacement were his predecessor resurrected.[42]

The risk of replacement dynamics appears to lessen when the dead child did not survive infancy. Perhaps that is because the parents have fewer and less distinctive experiences with such a child and, as result, are less likely to have a clear mental image of that child to impress upon the next. Another reason may be that the subsequent child passes the crucial age of the earlier child's death unaware of that fact of the tragedy and so would not be able to identify with it. However, the child must still contend with his family's continuing grief, along with the hopes and expectations they place on him that derive from their earlier loss.[43] Perhaps this accounts for the detached and aloof quality of replacement children, who are also said to appear like condemned souls, awaiting catastrophe.[44] In addition, when a replacement child lives beyond the age to which the dead child survived, it becomes much harder for parents to imagine the child as the resurrection of the one they lost. If they had not already, they may then turn against the child who is undeniably becoming a different person than the one they knew their lost child to have been.[45]

Once the child learns that there was a prior child who died, the subsequent child may feel guilty for being alive.[46] The replacement child also derides his own achievements compared with those of his dead predecessor. He is burdened by "guilt laden with inexpressible rage" provoked by constant comparisons to his predecessor, whom he is also expected to mourn and idealize.[47]

Replacement children may also have profound identity problems. Cain and Cain comment that these children "could barely breathe as individuals with their own characteristics and identity. Their parents compelled them to be like their dead siblings, to be identical with them, yet made it clear that they never would be accepted as 'the same,' and could never be as good." They were "burdened with unbelievably detailed knowledge of their dead sibling, constant comparisons and imposed identifications."[48] Replacement children are also thought to have serious difficulties with both separation-individuation and adolescence, stages of development critical to the development of an autonomous self and secure identity.[49]

The plight of the "replacement" has been contrasted to that of the "exception," with the replacement child carrying the burden of "not having been chosen."[50] However, at the same time, every replacement child is chosen, if only as a consolation. Any child cloned to be a replacement would also have been chosen, with the hope that he or she will be the closest possible approximation, if not the recreation or resurrection, of the original.

It might be that a replacement child might be prone to what has been called a "twinning reaction." Just as some twins fail to completely differentiate psychologically from each other, a replacement child may feel ambivalently tied to the dead child. We explored the "twinning reaction" in the Identical Twin Model and also looked at some unique difficulties identical twins may face when they lose a co-twin. Clones, as replacements for dead children, might be inclined to some of the same responses.

Disturbed Patterns of Interaction between the Replacement Child and His Mother

Almost certainly, a mother's maternal responsiveness to her replacement child would be impaired starting in infancy. Both parents, and perhaps especially the mother, would certainly register discomfort or worse at the lack of congruence between their image of the lost child and the new infant. The mismatch might well lead mothers of replacement children to have "difficulty responding to the rhythms of the new infant." Such a mother would be inclined to "try to impose the pattern of the dead infant" on him and their interactions. As a result, "the reciprocity built up between mother and infant may lack the spontaneity of a unique relationship."[51]

Although a replacement child's mother may be reliably present, she is not present for him. Instead, she directs her parenting toward the dead child—her memory of which is embodied in her perceptions of the new one. She does things "for him and to him as if he were another child. It is not his self but another that is cherished." Important people in his life respond to him "as if" he were his predecessor. As a result, he starts to act "as if" he were: "A reciprocal reinforcement begins with parents and child locked into a pseudoreality." This puts the replacement child at risk for the sort of inauthentic experience of self that is typical of people with "as if" personalities.[52] A clone would probably run the same risk for the same reasons.

In addition, perhaps some of the difficulty a replacement child has in being her own person might be related to the fact that she has never really been alone with her mother. It is important for a child to develop an autonomous self for him to be able to be alone in the presence of his mother and have her full attention.[53] A clone would probably have the same problem.

BROADENING THE ORIGINAL CONCEPT OF THE REPLACEMENT CHILD

Michelle

Michelle had an older brother, Michael, who died as a toddler. When his parents attempted to replace him with another boy, they failed when a girl baby was born. When the next child was also a girl, they elected her

Michael's replacement and named her Michelle. Not surprisingly, Michelle felt that she had been born the wrong sex. Her parents reinforced that feeling every time she saw them gaze admiringly at a neighbors' child, also named Michael.[54]

This case illustrates that the original concept of replacement child can be legitimately broadened. It shows that a replacement child need not be the first child born after a sibling's death and may also be of a different gender.

Ed

Ed was a psychology graduate student in his late twenties. He presented for treatment of anxiety and a chronic low-level depression. After a few years of intensive psychotherapy, an assignment in a family therapy class helped to accelerate his progress. Instructed to draw a "genogram," or family tree, Ed was faced with the fact that his mother had a brother who had died in childhood. He had known this but had not thought about it in years. While sitting at his mother's side as a boy, she had wistfully and tearfully viewed photos of her younger brother, Benny, in the family album, and Ed had futilely tried to comfort her. His mother had told him that Benny was a "little angel." "Don't you see it in the pictures?" Indeed, in the old sepia-tone photos, Benny seemed to have a serene, angelic aura of goodness about him. Ed had wondered if he looked anything like this "angel," but the aura that seemed to emanate from Benny's image nearly blinded him from being able to clearly see whether there was a resemblance.

To complete the genogram, fully grown Ed called his mother to try to get the facts. When Ed's mother was twelve and Benny was six, he was hit by a car and died. She had felt protective of her little brother and could also imagine him as her own son. She had felt terribly guilty when he died, and decades later she continued to believe that she ought to have been able to prevent it. Ed knew that his mother had been treated for depression several times and figured that losing her little brother must have been part of that picture. But, more personally, Ed had come to understand what must have been an important source of his mother's anxious, overprotective attitude toward him. She had been overly attentive to his health and safety but seemed inattentive to him as a person, not remembering important details of his life, interests, likes, and dislikes. He had felt nearly invisible to her in those important ways.

As a child, Ed had felt that his mother did not really know him and did not sincerely seem to want to. In therapy, he came to see these as deriving from her psychological vulnerability and preoccupation with Benny. Aside from those photo album sessions, his mother never spoke of Benny or overtly compared Ed with him. In fact, she had given Ed much adulation, but her compliments seemed overblown, hollow, and disingenuous. At a most basic level, he felt she had seldom given *him* her complete attention.

It all began to make sense to Ed when he concluded that Benny was the invisible child his mother loved best. His mother had adored not him but, rather, an idealized image of Benny. She must have been both distracted by thoughts of Benny and disappointed that Eddy was not him, loving Benny more than she could ever love him, his sister, or even his father. It seemed more than a strange coincidence that his father's name was Ben.

The family therapy assignment catalyzed important work in Ed's therapy, helping him make sense of experiences with his mother. He realized that in his mother's psychological reality he probably was not really his mother's first and only son but, in fact, her second. Though not a replacement child in the narrow definition, Ed certainly was one in essence. He realized that in his mother's heart, he never could have replaced her dead little brother.

Ed's replacement status also complicated the resolution of his oedipal issues. In the normal resolution of the oedipal situation, the boy acknowledges that his father has won the boy's mother. This leads the boy to identify with his father, sharing in a victory he could not himself achieve. As a boy, the other male who had Ed's mother's heart was not so much his father, her grown living husband, but an angelic and dead little boy. This explained Ed's childhood sense that his father was not much of a rival for her. It was not that little Eddy felt he had prevailed over his father but, rather, that neither had won. This interfered with his ability to identify with his father. Perhaps Ed's mother had been drawn initially to his father because he was also named Ben and as such was a disguised replacement, later to become her husband. Ed made good use of these insights. He became more confident with women and less fearful of intimate relationships with them. Years after concluding his psychotherapy, he is now happily married and the father of three children.

In therapy, Ed several times used the biblical story of Jacob wrestling with but unable to prevail over an angel as a metaphor for how he felt that, despite his best efforts, he was destined not to get whatever it was that he most wanted. He identified with Jacob and felt that Jacob's dream could have been his own. With the insight that he was a replacement child, Ed realized that the "angel" with whom he was struggling was Benny. We might even call the commonly described struggle of replacement children with the "angels" they were meant to replace a "Jacobean complex."[55]

∽

Ed's case demonstrates that, in its essence, a child replacement need not replace an earlier child of the mother. It also illustrates that parents need not intentionally and consciously conceive or choose to have a child as a replacement for the child to be one for them. Unconscious motivation is just as powerful as conscious motivation in its effects on behavior, often more so. Thus, even when a child is not intentionally conceived as a replacement, if the child is perceived and treated as if he were one, then he effectively is one. However, his status as a replacement is probably more likely to remain a mystery to him than it would for the "classic" replacement child who had a sibling who died before the new child was born. Ed's case illustrates the clinical utility of broadening the replacement child concept and also illustrates some complications that, to my knowledge, have not previously been pointed out.

Although typically, a replacement is desired for a beloved lost other, a replacement could also be of another person about whom the mother has unrecognized feelings of hurt, anger, disappointment, or guilt. In a kind of "repetition compulsion," parents may try to master a troubled, ambivalent relationship, perhaps most often one with their own mother or father, hoping to get it right this time, now that they are in authority. Ed's mother's overprotectiveness was her attempt to prevent him from dying as Benny had.

Ed's mother is one of many people with unresolved grief and separation issues who may want another child to try to undo earlier losses and deny current separations. Even before their child is born, such parents unconsciously identify him with the memory of a person from whom the parent has been separated, though not necessarily by death.[56]

Ed's mother experienced him as if he was her beloved little brother. A woman may also experience her unborn child as representing her parent or even herself.[57] It is said that the pregnant woman embodies three generations and that "the mother-to-be recapitulates her earlier stages of separation-individuation from her own mother in the growth and impending delivery of her child."[58] It is necessary for a mother to reconcile her ambivalence toward her own mother to be able to respond with optimal empathy to her own child.[59] Although it seems likely that Ed's mother was hampered in being genuinely empathic with Ed because she perceived him as a replacement for Benny, to some extent she might also have been impaired by experiencing him as a replacement for her own mother. We have seen in the chapter on parent–child resemblance that narcissistic mothers frequently relate to their children in this way. The children of narcissistic parents are typically "replacements" in this way and are also often replacements of the parent's own self.

Although the narrow definition of *replacement child* is a child born to parents who have lost a child to death, with the subsequent child conceived to fill the void, similar "configurations" may be found in the adoption of a child by parents experiencing the loss of being unable to conceive.[60] Cain and Cain in their original formulation of the concept mention that bereaved parents need not sexually conceive a replacement child but may adopt one for the same purpose. We have already noted one instance in which the parents made every effort to adopt a child who would closely resemble the child they lost.

In our broader conception, a "replacement" may also coexist with the original. A child replaced need not have died but may have been a deep disappointment to his parents. He may not have lived up to what his parents had hoped for him, having defied his parents' wishes for his goals or, having fallen ill, been incapable of achieving them.

One might object that in generalizing the concept of replacement, we have blurred the line between it and the more general concept of transference, but this is intentional. We might think of replacement phenomena as one end of the transference spectrum. Adam Phillips conceptualizes transference as the process of mental cloning.[61]

The philosopher Franz Brentano has written that mental contents have "intentional inexistence," that is, they exist but only in our minds.[62] Nevertheless, the nonmaterial inexistence of the contents of our imaginations, fantasies, and dreams is a reality. We might conceive of transference as creating, unconsciously and unintentionally, "inexistent" people in our minds based more or less on important people in our early lives. We unconsciously superimpose these internal representations onto real people in our lives, a process that often interferes with our ability to see the real people for who they really are. Replacement children, whether of the classic, intentionally conceived type or of the unconsciously begotten type, are real embodiments of transference regarding the person they are meant to replace.

The child conceived as a replacement is "treated more as the embodiment of a memory than as a person in its own right."[63] *The replacement child is an embodied enactment of transference.* A clone's transference identity for his parents would be "plain as the nose on his face"—as well as in all the other features and traits he would share with his progenitor. But even before such resemblance is apparent, *his transference identity would be concretely embodied in a genome identical to the progenitor his parents selected.* However, this would not exclude others that might coexist with it.

NAMING A NEW LIFE FOR A PREVIOUS ONE:
A POSSIBLE SIGN OF REPLACEMENT STATUS

Vincent van Gogh

While Vincent van Gogh's mental illness may have had important biological causes, being a replacement child might also have played a key role in his misery.[64] He began and lived his life in "the shadow of death"—that of his stillborn brother, his mother's first child, also named Vincent.[65] He deeply identified with his dead brother, and that identification was reinforced by his relationship with his depressed and grieving mother. It would appear that she was so preoccupied with her dead firstborn son that she could not provide her living second son with the empathic loving relationship he needed.

He might have come to feel that to be loved, accepted, and valued he had to suffer and die.[66] To put it somewhat differently, there was only one way he could both be like the original, idealized Vincent and be himself as a unique individual, and that was to die by his own hand.

Erik Erikson described children of mothers who, unable to attach to them because of unresolved grief over a first child, come to believe that the only way they can be "recognized" is to be sick or dead.[67] Van Gogh may have been both terrified of death and drawn to it through his identification with his brother. Death would be the accepting "warm embrace" from his mother that he had so long sought.[68] We do not know how many other replacement children eventually commit suicide, but psychological postmortems of suicide victims might start to ask that particular question. Perhaps being a namesake of a dead child might indicate that the person is also a replacement child. We will further examine Vincent van Gogh in the next chapter, where we will focus more on him having also been a namesake.

CONCLUSION

In this chapter we explored the concept of replacement child and how the wish for a replacement arises out of unresolved parental grief. By examining motivations for parenthood and prospective parents' hopes and expectations for their children, we were able to put parents' grief at losing a child in a more complete context. This allowed us to better understand why many parents wish to have a child subsequent to losing one and how some parents desperately want that child to replace the one lost.

We have found that the clinical and scientific literature suggests that bereaved parents commonly hope that the next child will strongly resemble the one lost and at least be of the same sex. Parents coping with their loss in such a manner might prefer cloning to sexual reproduction because cloning promises the most resemblance one could ever expect, resemblance based on genetic identity.

The process of grieving is a dynamic evolution in letting go versus holding on. Unresolved grief is a way of not letting go but of holding onto the dead child in a way that attempts to deny his death. The parent who creates a replacement child copes with her loss by holding on in a particular way that interferes in her being able to relate to the new child as a separate and unique person. An ideal prototype exists before the child's birth, and the new child is doomed to fail to match it.

We have looked at possible adverse psychological consequences of replacement. Those consequences may begin at the outset of the child's life outside the womb, with impairment in the bond of mother to infant. The parents, anxious and fearful that this child, too, will die, raise an existentially insecure child who naturally has difficulty with separation-individuation. The ghost of the dead child gets more attention and devotion than his replacement, and this adversely affects the new child's self-esteem. Also, given the parents' hopes and expectations that the new child will be "as if" the one lost, the replacement child typically has difficulties in developing his own identity.

We have seen how for his parents, the replacement child is the embodiment of an idealized memory of the dead. Individuals who decide to clone a replacement might very well be enacting their transference to the person lost by embodying it in the clone. However, even when death has been accepted, and when transference propensities are under control, a clone might be at risk for thinking of himself as a replacement and of becoming a living memorial to the dead.

Replacement children have not been studied into middle age and late adulthood, but my clinical experience suggests that they do not age out of their early difficulties. Rather, as a result of personal development and maturation in other areas, such difficulties simply manifest and express themselves somewhat differently. For example, oedipal complications related to being a replacement may have consequences for intimate relationships in adulthood. Most fundamentally, a replacement child grown up, though less fearful, tends to feel just as lonely, angry, and guilty but more hopeless. He may start to heal when he realizes that the life he has lived has not been fully his.

We have also seen that parents desiring a replacement child want that child to highly resemble the child lost. Indeed, replacement and resemblance come together inextricably like two rivers flowing into a confluence. For example, Ed's mother might have chosen to clone her beloved little brother, Benny, if cloning and a suitably preserved tissue sample had been available. Whether or not that hypothetical boy was to learn that he was a clone of his mother's deceased brother, his difficulties would probably be similar to but greater than those experienced by Ed.

Because parents wishing for a replacement want the subsequent child to closely resemble the one lost, they may be emotionally unprepared for a baby. Unconsciously, they expect the child they lost at the age they lost him to death or began to lose him to illness. Like parents of a replacement child, parents of a clone, regardless of their intelligence and rationality in other areas, might be similarly disappointed.

Because a clone would be an exact genetic copy of a prototype that the parents consider ideal in some way, the parents might have even greater expectations that he will reflect that ideal than they would for a sexually conceived replacement. But parental disappointment in the replacement child, clone or not, is apt to begin early in the child's life and produce the same psychologically harmful consequences.

As a clone approaches the progenitor's age at death, he would probably increasingly resemble the deceased, and the high degree of resemblance between him and his dead predecessor might reinforce the clone's psychological and social status as a replacement.

If a child were to know that he was a clone of a deceased person, especially a child, he might overidentify with that person. Although this might be similar to how identical twins may sometimes overidentify with one another, without the reality of a living progenitor to constrain it, we would expect such fantasies to take unusual forms. The clone might temporarily represent the dead progenitor in his mind as an imaginary companion or even an alter personality if he had Dissociative Identity Disorder. Recall that these kinds of reactions may occur in noncloned replacement children. By analogy to the findings with replacement children, we might expect a clone of a dead child to feel like a twin or double of it or even less substantial and alive than the deceased.

What we think we know about replacement children derives mainly from clinical case studies. Research relevant to the replacement child with nonclinical populations indicates that most parents having children subsequent to the death of a child are very different from those parents described in the clinical literature whose subsequent children became replacement children. That is reassuring. However, the same research reports findings that do suggest that a new child would run a risk of being viewed as a replacement. Thus, those dynamics are not found only in emotionally disturbed parents and children like those in clinical case studies. In fact, it is typical for parents who have lost a child to be prone to view the subsequent child as a replacement in some ways.

We have seen that it makes sense to broaden the concept of replacement child to include siblings, parents, and even the self and living children. Replacing an individual other than a parent's own dead child may also create a replacement child. And the loss need not be of a dead other. It may also be a sense of loss or lack in a parent's self or relationship to an important other. Lastly, a parent confronting the ultimate existential issue, death, might view his child as his own replacement.

Just as we have usefully extended the concept of replacement child, we can also further broaden it to include clones. In this view, any clone as a replacement for a previously existing individual is at high risk to be a replacement child. To understand how it might manifest, we would need to know what sort of relationship the rearing parents had with the progenitor of the clone they are raising as their child. Was the progenitor one of them or one of their parents, a sibling, a spouse, or a friend?

At various points in this chapter, we have seen the need for further psychological research. Longitudinal studies of children born subsequent to the death of a sibling are clearly needed. As we have seen in other areas, interviewing parents about their attitudes regarding their children can only give us an inkling into the parent–child relationship and the psychological outcome for the children. Also, many psychological problems based primarily on faulty child rearing are not readily apparent early in life but manifest in adolescence, young adulthood, or even later.

To better understand parental motivation for replacement children, we might want to use an "animal model" with research on bereaved pet owners who elect to acquire new pets as replacements. Such research could not model the psychological consequences of replacement status on a child, but it might shed light on parental psychology regarding cloning. Relevant by analogy to human cloning, it would also be pertinent because some people, in fact, do want to clone their much-loved and now deceased pets.[69]

It may be useful for psychological research on children born subsequent to the death of a sibling to look for a relationship between their name and that of the deceased. Perhaps being similarly or identically named might be a marker of replacement status, though neither necessary nor sufficient in this regard. As we shall see in the next chapter, namesaking itself is useful as a model of cloning. A replacement child who is also a namesake, like Vincent van Gogh, exemplifies two models of cloning. Not making a clone a namesake might not reduce the psychological pressure on him to be a replacement, but making him one would probably reinforce his replacement status. Suicide reflects the worst outcome of psychological trauma. From what we have seen in this chapter, perhaps "psychological autopsies" of suicide victims might do well to try to determine whether the victim was born subsequent to the death of a sibling and, if so, if he was the dead child's namesake.

The Parent–Child Resemblance Model can help us to anticipate expectations and pressures a clone might experience as a replacement—whether of an earlier child, some other person important to his parents, or even one of his parents. In addition to whatever else it might be for the parents psychologically, cloning would also be an attempt to give a kind of immortality to the progenitor. But the mortal reality of the clone would undermine that wish in two ways. The anxious overprotectiveness of a parent toward her replacement child reflects an excruciating awareness that the replaced person is, in fact, dead and that his replacement is not an immortal resurrected presence but, rather, a vulnerable human being who could meet the same fate. Such parental responses might be even more likely for a clone. Like any other replacement, a clone would not live forever.

Another way the mortal reality of a clone would undermine his parents' wish to immortalize his progenitor would derive not from the clone's lack of everlasting life but from him being incapable, like any replacement child, of immortalizing his parents' ideal conception of the person they intended him to replace. Simply by being mortal, the clone would also be incapable of satisfying his parent's idealized expectations.

Replacement child issues might be magnified for a clone who is both a replacement for and also a clone of the dead child. Presumably, most replacement clone children would be of this type, but if there were no viable DNA from the dead child, parents might elect to clone another person. In those circumstances, replacement status would be dissociated from clonal identity. Such a child would be a double replacement—for the dead child who preceded it and also for the progenitor. Both preexisting parental expectations and the reality of heightened resemblance of clone to progenitor might well psychologically and socially magnify the child's status as a replacement.

Not all children born subsequent to an earlier child's death are replacement children, but the risk might be high for a child whose parents choose to clone their dead child, or any other person, when other options exist to replace their loss. Through unassisted or assisted sexual reproduction or adoption, it appears that parents often hope to replace not only their loss but also the person lost. *Psychologically, a "cloned child syndrome" is likely to share many of the features of the "replacement child"* if only because the clone would be a replacement in one way or another.

The classic replacement child is an example of the intergenerational pathology of grief visited on a subsequent child. There is a danger that, despite the best of conscious intentions, parents of a child clone would experience great unconscious psychic pressure to view and treat the child clone as a replacement. Also, if others were to know the child is a clone, then wider family and social expectations would reinforce his replacement status.

We may hope that by helping bereaved parents psychologically, their next child will avoid the fate of the replacement child. Likewise, appropriate psychological assessment and treatment of individuals seeking to parent a clone, whether of a dead child, another person, or a self-clone, might help to prevent the clone from suffering the replacement child's fate. It might even lead some prospective parents to reconsider cloning in the first place.

As another model for thinking about psychological and social aspects of cloning, the next model, on the namesake, evolves naturally from this one on the replacement child and also the previous one on parent–child resemblance. In the Namesake Model, we will consider parents' desire to clone a name and perhaps by assigning that bit of identity to their child, create a person they hope will resemble or replace his namesake.

8

The Namesake Model

A person's name helps to identify that person as a unique individual. A last name identifies an individual as part of a particular family and often suggests ethnic or national background, but first names symbolize personal identity.[1] In fact, the name given us at birth constitutes "one of the most fundamental pillars of selfhood."[2] Our names—first, middle, if any, and last, as well as any suffix, whether "Jr." or numerical—help to define who we are. Like race, ethnicity, or adoptive status, one's name is an "assigned" piece of identity. Like other assigned aspects of identity, it can be a potent psychological factor in a person's life, especially in infancy and childhood.[3] Just as an individual's name is a critically important "anchor point" in his or her identity, a progenitor's identity would be an additional one in the identity of a cloned person.[4]

In previous chapters we review other factors importantly influencing individual identity. Among them, we have noted that the name parents give their child sometimes is a reflection of parental narcissism or wishes for a replacement child. In the Parent–Child Resemblance and Replacement Child Models, we explore how making the child a namesake of oneself or a lost other may be an additional vehicle to express parental psychological issues.

Namesaking is not only analogous to cloning; it is conceptually equivalent, albeit linguistically. The parallels between cloning and namesaking make the history, psychology, sociology, and cultural anthropology of namesakes and namesaking practices relevant to anticipating the psychological and social consequences of reproductive cloning. Let us see what they suggest.

NAMING ONE'S CHILD FOR ANOTHER

The Namesake as Replacement for a Dead Sibling of the Child

Vincent van Gogh, Again

The renowned artist, Vincent van Gogh was born March 30, 1853, exactly one year after his brother, also named Vincent, who died in infancy. Both Vincents were named after two grandfathers and a paternal uncle.[5] The first Vincent was buried near the entrance of his father's chapel. The second might well have seen his brother's grave every Sunday and read the name they shared on its headstone.[6] As discussed in the previous chapter, the fact that the second Vincent was given his dead brother's name may indicate that he was a replacement child.

While working on the foreboding painting *The Reaper*, Vincent learned that his sister-in-law, the wife of his beloved brother Theo, was pregnant and that if the baby was a boy the couple planned to name the child after Vincent. Vincent had a recurrence of severe depression with suicidal ideas soon afterward.[7] The child was a boy, and his parents did indeed name him Vincent. His Uncle Vincent committed suicide a few months later.[8]

Naming a child for another person can be a way of trying to replace that person. Ideally, a replacement should be identical to the original, but absolute resemblance can be guaranteed only in name. It seems that the parents of the second Vincent van Gogh intended him to replace their first son and assigned him the same name in that hope.

An existentially vulnerable person might feel replaced by the birth of a child named for him. It was Theo's intention to name his son Vincent to honor his gifted brother, but, in his depression, Vincent might have felt not so much honored as replaced. Of course, this is highly speculative, but if he had, then this could have been a factor in worsening his depression. Because the process of replacement may involve discontinuing the earlier version, he might have felt that suicide was obligatory. He would make room on this earth for his replacement, just as his brother Vincent had made room for him by dying a year before his birth.

Patterns of Naming for Dead Siblings: Namesakes and Name Taboos

Before we go too far in trying to analyze the phenomenon of namesaking psychologically, we should place it in historical and sociocultural context. When it comes to siblings sharing the same name, there appear to have been three historical stages since medieval times:[9] In the medieval period, living siblings, especially if named after a parent, could have identical first names and were distinguished by birth order or other modifiers. Beginning in the early modern period, we find a different pattern: with only rare exceptions, living siblings were not identically named. However, it was typical for a dead child's name to be given to the subsequent child of the same sex. In Hingham, Massachusetts, in the seventeenth and eighteenth centuries, this pattern was accentuated when the child who had died had been a namesake of a parent. Naming for the dead child was also more likely the longer the child had lived before he died.

Although we can only conjecture on parental psychology at that time to account for these facts, it would appear that the name of the dead child was often reused because the parents wished to re-create the unique personality of the longer-lived child in his namesake.[10] Twentieth-century research and case studies on the replacement child phenomena, reviewed in the last chapter, are consistent with this idea.

In contrast to the medieval and early modern patterns, the modern pattern of naming siblings finds no two being given the same name, even if the older one has died. But, as we know from the last chapter, there are exceptions to that rule.

This historical analysis of sociocultural epochs in the naming of siblings suggests that the parents of the replacement children Vincent van Gogh and Salvador Dali followed an early modern pattern that was more typical of Beethoven's time. Beethoven was also given the name of a dead sibling.[11] It would appear that there has been a historical evolution in Western cultural in which name uniqueness reflects respect for the individual uniqueness of the child.

Given this historical trend away from naming children for dead siblings, what does the persistence of the practice mean today? In the prior chapter we saw that the cases reported in the clinical literature suggest that replacement children are often named for their dead siblings. This stands in contrast to studies of parents with children born subsequent to the death of a sibling. None of those parents planned to name the new child after the one who had died. The clear historical evidence of cultural evolution away from naming a new child after a dead sibling of the same sex puts the clinical case studies on the replacement child in a new context.[12] It would appear that in our time, when parents name a child for a dead sibling, the parents, consciously or not, probably intend the living child to replace the dead one.

Twins and Their Names: Replacements and Replaceable?

The way parents name identical twins might be related to the phenomenon of naming a child born subsequent to the death of a sibling for the dead child. In some ways, this is approximated by the instance of a man who wished to change his name to that of his dead twin brother, making himself a replacement.[13]

While not the same as naming a child for a dead sibling, sometimes identical twins bereft of a co-twin name one of their children for their dead twin.[14] This not only commemorates the dead loved one; but because twins are genetically identical, in theory, the child of the living twin could have been a child of the dead one.

It has been argued that reproductive cloning would produce unique individuals because clones by embryo splitting, or identical twins, are distinct individuals by virtue of the fact that they are given different names.[15] Actually, however, identical twins are often given similar names and, as mentioned above, more often than not share first names with the same initial letter. Parents give twins names with the same first initial 62 percent of the time and 67 percent of the time if they are identical, making them partial namesakes of each other.[16] In doing so, are the parents simply succumbing to a desire for alliteration in pronouncing the twins' names together? Might this practice hint at a sense that, as exact genetic copies of each other, one twin is replaceable by the other in the sad event that one child dies? Might name similarity influence how identical twins are perceived and their development of individual identities?

Namesakes of Nonsibling Dead

Let us look more generally at namesakes of the deceased, beyond siblings and the special case of twins. Although perhaps replacement status is most clearly indicated when a child is named for a dead sibling, a more general pattern may hold. When a person is named for another person, we might wonder whether the new bearer of the name is intended as a replacement. Being a namesake might be as much a "marker" of psychological risk for replacement status as certain genetic markers may indicate risk for specific medical illnesses.

Though the person namesaked could be dead or alive, let us focus on the former. Societies have one of two general responses to death, and the attitude toward the names of the dead may reflect these. One response is to try to keep memories of the dead alive. In addition to ancestor worship or treating children as if they were reincarnations of the dead, the names of the dead may also be given to new life. Other societies try to forget the dead and taboo their names.

Predominant attitudes in different societies, when they are at variance with our own, may give us insights into our own repressed or latent ambivalences. When it comes to attitudes about the dead and their names, we might do well to look to these other societies for what they may tell us about universal human attitudes. While certain ideas and attitudes may not predominate in our society and might seem quite alien or strange to us, they are still human. As such, they may lie dormant in the subconscious minds of individuals constituting our society, existing as repressed trends for the society as a whole, with the potential to emerge under the right conditions. In my view, the prospect of human reproductive cloning is such a condition. Both conscious and subconscious individual and societal attitudes about the dead may help to shape attitudes toward cloning and, in particular, cloning the dead. I suspect that some arguments against cloning drawing on science fiction horror have their origin and appeal at this level. If we learn more about the variety of attitudes toward death and their effect on namesaking practices, then we may gain insights into psychological, social, and cultural aspects of reproductive cloning.

Perhaps relevant in this regard are those societies that encourage their members to forget the dead. Name taboos may apply to everyone or only to relatives and close friends. The taboo may last forever or only for a set period of time.[17] The sociologist Richard Alford finds that names of the dead are avoided in about one-fourth of a large number of societies studied. The names of the dead tend to be tabooed in smaller and simpler societies, typically societies where there is subsistence hunting, gathering and fishing, and lack of a written language.[18] People in these societies fear that using the dead person's name, especially if it is unique, risks calling forth that person from the dead.

In more complex societies, greater importance tends to be given to the dead, signs of which are ancestor worship and beliefs in reincarnation.[19] Sharing the name of a dead relative sometimes reflects the belief that the relative has been reincarnated in his or her namesake. The namesake may be considered *to be the same person*, including occupying the same position in a kinship network. In some instances, "name sharing reflects a belief that namesakes should or will share characteristics or attributes."[20]

The way in which people in these simpler societies think about names and personal identity resembles the thinking of the developmentally primitive young mind of children everywhere. They understand the same name to indicate identity. This way of thinking has been described by the developmental psychologist Jean Piaget. He found that children younger than eight often equate a thing or person with its name and believe that it cannot exist without it. Piaget called this "nominal realism."[21] The same kind of thought process that produces nominal realism also equates a person's identity with his or her name. Piaget called this "nominal identity."

Silvano Arieti coined the term *paleologic* for that kind of thought process that underlies nominal identity. Paleologic is based largely on "Von Domarus's principle." According to that principle, a fully rational person accepts identity on the basis of identical subjects, whereas a person employing paleologic accepts identity on the basis of identical predicates. In

addition to adults whose thinking has been disordered by psychotic illness, children aged one and a half to three and a half or four have a
propensity to think paleologically, as do people in primitive societies.[22]

To illustrate, let's apply Von Domarus's principle to a specific statement
that I can correctly make about myself: "I am Steve." In this sentence, "I" is
the subject. The rest of the sentence, the predicate, makes a statement about
me, the subject. "Steve" is not the subject of the sentence, even though it
could be a subject of many others and even though the sentence indicates
that "I" and "Steve" are one and the same. That identification, however,
holds only for "I," the subject of that sentence. About another person also
named "Steve," I could correctly state, "You are Steve," but if I were to identify "You" and "I" as the same on the basis of the same name, "Steve," paleologic would be guiding me to an incorrect conclusion to the syllogism.

Symbolization in dreams follows the rules of paleologic, as do delusions in
psychotic people.[23] However, evidence that paleologic is at work need not be
so blatant. Indeed, the operation of paleological thinking may only be subtly
and implicitly evident. None of us should believe that it does not influence
our thoughts and feelings, though hopefully we are able to edit it rationally
so that it has minimal effect on our behavior. Most relevant to our discussion,
evidence of it can be seen in attitudes and behavior that make it seem as if a
parent is relating to the child as if the child were an extension of himself. If a
parent prone to this kind of thinking assigns a predicate identical to one he
holds himself—his own name—to his child, perhaps that parent might have
more trouble seeing himself and his child as separate subjects.

Lest I leave you with the impression that this kind of thought process is
the sole province of the developmentally immature, severely mentally af-
flicted, or societally "primitive," I should add that the normal process of
identification is not completely rational and utilizes more primitive
modes of thought. For example, recall from the chapter on parent–child
resemblance that twinship transference can predominate as a narcissistic
transference, but it also manifests to a much lesser extent in normal people. I suspect that it is operative in the fascination and immediate sense of
identification a person may have in meeting another unrelated person
with the same name.[24] In addition, a pathologically narcissistic parent
may usually be quite rational, but situations that challenge her inflated
sense of self may bring forth narcissistic defenses, aided, abetted, and subtly contaminated by paleologic.

Cloning, as a way to produce an offspring genetically identical to the
progenitor, though an achievement of an "advanced" society, might evoke
in our minds the same sort of psychologically and socially primitive understandings of identity as seen in so-called primitive societies. The phenomenon of namesaking may model the likely psychological and social
reactions to the genetic identity of progenitor and clone. A mental process

parallel to the equating of nominal identity and personal identity might well infect what we could call "clonal identity." *Clonal identity* reflects the fact that clone and progenitor would be genetically the same. But when the concrete thinking represented by "nominal realism" contaminates logical attempts at understanding, *the distinction among "clonal identity," "nominal identity," and personal identity* may become confused, eroded, or lost in a paleological haze.

Namesaking for Those Closer to Death but Still Alive

Sometimes the person for whom a child is named is not dead but, because of age or illness, is closer to death than most people. For example, the Inuit name their children after aging or ill individuals, usually family members who want the children to "live as an extension of themselves." The child in that society is so strongly identified with its dead namesake that a girl named after a maternal grandfather "is" her grandfather, and her mother addresses her as "father": "The continuity of the name denies death."[25]

In addition, the Iroquois believe that "names possess spiritual forces, and that these forces are transferred to the new namesake." For them, particular names suggest particular, traditional personalities, and the bestowal of names is a way to suggest to their bearers what kind of behavior is expected of them.[26] Naming for kin "provides parents with a means to honor name sources, express and display family continuity (and their transient status), encourage the child's identification with the family, and/or encourage name-source-like behavior."[27]

The motivations for kin cloning may be similar to those for kin naming. Cloning the dead, or even intergenerational cloning of the living, would be the ultimate expression of remembering the dead or honoring living progenitors by the most complex of societies.

Other Than the Names of Parents, Whose Names Are Given to Whom in Our Society?

Although aesthetic preference governs the naming of the majority of children at this time in our society, when children are named for another person, that other person is typically a parent, and most commonly the father, regardless of the child's sex.[28] We will focus on self-namesaking by a parent in the section that follows this one. Here, let us look at which other people are commonly namesaked.

Next to the child's parents, Alford finds that children are most commonly named for friends, neighbors, and grandparents.[29] Less commonly, they are named for great-grandparents and more remote lineal relations and for famous people.

A mother who names her baby girl for her own mother may sometimes be complying with the wishes of a narcissistic mother who has made clear her desire that any granddaughter be named for her. Recall the case of Julie in the chapter on parent–child resemblance. Let me add one detail to it now: Julie's mother on several occasions expressed the strong wish for a granddaughter who shares her name. Julie has postponed motherhood for a variety of reasons, but one factor is reluctance to face defying her mother's wish that she name the baby for her. To Julie, this would be tantamount to symbolically giving her child over to her mother.

In an Oklahoma sample, 2.6 percent of boys and 3.4 percent of girls were named after a famous person.[30] Given the parallel we have drawn between namesaking and cloning, we might expect that a possibly similar small percentage of those seeking to parent a cloned child would want to adopt and parent a clone of a famous person.

WHEN A PARENT NAMES HIS CHILD FOR HIMSELF

Parental namesakes share not only the last or family name but also the first name with a parent.[31] The three cases that follow illustrate some different ways in which having been named for a parent may be an important psychological and social factor in a person's life.

Case 1—Joe Jr.

Joe Jr. is a single man in his forties with a psychotic disorder. Throughout most of his life, he was the meek and obedient son to his dominant, powerful father, a former professional athlete and self-made man. While his father loved Joe, he could also be emotionally and sometimes physically brutal as he compelled his son to follow his directives. Joe Jr. said that his father "used to beat the shit out of me for not being him." Joe Sr. has been dead for several years, but Joe Jr. is not spared the verbal abuse he used to receive. Now, he hallucinates "voices" calling him "stupid," along with other "names." One "name" he hears is a surname similar to his own. If Joe's last name were Spitzer (it is not), then the voices would call him Joe "Shitster." Joe's illness is under enough control that he is able to play a peripheral role with the company that his father started and also named for himself—which Joe's younger brother now directs.

Case 2—Abe Jr., Now Known Simply as Abe

Abe is a mental health professional, friend, and colleague. I had no idea that he was a "junior" because he does not use the name suffix. Abe says he only uses "Jr." on formal documents, having stopped using it otherwise when he graduated high school and left home.

Abe's thoughts about being a "junior" and about why he stopped using the suffix are worth repeating here. He said that his father, now deceased, "wanted to mold me in a certain way, to be like him—or like the person he wanted to be, or the way he thought he should've been." Abe stopped using "Jr." in order "to be my own person," "to let go of something that is no longer me." Letting go of it was "part of a separation process"; "'Jr.' is still part of me, but in a developmental way, not something I display proudly or rely on in my self-image."

As a "junior," Abe felt "special status" as the oldest boy, a boy with his father's name. It seemed fitting to share his father's name because sometimes as the oldest child Abe functioned as a surrogate father to his younger siblings. Being his father's namesake got him "respect" from siblings and cousins, and a sense of "responsibility" came with it.

The "Jr." both made him like and also distinguished him from his father: "It was 'Big Abe' and 'Little Abe'"; "When I was younger, it felt good to be called 'Little Abe.' I was like my father. I felt bigger." This continued until puberty when he told his family that he did not like being called "Little Abe." At that point, being a "junior" and "little Abe" made him feel smaller.

However, being named for his father is something about which Abe is still proud. It is in the tradition of his family and ethnic group to name the firstborn son after the father and assign him the suffix "Jr." In fact, Abe's four brothers all gave their firstborn sons their names and the suffix "Jr." *Abe* did not: "I didn't want to impose a 'III' on him," not wanting "to interfere with his developing his own identity."

Case 3—Charles Richard IV, Always Known to Me as Rick

Rick, a childhood friend, carried the Roman numeral "IV" after his name. He was never called "IV," just as his father had never been referred to as "III," and he was spared the diminutive term of address "junior."[32] Rick came from a family whose roots in the area go back almost 200 years, and his name enhanced his sense of belonging to a long line of accomplished people. Interestingly, though his first, middle, and last names were all identical to his father's, his parents always addressed him as Rick, a nickname for his middle name, Richard. In contrast, his father always used the first name, Charles. Rick's parents intentionally addressed him in a way that distinguished him from his father: "They didn't want to belittle me by calling my father 'big Charles' and me 'little Charles.' They said they wanted me to be my own person."

The only time anyone knew that Rick's first given name was not Richard and that he was a "IVth" was when teachers read his name on their class list the first day of school each year. While some classmates

gently teased Rick for his suffix, I recall finding it rather exotic and confusing, an early experience of a culture and tradition different from my
own.[33]

Like my friend and colleague who is a "junior," Rick does not use his
name suffix now, even though he lives in a different part of the country,
one where numerical suffixes are not as unusual: "After I left the area, it
didn't mean anything. It only has meaning if you know the father or the
grandfather. It would sound a little pretentious to use it." Rick decided
not to make his son a "Vth."

What can we conclude from these three cases of paternal namesakes? All
three demonstrate the importance of parental attitudes toward the
parental namesake. We may think of them as falling along a spectrum
from extreme lack of respect for the namesaked child as a separate,
unique, and autonomous person, to a more mixed picture, to clear respect
for the child's separate unique self *despite* being his father's namesake.

At one end of the spectrum is Joe Jr., whose personal autonomy and
self-identity are quite impaired. Perhaps being a "junior" and a parental
namesake contributed to this, or maybe it is simply a marker of his father's narcissistic motivations for fatherhood. Joe Jr.'s father had no respect for him as an autonomous person, as evidenced by the verbal and
physical abuse Joe Jr. endured from him.

In the middle of the spectrum is Abe. His father very much wanted Abe
to "be a certain way, like him, or the way he thought he should have
been." However, Abe's father was not abusive. Furthermore, Abe felt that
being named for his father garnered him special status in the family and
helped him feel big when he was little.

Rick is at the other end of the spectrum from Joe Jr. His parents appear
to have had an ambivalent relationship with his name. They made him a
namesake of his father, grandfather, and great-grandfather but, by addressing him by his middle name, seem to have tried to undo it for most
practical purposes.[34] Presumably, the implicit message in making Rick a
namesake was that they hoped he would highly resemble other bearers of
his name. However, the explicit message was that they wanted him to be
his own person. To linguistically convey their respect for his separate personhood, his parents called him by his middle name, a name his father
also held but did not use. They also never applied any diminutive forms
of the name his father used to Rick, not wanting to "belittle" him.

Rick's case exemplifies that being a namesake need not be the kind of
burden it can be for namesakes who are also designated with the suffix

"Jr." Rick's Roman numeral suffix signified high social status for his family, and I suspect some of the teasing he endured arose from jealousy over this fact. Though he never drew attention to his numerical suffix, Rick found that its relevance was limited to the region where we grew up; for it was only there that his family name was well known.

That Abe and Rick stopped using their name suffixes in early adulthood may indicate a reasonably well established sense of autonomous identity and a willingness to claim more. In contrast, Joe Jr.'s sense of autonomy is undermined in a most fundamental way—by hostile and insulting hallucinated "voices." I do not believe he ever felt he was anything other than "Jr.," and so he could never even have considered dropping that suffix.

Let us now venture beyond these individual cases, by first constructing a historical and sociological frame for parental namesaking and then viewing within this frame possible parental motivations. We will then proceed to review psychological theory and research on the subject of children named for their parents. Joe Jr., Abe, and Rick serve as individuals illustrating more general findings and ideas.

SOCIAL AND CULTURAL INFLUENCES
ON PARENTS NAMING
CHILDREN FOR THEMSELVES

Before the advent of surnames, people's names were modified by designation of profession, locale, or lineal relationship. In patrilineal societies, for example, Matthew might be known as Matthew the son of John. The many surnames that end in "son" are the result. In Hebrew, the phrase "son of" is *Ben*. You may be familiar with it from the movie classic *Ben Hur*. The use of "Jr." and numerical suffixes would appear to be another example of the practice of patrilineal designation.

Historically, in North America, naming first sons for the father and first daughters for the mother was the predominant mode of naming children in the 1600s. Wealthy parents were more likely to name children after themselves and to name for individuals on the father's side of the family.[35]

In the seventeenth century, over five-eighths of first sons and nearly three-quarters of first daughters were given the same names as their parents. By 1840, this pattern began to change, and by 1861–1880, the frequency had declined substantially, to two-fifths and one-sixth. Thus, by the middle of the nineteenth century, there had been a shift away from the "lineal" tradition of naming children toward the modern conception of children as individuals. This involved less naming for parents or dead siblings and a move toward using middle names to differentiate them.[36]

In the twentieth-century Oklahoma sample, 20 percent of boys were named for their fathers, and none, for their mothers; and 3 percent of girls were named for their mothers, and 5.1 percent, for their fathers.[37] The frequency of paternal namesaking varies regionally within the United States and seems to be greatest on the East Coast. In Richmond, Virginia, 31.6 percent of firstborn Caucasian boys born in 1968 were named for their fathers and given a suffix. That year, an almost identical 29 percent of African American firstborn boys were also paternal namesakes.[38]

The historical trend away from naming children for parents is less evident for families in the wealthier strata.[39] To this day, fathers in higher occupational groups are most likely to name the firstborn son for themselves.[40]

Perhaps these findings can be explained in part by what has been called the "prestige of decent." The name given as a namesake and name suffix may be a kind of "symbolic estate": "In the absence of hereditary and title aristocracy, the *suffix* became a substitute for the *prefix*." Upper and middle classes are thought to have a greater stake in maintaining the symbolic family estate to claim and maintain what they feel is their rightful place in the social hierarchy. As symbolic estate, the namesake name and the name suffix are a claim to status.[41]

There are also geographic, ethnic, and racial factors affecting the likelihood that a child will be made a parental namesake or assigned a suffix, and at least in the twentieth century, historical trends for blacks and for whites are not entirely congruent. For example, at a time when whites became less likely to name a firstborn son for his father, blacks became somewhat more likely to do so.[42] Any conclusions about the psychology of naming parents for children (see below) must take these factors into account or at least consider that they may complicate interpretation of research findings.

Progress toward individuation within the family, reflected in a reduced frequency of parental namesakes, has not gone so far as to divorce individuals in the early twenty-first century from their families of origin. Hence, surnames continue to be passed down. It is interesting to note that radical thinkers early in the American republic saw surnames as a threat to equality and democracy.[43] Perhaps that is because our last names make us partial namesakes of our fathers, and as such, they may also be a kind of hereditary and, hence, antidemocratic "symbolic estate."

INITIATORS, TRANSMITTERS, AND TERMINATORS OF NAMESAKE PRACTICES AND NAMESAKES WHO CHANGE THEIR NAMES

What leads a parent to decide to name a child after himself? What leads a child named for a parent to do the same with his own child? What leads such a parental namesake when grown up to terminate the parental namesaking? There has been no research on any of these questions, all of them relevant by analogy to cloning. Answering them might help us to predict who would be inclined to clone themselves and their motivations, why a clone may not want to clone himself, and whether we might expect a tradition of self-cloning to begin in families of some clones.[44]

There is also no information on the psychology of namesakes who drop their suffixes, though the cases of Abe and Rick suggest that doing so can be a symbolic act of self-redefinition. To my knowledge, namesakes tend not to change their given names but may drop the "Jr." or the Roman numeral suffix. This is analogous to the fact that a clone could not easily disavow that he is a clone of a given progenitor. Although genes cannot be discarded as readily as suffixes, understanding why someone decides to drop a suffix may help us understand why a clone might come to wish he was not one.

How do people who drop their "Jr." or Roman numeral suffix compare with those who do not in successfully negotiating Erikson's stages of psychological and psychosocial development, particularly the achievement of a stable and secure self-identity? Comparing Abe and Rick, who both dropped their suffix, with Joe Jr., we might speculate that dropping the suffix is a symbolic act that indicates that the person feels that he is already standing on his own as a unique and autonomous individual and that the name change may itself further the process. Obviously, Rick, Abe, and Joe Jr. are a small and not randomly selected group, and certainly there are many people assigned a suffix at birth who carry it through their lives without a sense of handicap to their personal identity. Only psychological research with a larger, randomly selected sample, controlling for possible confounding variables, can really answer the question.

Historically, as reviewed earlier, initiation of self-namesaking has decreased since the Middle Ages, and that overall trend continued into the twentieth century, dropping to about 30 percent or less by 1968 along the East Coast of the United States.[45] In a reverse trend, fathers who were themselves namesakes in Richmond became increasingly likely to transmit their name with a suffix to a son from 1930, to 1950, to 1968, the last date for which I can find published data. In 1968, 8.3 percent of white fathers who were themselves namesakes transmitted their own names to their firstborn sons. A countervailing trend over the same years sees an increasing percentage of parental namesakes terminating the practice. In 1968, 15.2 percent of white fathers who were themselves namesakes did not pass on their names to their own firstborn sons. To make sense of these opposite trends, we need to look at the ratio of namesake fathers terminating the practice to those continuing it with their own sons. In 1930, the ratio was about one to three; in 1950 it was about equal; and by 1968 it was about two to one, with those terminating the practice outnumbering those continuing it. Thus, it appears that there is an overall lessening in the popularity of parental namesaking, including that by those who are themselves parental namesakes. Occupational status and race interact with this overall trend but do not fundamentally alter it.[46]

We should note that the sociologist Rex Taylor, who looked at parental namesaking in Richmond, appears to have implicitly assumed that the fathers were the name-giving parents. Although the research tells us about paternal namesaking having been initiated, transmitted, or terminated, we do not know which parent made the name assignment. Richard Alford finds in a sample from Iowa that father and mother together decide the name of the firstborn boy about 60 percent of the time and that, when one parent is the primary decision maker in this regard, it is usually the mother who names the child.[47] When one parent is the primary decision maker regarding the naming of a child, we do not know if a parentally namesaked child is more likely to have been named by the namesake parent.

When it comes to changing one's first name, this was a rare occurrence in Hingham, Massachusetts, before 1790. The most frequent name changes sought were the adoption of a middle name and the dropping of the suffixes "Jr.," "III," "IV," and "V." Such changes are thought to reflect an emerging societal desire for individuation.[48] Other sorts of name changes may be indirectly relevant to the issue, but few have been studied.[49] To the extent that namesaking is psychologically and socially analogous to cloning, the latter practice would appear to run counter to a strong historical trend in our culture.

I have not heard of any individual not named for a parent making himself his parent's namesake by changing his name. Perhaps we should give the same degree of significance to this non-occurrence as we have given to patterns of occurrence in naming of children.[50] By way of analogy to cloning, we could conclude that no normal adult would wish that he were a clone of his father or mother.

"Jr." and numerical suffixes are avoided completely when the child is a parental namesake in all but his middle name. The George Bush family is a case in point. How do such nearly complete namesakes versus complete namesakes, suffix included, fare psychologically? A child's individuality might be better respected by the practice of addressing a namesaked child by his middle name or initial or a nickname to distinguish him from his parent. We may assume that many parents of children they have named for themselves, like Rick's parents, try to respect the children's individuality by calling them by a different name. This understanding finds indirect support in research showing that an elementary school boy who is a paternal namesake is more likely than a boy who is not to use a substantially different name in identifying himself.[51]

We cannot change who we are genetically, but we can change other things, including our names. Clones, if they were also namesakes of their progenitors, would be able to change their names, but they would not be able to change the fact that they are clones—and clones of a particular progenitor.

PSYCHOLOGICAL THEORY AND RESEARCH ON CHILDREN WHOSE PARENTS NAMED THEM FOR THEMSELVES

Three times as many "juniors" were found on a psychiatric outpatient clinic roster of patients than in the phonebook of the city where the clinic is located. At a Veterans Hospital in the same city, there were twice as many people with the "Jr." suffix among psychiatric than among medical-surgical patients.[52] The researcher makes no claims of statistical significance, but the results suggest a higher incidence of emotional problems among "juniors." Disturbance in conduct also appears to be more likely for "juniors" than for adolescent boys not named for their fathers. When adolescent boys in a home for delinquents are compared with boys in local high schools, delinquent boys are twice as likely to be "juniors." When boys who are also paternal namesakes but do not use the "Jr." suffix are added to these, the delinquent boys were more than twice as likely to have been named for their fathers. These findings hold regardless of race and ethnicity.

These paternal namesakes were twice as likely to have been abused by their fathers than delinquent boys who had not been so named. Although it is commonly believed that "juniors" are likely to be firstborn sons, this was not always the case at the home for delinquent boys. Regardless of birth order, paternal namesakes were more likely to have been subject to paternal abuse.[53] Even boys who shared a middle name or initial with their fathers were more likely to have been abused by them than were boys who shared no name or initial. It appears that the likelihood of having been abused was roughly proportional to the extent to which the boys shared their fathers' names.[54]

While I want to emphasize the relationship between abuse and being a paternal namesake, the statistic was influenced by racial variables: Having been named for one's father correlated most strongly with having been abused for Hispanic boys and less so for non-Hispanic whites. For blacks, there was no relationship between whether a boy was a paternal namesake and whether he had been subjected to parental abuse.

We should not be surprised by the history of prior abuse among these boys, as delinquent behavior is often associated with it. When parents have unrealistic expectations of a child and perceive him as different, they are more likely to abuse that child. Naming a child for a parent may mark that child as different, as a focus of excessive parental expectations, and an increased likelihood of abuse may be the consequence.[55] But perhaps the namesaked child is different in a very specific way that puts him at risk for abuse: He is bound to be different from the hopes and expectations of the parents that he be the same as or closely resemble his father. Think back to the chapter on parent–child resemblance in this regard.

Robert Plank, a pioneer in the psychology of personal names, thinks that "a boy is named 'Junior' in deference to the father's wish to perpetuate himself in his son, to have his son emulate him, and at the same time to retain a clearly superior role himself." Plank expects that this would aggravate the "natural conflict that may develop between father and son."[56]

It may also be that "a son bearing his father's name may mirror that parent's own failures and evoke abuse, or (with breakup) he can become a scapegoat by his mother [or stepfather?]."[57] It is important to note that the abuse described in this study is "parental," not necessarily paternal, though it is reasonable to assume that the father plays an important role in it.

As we have seen in other chapters, psychological research typically starts with a specific population with mental health problems and examines whether a given variable is more commonly found in that population than in an otherwise comparable group. Then a larger population is studied to see whether that variable distributes itself in a way that suggests that a mental health problem might be associated with it. In previous chapters, being an identical twin or an adopted person was one such variable. In this chapter, it is being a parental namesake.

A population of parental namesakes not selected for mental health problems was identified from all entering freshman at Wesleyan University between 1966 and 1973. All entering freshman were given the California Psychological Inventory. The results were compared for students with and without the name suffixes "Jr.," "II," "III," and "IV"—clear indicators that they were parental namesakes.

The most powerful finding is one that we often do not think about, probably because we assume it as members immersed in our culture: All names with suffixes were male.[58] This does not necessarily mean that none of the female students was named for her mother, but we can assume that the incidence was much lower than that for boys being named for their fathers. In this context, recall that in the Oklahoma sample, 20 percent of boys were named for their fathers, and none, for their mothers, while 3.3 percent of girls were named for their mothers, and 5.1 percent, for their fathers. When parental namesaking occurs, it is more likely to be for the father, even when the child is a girl.[59]

That all names of the Wesleyan University students with suffixes were male and that many times as many boys as girls are named for the same-sex parent reflect a powerful gender-related disparity in an important cultural practice. *Applied by analogy to cloning, more males than females are likely to be cloned.*

Beyond the dramatic gender difference in Wesleyan students being parental namesakes, interesting findings emerge when the psychological profiles of individuals with the name suffix "Jr." are compared with individuals with the suffixes "II," "III," and "IV." Those with the Roman numerals score higher than those with a "Jr." appended to their names on eight scales: capacity for status, well-being, responsibility, socialization, self-control, tolerance, achievement via conformance, and intellectual efficiency.[60]

The researchers speculate that "possessors of a 'Jr.' may feel as though they are the lesser or smaller versions of their fathers," a feeling that Abe in adolescence came to have about his "Jr." suffix.[61] The "'Jr.' title reminds everyone, and the namesake that he is younger, most likely smaller [in his earlier years] . . . and lesser in status *than the real thing.*"[62] For clones, their progenitors would be the "real thing."

In contrast, the Roman numeral suffix appears to have quite different meaning to the individuals carrying it, who may feel that they are part of a long line of respected individuals of the same name. As a "IVth," Rick owes some of his apparently good self-esteem to being part of a locally distinguished line. The Roman numeral suffix reminds him of his lineage and heritage. Good self-esteem can be inferred from the profiles of those with Roman numeral suffixes on the California Psychological Inventory. In this respect, it is interesting to note that "IIs," identical to "juniors" in having been the first generation to be named for their fathers, scored higher than "juniors" on capacity for status, well-being, achievement via conformance, and intellectual efficiency.[63]

Regarding achievement through conformance, we might conclude that a "II" compared with a "Jr." has less reason to rebel, not having to assert that he is not a lesser version of a "senior." Those with other Roman numeral suffixes also scored higher on this scale. Might a person carrying such a suffix be more likely to conform to his parents' wishes and expectations and so be at greater risk for the "foreclosed identity" status discussed in chapter 4?

In my view, while both "Jr." and "II" indicate that a person is the second generation bearing the same name, the "II" suffix invites the bearer to continue the tradition, joining the original name bearer in making himself senior to those in succeeding generations. A father namesaking a son and assigning a suffix of "III" or higher is calling attention not so much to himself but to his forebears. On the other hand, perhaps a father who names his son for himself and assigns him the "Jr." suffix is more likely to also call attention to himself as the more powerful "senior," possibly viewing his son as a narcissistic extension of his enlarged self.

These psychological interpretations have never been researched, but they are consistent with the data and are relevant by analogy to the attitudes of a progenitor parent toward his child clone. Whether a clone is to his progenitor as a "Jr." or as a "II" namesake is to his father would almost certainly be influenced by the attitude of the parent toward his clone.

There was one scale on the California Psychological Inventory that did not differentiate "juniors" from those with numerical suffixes but did differentiate all of the namesakes from those students without any name suffix. I refer to the "femininity" scale. On this scale, those with name suffixes scored lower. Perhaps being a namesake of one's father enhances identification with him as a man. But, we may ask, *what is the optimal degree of identification*? For some of these young men, might the low femininity score possibly reflect the influence of an overbearing father who insisted on a degree of identification that demanded denial of any feminine aspects of self?

We have already seen evidence from the earlier mentioned studies on naming a child for oneself that doing so can produce "heightened identification, frustration, conflict, and, sometimes, aggression" in the parent.[64] The association among being named for a parent, parental abuse of that child, and childhood delinquency also suggests that the heightened identification, frustration, conflict, and aggression may be present *in both the child and the parent who made that child his namesake*. The child may feel favored at first by being the father's namesake, as Abe felt, but also feel overwhelming expectations to strongly identify with his father—to be like him or else.

The disparities in power and dependency of the parent–child relationship, to which I have referred in previous chapters, may be reinforced and symbolically magnified when a parent names a child for himself, especially if he also assigns the child the suffix "Jr." As we have already seen, psychological issues relevant to resemblance and identity appear to be intensified by the inequalities inherent to the parent–child relationship, and so it is not surprising to find them coming into sharp focus with parental namesakes. That focus becomes even clearer when famous parents name their children for themselves.

WHEN A NAMESAKE IS ALSO A CHILD OF A FAMOUS PERSON

While nothing is to stop a child unrelated to a famous person from being named for her, the child invariably retains her own family's last name, so these children are at best partial namesakes. More interesting and relevant to cloning is when the child is named for one of her parents and that parent happens to be a famous person. Famous last names can come to be seen as brands, and the brand effect is maximized if all of the person's names, but especially first and last, are identical to his predecessor's.[65]

In an example of both the Child of the Famous and the Namesake Models, Andre Dubus III, a short story writer like his father, says that he has often felt like "Elvis Jr.": "People would see a story by Andre Dubus—it probably got my name to the top editor before it got rejected."[66] Another example is that of George P. Bush, a nephew of George W. Bush, son of Jeb Bush, and grandson of George H. W. Bush. George P. Bush admitted that "his instantly recognizable name had been a passport to extraordinary experiences" but in some ways had also been a burden and a responsibility: "'It's kind of an unwritten law, I think—carrying the name—George Bush' he said. 'At times I ask myself, Am I going to be able to live my own life the way I want to?'"[67] In the Child of the Famous Model we briefly consider another member of that family, President George W. Bush. Perhaps sharing both his father's first and last names might have added to his status as a child of a famous man, but it must also be an assigned component of identity with which he has had to contend.

Recall that the identical twins Jerry and George, whom we first met in the chapter on that model, have a brother, Sam Jr., named after their locally famous father. Jerry has told me that Sam Jr. believes that he has had a hard time in establishing himself as a person who respects himself and is respected by others for his unique individual self because he is both a namesake and a son of his famous father.

WHEN A PARENT IS NAMED FOR HIS OR HER CHILD

"Teknonym" is a form of naming in which the parent derives his or her name from a child. It is common in the Moslem world. A man might be known as Abou Daud ("father of David") or a woman as Um Daud ("mother of David").[68] As the father of a boy in elementary school, I have some familiarity with this form of address, known by more children as "Noah's dad" than by my own name. Loving and admiring my son as I do, I am proud to be known by this teknonym. It does not acknowledge my individual identity, but it does emphasize a piece of it that I feel very good about.

Fathers who self-namesake a son and assign him the suffix "Jr." sometimes also give themselves the suffix "Sr." The "seniors" have never been studied, but it would appear that these fathers are giving themselves a self-reflexive teknonym.

Teknonyms are relevant by analogy to people wishing to adopt and parent clones of famous people and celebrities. Such people might well share similar traits with people who name their children for famous people. They may be inclined to draw attention to the fact, hoping to bask in the glow of the progenitor's fame, and so may want to be known as the "father of" or "mother of" the famous progenitor's clone. The chapter "The Child of the Famous Model" discusses psychological and psychosocial risks that these children may also run.

The adoptive parents' of a clone of a famous person might try to make their status as his parents socially obvious by making the clone a partial namesake—melding the coveted identity of the progenitor with their own name. While copyright law protects the names of companies and products, the names of individuals are not so protected. The adoptive parents of a clone of a famous person might feel justified in appropriating the famous progenitor's name for their child and, by extension, for themselves in a teknonym.

To illustrate, let us imagine that the brilliant theoretical astrophysicist Stephen Hawking was willing to make himself available for cloning, if, by way of genetic modification, his clone could be spared the neurological disease that afflicts him. Let us also imagine that a couple admiring him, such as my wife and me, wanted to adopt and rear his clone. My wife could elect to carry the pregnancy, or we could hire a surrogate.

If we were very invested in the child being like his famous progenitor and wanted everyone to know both his identity and that we are his clone's adoptive parents, then we could name the child Stephen Hawking Levick. Every time his name was written or spoken in full, the fact that the child is a clone of that famous person would be advertised. Because Stephen is my first name, too, the boy would also be my namesake except for his middle name. At the same time, while not formally a teknonym, the name Stephen Hawking Levick functions like one. It simultaneously emphasizes that *I* am the adoptive *father of* this clone of a *famous person* and that *he is my son*.

More generally, making the child a namesake may serve a teknonymic function for a parent wishing to take credit for a child's accomplishments, as narcissistic parents are wont to do. Individuals and couples who choose cloning, as a route to both parenthood and enhanced self-esteem, might be inclined to select progenitors whose names have achieved positive near-stereotype or brand status.

We might expect that when a clone is named for his famous progenitor, being his namesake might have an even more profound impact on the child's identity than when a famous parent gives his name to a noncloned offspring.[69] Whatever psychological burdens a clone may shoulder, being named for one's progenitor, famous or not, would likely compound them. *Such a child would be a clone in both name and genes.*

CLAIMING A CHILD AS ONE'S OWN
AND ASSISGNING PATERNITY

One way in which parents claim a child as their own is to name him or her for one of them.[70] The child shares the parents' last name, but usually the first and middle name assigned to him acknowledges the child's individual identity within the family. However, regardless of the name the parents give to a child, simply the act of selecting a name for him claims him as theirs. In the biblical story of Genesis, God creates the world and all nonhuman life and only then creates man and gives him dominion over his other creations. *Naming* them was a key exercise of man's dominion.[71]

Naming a child for a parent claims a child doubly for the parent for whom he is named. He will not be able to do anything about it until much later and may drop a suffix, as my colleague Abe has done, rather than changing his name more fundamentally.

The notion of "naming as claiming" finds further support in research on name assignment by parents of adopted children: Adopted children are more likely than genetic children to be parental namesakes.[72] It would appear that lacking the genetic tie, adoptive parents are more likely to take an extra step in claiming the child as theirs by making him a namesake of one of them. In this regard, we should note that although nonadopted children are more likely to be given a name from the father's than the mother's side of the family, this is not true for adopted children. Here there is a nonsignificant trend in the other direction. Not having borne the child from her womb, the adoptive mother cannot claim the child from that experience and is more likely to claim him by other means. One of these would be to assign the child a name from her side of the family.

While adoptive parents actively seek to become parents, the parents of children born out of wedlock generally have not, and sometimes the paternity of the child is in question. How do patterns of naming these children reflect these facts?

It appears to be an advantage for children born out of wedlock to be named for their fathers, at least in their early years. Such children are more likely to receive support from and have contact with their fathers during their first five years. For children born out of wedlock, naming the child for the father appears to enhance the long-term relationship of father and son and the father's financial support. It also lessens the risk that these sons will have behavioral and cognitive problems. It appears that mothers bearing children out of wedlock may intentionally name the child for the father to strengthen the bond between father and child.[73]

These findings would appear consistent with those on naming and adoption—but with a twist. Where the adoptive parents appear to be *claiming the child as theirs by naming him after one of them*, the mother of the out-of-wed-lock child seems to be *asserting the claim that a particular man is the child's fa-ther*. By asserting this claim, she is *claiming the child for the father by proxy*, and the father appears to respond. She has assured the man of his paternity or has tried to compensate for any insecurity he may have in that regard.

These findings make it appear that naming an adopted child after a par-ent also may be an attempt to enhance bonding with the child by the par-ent namesaked. The fact that biological children are more likely to be pa-trilineally than matrilineally namesaked may reflect a lesser degree of confidence experienced by the father compared with the mother that the child is kin.[74] Evolutionary theory on sex-biased parental investment may explain the fact that nonadopted male children are more likely to be pa-trilineally namesaked than female children are.[75] It is an obvious and in-escapable fact that at the time of birth we only can be certain of the iden-tity of one parent, the mother.

Might concern about paternity be a source of problems experienced by namesakes? If a father who names his child for himself is less secure that the child is his, then he might be less likely to treat the child well. This seems a reasonable supposition, given the data for stepfathers. A stepfa-ther knows for a fact that he is not the biological father of his stepchild, and as we saw earlier, being a stepchild is a powerful risk factor for abuse by the stepparent. A narcissistically insecure man, though he might not seem it, at a deeper level is probably insecure sexually. We might, there-fore, expect him to have some degree of doubt about his children's pater-nity and to be more likely to assert vigorously that he is the father by nam-ing the child for himself.[76] However, such a father's underlying insecurity would remain, possibly manifesting in abuse of the child. Of course, such a child would also be subject to other deleterious consequences of narcis-sistic parenting, reviewed in the Parent–Child Resemblance Model. Just as self-namesaking may seem a balm to the narcissistically insecure man, so might self-cloning. However, the father would inevitably be frustrated that his clone was not living up to his idealized expectations and just as frustrated should his son outdo him.

Research on naming of adopted and out-of-wedlock children is relevant by analogy to cloning. It points to the likely need for parents adopting a clone from a progenitor other than either of them to claim that child as theirs. One way they might do this is to do what so many other adoptive parents do in naming the child for family members or for themselves. We may draw a related analogy between the alleged father of an out-of-wed-lock child and the nonprogenitor marital partner of a child clone of the other parent. As it would be absolutely clear to her that she is not a ge-netic parent to this child, the nonprogenitor partner might need to bolster her status as his rearing parent. While a mother might do this by carrying the pregnancy of a non-self-clone, a nonbiological way to highlight her status as the child's mother might be to name the child for herself. If pregnancy is not an option, then a nonprogenitor parent, whether male or female, might feel it to be more imperative to name the child either for him- or herself or for a "blood" relative.

Based on this, we may also anticipate some likely considerations in the naming of clones. If the purpose of namesaking is to claim the child where there is doubt of a genetic bond, then a man cloning himself should have no doubt and so be less motivated to name the child after himself. The child is clearly his. However, the deep insecurity of the narcissist would probably not be assuaged by this reality. Additionally, the desire for self-resemblance in the child, already expressed in the wish to self-clone, might also lead him to name the child for himself.

The wives or partners of men cloning themselves might be less likely to want to name the child for the father because, as progenitor, his paternity is indisputable. However, a father who is not primarily motivated by narcissism might wish to have his son bear a name from his wife's side of the family—as a way to claim the child for her, too. Doing so would symbolically proclaim her relevance to a progenitor–clone dyad that could threaten to sideline her. If a woman were to clone herself within a heterosexual marriage, then the same sort of issues would apply. If a self-cloning person were homosexual, his partner might very well want to name the child for himself to try through the nominal equivalent of cloning to assert the claim that this child is also his.

NAMESAKING AS AN ATTEMPT TO HEAL A WOUNDED SELF

Case 1

A case is mentioned in the neonatal nursing literature of a father of a stillborn son who gave his dead child his name. The father had always disliked his name but felt that it was all right to give it to this firstborn son, knowing that his dead son would never experience any negative consequences in carrying it.[77]

In this case, we may speculate that this father named his stillborn son for himself to try to free himself symbolically from the name he had negatively experienced as a child. By giving the child a name he himself had no choice in receiving, the boy's father may have been striving for a sense of mastery and control in the naming process. In addition, his son could replace him as the child carrying the name and, because the boy was dead, would not be harmed by the name, as the father felt he had been. Further, because the child was dead, the father could bury with his namesake painful childhood memories associated with his name. In naming his stillborn son for himself, the father would also get to both mourn himself and send the painful memories to the grave.

There was nothing this father could do about his son's death, but he was able to capitalize on it psychologically. Similarly, some might not only mourn a failed cloning attempt but also use it as an opportunity to try to symbolically shed unwanted aspects of the self.

But a child need not be dead for the parent to self-namesake her to try to heal a wounded self. Any self-namesaking could have this as a major or subsidiary motivation, and quite likely the motivation would not be conscious. Refer to the Parent–Child Resemblance Model and the discussion of attempts by the narcissistically injured parent to repair his wounded self through his child. Recall how the child is expected to be everything the parent wanted himself to be, and we can see how any self-namesaking may be an attempt at healing the wounded self of the parent. As such, taking ethnic heritage and tradition into account, it may be a possible marker of parental narcissism. By analogy, similar motives might be at work, probably unconsciously, in self-cloning.

Case 2

Erik Erikson, the renowned psychological expert on identity, was never told his biological father's name and was adopted at an early age by his mother's second husband, Theodor Hamburg. Erikson said that "we create ourselves in a way. When I became an American citizen, I realized that I could take a name of my own choosing. I made myself Erik's son. It is better to be your own originator."[78] By becoming his own namesake, Erik Erikson symbolically declared his independence from a biological father he never knew, from a mother who would not disclose his father's identity, and also from an adoptive father he did not fully accept.

Because a person's last name is typically an assigned piece of identity identical to that of one or both parents, most of us are *partial namesakes*. One might object that using the term *namesake* this way has little distinguishing meaning, but when the father's identity is not known or disclosed to the child, it is relevant. Such was the case for Erikson, who, not knowing his biological father, was denied the normal partial namesaking of bearing his biological father's last name. Erik Erikson tried to solve that problem by making himself the father of himself. In linguistic self-reflexivity, he made himself his own namesake, symbolically trying to make himself, as he put it, his "own originator." As applied to cloning, progenitors might mistakenly feel as though they are their "own originators," and if they treat their clones as part of themselves, they will be denying their offspring the opportunity to create themselves as individuals.

CONCLUSION

In condensed symbolic form, namesaking integrates the parents' desire that their child resemble the person for whom she is named with their wish that the child also replace or perpetuate that person. In the case of naming a child for someone other than a parent, namesaking of the deceased is a way to honor the memory of that person and partially embody that memory in the child by making him a replacement in name of the lost loved one. In some instances, such namesaking may be a marker of replacement child status. Self-namesaking may reflect a parent's wish that the child resemble him perfectly or at least perfectly resemble his ideal. As a form of self-replacement by self-perpetuation, self-namesaking may be an attempt to deny death and achieve immortality.

The developmental and psychopathological concept of paleologic helps us make sense of how some simple or "primitive" societies conflate a person's name with identity. Such thinking may also contaminate the thinking of normal people in advanced societies, leading them to misperceive namesakes "as if" they were the same people. By analogy, the same risk might apply to how some would perceive clones.

Although parental namesaking was once the norm in Western and, more specifically, American society, parents have become less inclined to name their children for themselves. Granted, birth order, race, social class, geography, and uncertain paternity all appear to be separate, sometimes interacting social, and in some instances possibly evolutionary psychological factors influencing the incidence of parental namesaking. However, in a broad historical context, these factors produce only interesting eddies in what is clearly an overall current toward less paternal namesaking, reflecting greater respect for the individual. Both literally and figuratively, we might understand the practice of namesaking as a nominal artifact of earlier sociocultural epochs. Cloning would be the reproductive analogue of namesaking, and although thoroughly modern and radically new scientifically, it would appear to be a regression or throwback in cultural evolution to a few hundred years ago, if not the Middle Ages.

Just as parental namesaking may be a sign of psychological and social forces countercurrent to historical progress toward human liberation, the same might be said of reproductive cloning. Both might reflect and also interfere with the liberating effects of perceiving and treating children as separate beings. While *of* their parents and in need of their guidance, children do not *belong* to their parents but, rather, are individuals with their own desires, thoughts, feelings, and selves.

We might suppose that as succeeding generations of parents move in this direction in rearing their children, those parents in current and future generations who do not will be more likely to be narcissistically motivated.

We have inferred from psychological research on namesakes that men who name their sons for themselves are more likely to be narcissistic and have surmised that, for these men, self-namesaking might be an attempt to compensate for sexual and procreative insecurity. Likewise, self-cloning might also be an attempt to eliminate concerns about paternity and sexual insecurity, more reasons to make it particularly appealing to a narcissistic man.

The psychological and social costs for a child having parents who feel that he belongs to them is suggested by psychological theory and research on children whose parents named them for themselves. We have seen that there appear to be more "juniors" with psychiatric and delinquency problems and that paternal namesakes are more likely to have been abused by their parents, a possible cause for their problems. We have reasoned that narcissistic parents, their expectations having been disappointed, would be more prone to react in an abusive manner with their child namesakes.

On the one hand, these findings might seem counterintuitive: Being a paternal namesake might benefit a boy in his early years, strengthening a father's bonding and identification with his son, helping to ensure financial and emotional support. It might also help the boy identify with his father, strengthening his masculine identity.

But what can be a plus at one developmental phase may become a detriment at another. When the child parental namesake enters adolescence, a time when an individual attempts to develop his own unique personal identity, he reworks earlier identifications, and a degree of rebellion is normal and healthy. A parentally namesaked child, particularly a "junior," has more to rebel against to assert his autonomy and develop independence. Recall how Abe felt "bigger" as a boy, being a "Jr." to his father, but in adolescence, bearing that suffix made him feel "smaller."

Apparently, boys carrying a Roman numeral suffix feel that they have less to rebel against than do "juniors," and their self-esteem is better.[79] A strong positive identification with their fathers may account for their greater ability to succeed through conformance, but it might also put these boys at risk to develop a "foreclosed identity."

Would a clone be more likely to be like a child with a "Jr." or a Roman numeral suffix? It certainly would depend on how the child is reared. In theory, cloning need not be psychologically detrimental for the child, even if the clone were of one of the parents. Some of the potential psychological pitfalls of cloning might be avoided or greatly ameliorated if parents raise their child clone in a way that emphasizes his unique individuality. That said, however, the very fact that the parents decided to become parents through cloning might well indicate that they have certain psychological issues that make it more likely that they would rear the child in ways that might lead to difficulties. Even in the unlikely event that she was not an embodied expression of parental transference, the clone might still not be free from the pressure to conform to familial and sometimes social expectations that she be like, if not "as if," her progenitor.

Self-namesaking a child is also relevant to Richard Dawkins's concept of "memic immortality."[80] Dawkins sees two forms of immortality: genetic and memic. Memic immortality refers to a person living on through her ideas and achievements. In my view, self-namesaking one's child may be an attempt at double self-perpetuation, both genetic and memic. A person who would self-clone might be trying to attempt by way of absolute genetic self-perpetuation the same thing a person naming his child for himself may be attempting to accomplish memically.[81]

But a name is the simplest of memes, the memic analogue of a virus. Just as a virus can do nothing without a host organism, by itself a name means nothing. It is what one associates with a name that gives it meaning. What a person's name means is determined by what she has done and how she has affected others—in other words, what sort of contribution she has made to society. As we have seen in the Child of the Famous Model, a child with a famous name has a name with meaning he has not shaped. He must struggle with that fact in trying to create his own autonomous identity. Will he be able to have his name come to mean something apart from his famous namesake, even if only to himself?

It appears that the Namesake Model reflects many psychological and social aspects of cloning better than other models that are clearly more biologically relevant. The Namesake Model also suggests research that can be done now. There are tens of millions of living namesakes in all stages of the life cycle. Psychological research with some might help to answer the kinds of questions raised here. By analogy, such research might also help us be better able to anticipate the psychological consequences of cloning.

Research with namesaking parents and their namesake children is also essential. It could test some of our inferences and conjectures regarding possible parental motivations for namesaking and, by analogy, cloning. It could also help to put in a developmental perspective any initial indications that clones might not be harmed psychologically and might even derive psychological benefit from being clones. Through further study of namesaking practices throughout the world and over the centuries, we might also gain insights about the social, cultural, and political implications of cloning for future generations.

But with all that we have looked at and analyzed in this chapter on the namesake and what it implies for clones and cloning, isn't the fact that a name is purely symbolic still a weakness of the model? Yes and no—if a

name as a symbol of identity can produce some fairly powerful psychological effects, then imagine how much greater the consequences might be
for a clone as a genetic namesake. We will see more fully in the next chapter, where we integrate the models, that sometimes a model's apparent
weakness can be among its greatest strengths.

9

The Models Integrated

It is time to try to put it all together. We have looked at eight models relevant by analogy to possible psychological and social consequences of cloning.[1] From the outset, it was clear that no single model would perfectly capture the psychological and social essence of cloning, but that is in the very nature of what it means to be a model. Within the domain of each of the eight models presented here we have analyzed scenarios and phenomena conceptually relevant by analogy to the biological, psychological, and social aspects of cloning. We have found the theoretical scaffolding of each model buttressed at points by clinical and empirical data and have discovered other points that future research might usefully address.

Standing alone, each of the eight disparate models makes its own contribution to anticipating possible psychological and social consequences of cloning, but by integrating the models we shall find areas of convergence and complementarity. We can have the most confidence in the validity of predictions in those areas.

In this chapter, we will synthesize the models to see what we may reasonably conclude. Then we'll look at cloning from key perspectives informed by the models and their integration. Foremost, we shall take the perspective of the clone and then that of his progenitor and parents. Finally, we will briefly look at cloning from the perspectives of siblings of a clone and those promoting or facilitating cloning.

Once again, it is important to distinguish between practices to which each model refers (e.g., adoption, namesaking, etc.) and the practices themselves. With rare exceptions, the psychological and social risks resulting from any of these existing practices are insufficient reason to judge any of them unacceptable. However, it is quite another matter to use any given practice as a model relevant by analogy to analyze a new practice, reproductive cloning.

RATIONALE FOR
A CONCEPTUAL INTEGRATION OF THE MODELS

The first step in coming to conclusions about which we can have the greatest degree of confidence is to integrate the models conceptually. Each model has its specific conceptual or empirical strengths and weaknesses. However, in several instances, we have logically construed a given model's weakness as strength. That is because the nature of the weakness is such that even weak findings within those models might be greatly amplified as applied to cloning.

But we need not rely solely on this sort of reasoning because we have eight models that relate to each other interdependently in many areas. Where one model is weak, another model is strong. Not only does one model's strength shore up another model's area of weakness, but also a model strong in one area may be weak in another and a model weak in one facet may be strong in another. The integrated architecture of the models is strong because of the conceptual complementarity among models. This feature should lead us to attend to any redundancy in concerns raised by the very different models. Typically, arriving at the same answer in more than one way makes it more likely that it is correct. It is likely that where the models of cloning converge with parallel or triangulating evidence, we are encountering valid core issues and sound predictions for the consequences of cloning. Let us begin to synthesize the eight models by taking each in turn, analyzing their conceptual strengths and weaknesses, looking for how they interrelate, and noting convergences with other models in the predictions they make for cloning.

THE MODELS CONCEPTUALLY INTEGRATED

The Identical Twin Model

Conceptually, identical twins help us to understand a particular kind of cloning that occurs naturally, by embryo splitting. However, it is far from a perfect model to understand cloning by somatic cell nuclear transfer. Nonetheless, the model is strong in showing how resemblance and sameness may be confused. Given the fact that an age difference would lessen resemblance between clone and progenitor, we might think that the typically strong resemblance of identical twins is too strong to model cloning. Nonetheless, the same degree of resemblance would exist between clone and progenitor, not contemporaneously, but between the clone and the progenitor when he was of the same age as the clone. Knowledge of how the progenitor was at that age might create as great or greater expectations that the clone should be like his progenitor, compared with expectations of identity in identical twins. And not only would the expectations be greater; they would all be in one direction. With identical twins, we have no greater expectation that Joshua be like Jack, than Jack be like Joshua. However, we would be much more inclined to expect a self-clone to be like his progenitor than to expect a progenitor to be like his self-clone.

We have reviewed clinical literature in the Identical Twin Model that raises concerns that twins might be more prone to problems related to attachment, separation-individuation, and development of a separate self-identity. However, research with nonclinical populations of identical twins does not find them to have a greater risk of problems in those areas. Here, we should not be reassured by the lack of strong support for those concerns in twins to conclude that these same issues would not be problematic for clones. That is because the Identical Twin Model is clearly deficient in a key feature likely to be found in most instances of cloning. While identical twins are a genetically identical pair of the same age, a clone produced by somatic cell nuclear transfer would not be the same age as his or her progenitor. Furthermore, identical twins have a sibling relationship, whereas the relationship of progenitor and clone would most likely be that of parent and child.

This weakness of the model makes the profound and lasting grief following the death of a twin even more concerning for clones. It implies that grief at the loss of a clone or of a progenitor might be accentuated. We have also inferred that the special bond between twins might reciprocally decrease their ability to bond with others in an intimate way. If true, then this would imply that a clone might have a diminished ability to become close with people other than the progenitor.

That the relative dominance versus submissiveness in the twin pair seems to have no lasting detrimental effects may be explained by this key weakness in the model—it does not capture the unmatched asymmetry in dominance and dependency characteristic of the relationship of parent and child. At its worst, this is the relationship of master and slave. This deficiency in the Identical Twin Model is a good example of where the weakness of a model in one area becomes a strength, providing a logical basis for our dismissal of a conclusion as not relevant to cloning.

The social experience of identical twins might make it a relatively strong model of that dimension for clones. However, lacking the feature of significant precedence of one person genetically identical with another weakens it. In this regard, the Parent–Child Resemblance and, more particularly, the Child of the Famous Models are stronger. The Namesake Model converges with these others in suggesting that there would be significant social expectations that the clone be very much like her progenitor.

The Assisted Reproductive Technologies and Arrangements Model

Cloning is itself an assisted reproductive technology, a conceptual strength of the model that carries that name. It is likely that a sizeable proportion of people seeking cloning would do so because of infertility problems. Another important strength of the Assisted Reproductive Technologies and Arrangements Model is the fact that some types (donor insemination and ovum donation) produce children linked genetically to one parent and not the other.

We have seen that when assisted reproductive technology makes it possible for the mother to not be linked genetically to her baby, there may be less maternal warmth and greater mother–child conflict. This finding should lead us to interpret cautiously the lack of demonstrated problems with children born through such assisted reproductive technologies. Also, the fact that the children have not been studied into adolescence, when many psychological problems can emerge, forces us to turn to other models wherein individuals have been studied at this and later stages.

In the Assisted Reproductive Technologies and Arrangements Model, we see powerful social and cultural influences exerting themselves through sex-selective abortion and sex selection techniques in artificial insemination. The issues of disclosure and secrecy are well illustrated in that model, as well as in the Adoption Model.

The Stepchild Model

The Stepchild Model shows us that the stepchild is at risk of abuse by the stepparent. The model is weak in that although a stepchild has another genetic parent, the clone has no genetic "parent" other than the progenitor. Furthermore, the couples deciding to have a child through cloning are likely to decide jointly to become parents in this way. When it comes to self-cloning, a parent might sometimes wish to clone not herself but, rather, her partner to encourage the progenitor parent to attach to and be responsible for the child, perhaps at the same time hoping to strengthen the couple's bond.

The weaknesses of the Stepchild Model are partially compensated for by the Adoption Model in which both parents decide together to adopt a child, choosing a child to whom neither is genetically linked. Nonetheless, parents typically want their child to resemble them as much as possible, helping them feel like he is "one of us." This wish could literally come true with a parent cloning himself.

The Adoption Model

While paradoxical on the face of it, as an extreme opposite to cloning in some key respects, adoption demonstrates problems that might occur with cloning. In this sense, the model's conceptual opposition to cloning turns its weakness into strength. For example, the model might appear weak in the sense that adoptive parents lack a genetic claim to be entitled to control and regulate the child's behavior, with the courts needing to provide one through the legal mechanism of adoption. However, this weakness of the model becomes a strength in demonstrating that parents may feel entitled to have a child who resembles them and fulfills their expectations, even when there is no genetic linkage between parents and child. Complete genetic concordance between progenitor and clone would only magnify these feelings.

The lack of genetic linkage of the adopted child to her rearing parents may also contribute to problems of genealogical bewilderment, concerns about incest, and complications in contending with the oedipal conflict and the family romance fantasy. We have seen how the total genetic linkage of a clone with one rearing parent and not the other could also create complications, sometimes parallel and sometimes opposite to those of adoptees. Furthermore, even noncloned children would be likely to manifest these problems and complications in a world where clonal origin is a possibility.

The development of identity is another area that the Adoption Model addresses by being opposite conceptually to cloning in representing a complete lack of genetic linkage between parents and child. An adopted child may not know his genetic parents, whereas the child clone may know that she has only one genetic "parent"—her progenitor. Early on, identification with that parent might be enhanced by this knowledge, and with that may come a greater sense of security and self-confidence. However, if the progenitor is to serve as a mental template of parental expectations for the child, development of individual identity could be severely compromised. In both adoption and cloning, parental expectations could overwhelm the child's efforts to develop his own identity. The consequence may be "foreclosed identity" or rebellion against it, perhaps going to antisocial extremes.

Disclosure in adoption finds the adoptee often knowing too little about his origins; but the clone would inevitably know too much. Disclosure versus secrecy in assisted reproductive technologies and arrangements might even better model this issue. It seems that a critical variable is protecting the illusion that the rearing father is also the genetic father, a value often held to be more important than disclosing the truth to the child. The heightened father–son resemblance one would expect with cloning should assure the father and everyone else that this boy is his son. However, the father might be ambivalent about disclosing that he cloned himself, fearing social stigma, including the suspicion that he did so because he is sexually deficient, fertility often being lumped in with sexual potency.

Adoptees, because they have been "chosen," may also have a sense that they are replaceable. We might expect similar feelings in clones because they would also be chosen, though in a different way, the parents choosing a progenitor to clone, whether one of them or someone else. Compared with adoptees, a clone might have an even stronger sense of being chosen but also replaceable. As long as DNA from the progenitor is obtainable (and that includes from the clone himself), yet another clone could be created.

The Parent–Child Resemblance Model

The Parent–Child Resemblance Model, embodying intergenerational difference and inherent disparities in power and dependency in the parent–child relationship, is strong where the Identical Twin Model is weak. In this model, we see the parent's desire that the child resemble him as possibly having evolutionary roots. However, this does not preclude other understandings. We have analyzed it as a manifestation of normal psychological processes of attachment and identification but have emphasized that when excessive, it is the result of the aberrant psychological processes of pathological narcissism.

To the extent that some individuals might wish to clone themselves or a famous person for narcissistic reasons, a clone could suffer the same deleterious consequences of empathic failure risked by any child with narcissistic parents. In such instances, the parent expects the child to be just like him or his ideal—a perfect "mini me." The heightened degree of parent–child resemblance we would expect between a parent and her self-clone would only enhance the power of these "twinship" narcissistic transferences. Hence, a clone might have a greater risk of developing a "false self."

Even when narcissistic motives are not prominent in self-cloning, the parent's perception of heightened self-similarity might be more likely to evoke "twinship" transferences in the parent. While perceived dissimilarity may be a hindrance in a parent attaching to a child and in a child healthily identifying with a parent, too much perceived similarity might interfere with accurate empathy by the parent and engender unrealistic expectations of the child.

The Child of the Famous Model

The Child of the Famous Model is a special case of the Parent–Child Resemblance Model and is particularly strong in demonstrating its social dimension. Some people might seek DNA from famous people so that they could parent a clone of a person they idolize. The Child of the Famous Model is strong in modeling the desire for fame—for oneself, one's parents, and one's offspring. It also shows that the wish to have famous offspring probably derives from a reactivation of a normal childhood fantasy, the family romance fantasy—the wish to have parents other than one's own, famous and important ones. We also have seen that denial of death and an attempt to enhance self-esteem or counteract feelings of inadequacy may be important in the wish to be famous oneself. Rearing a clone of a famous person might be motivated by all of these. The Namesake Model reinforces some of these notions.

The Replacement Child Model

The Replacement Child Model is strong in specifically modeling a circumstance that might motivate many instances of cloning—grief at the loss of a child. A child born subsequent to the death of a sibling may sometimes be viewed by her parents as a "replacement." A full-blown replacement child syndrome is not the typical result following such a loss. That is fortunate because such children are full of fear of death and illness, filled with guilt and rage provoked by comparisons to their dead predecessors, and are likely to experience problems with separation-individuation and the formation of self-identity. However, parents who would choose to clone a dead child would be choosing not only to have another child but to have a child as much like the one they lost as possible—a replacement. The Namesake Model suggests that cloning a dead child in name might be a marker for replacement status, and by analogy it also implies that a clone would be a replacement.

Conceptually, the Replacement Child Model makes a more general point—any clone would be a replacement and, as such, an object of transference specifically regarding the person he or she was chosen to replace. For this reason, a "cloned child syndrome" would incorporate some features of the replacement child syndrome.

The Namesake Model

Conceptually, the Namesake Model picks up where the Replacement Child Model leaves off. By naming a subsequent child for a dead sibling, parents appear to be expressing a wish that the new child strongly resemble his predecessor. The child's parents have cloned his predecessor in name, a sign that he is meant to be his clone in other respects. Although such a practice used to be commonplace, there has been a strong historical trend away from doing so. Hence, in contemporary times, naming a child for a dead sibling often marks replacement child status. Thus, the Namesake Model reinforces the Replacement Child Model and points back to that model to anticipate the likely consequences of being a clone of a dead sibling or other idealized person, a person who has been literally, if not nominally, cloned.

The two models also converge in developing a more general concept of replacement status. To be a replacement does not require that the person replaced is dead or that the person be one's child, just as a more general concept of "namesake" includes the idea of "partial namesake."

"Transference" is the key concept in broadening both models. The parents' transference toward the child plays a critical role in both the Namesake and Replacement Child Models. Both models make the same point—parental transference toward their child clone is likely to be overdetermined by the parents' choice of whom to clone. The Child of the Famous Model extends such transferences into the social sphere.

Developmentally primitive thought processes whereby a person conflates another's name with that person's identity are more likely where unconscious influences hold greatest sway. A namesake's name might be more likely to trigger these processes in those who named him. Likewise, the increasing degree of resemblance of clone to progenitor as the clone ages would be likely to trigger developmentally primitive modes of thought in the clone's parents and others. The clone's resemblance to his progenitor might well provoke transferential feelings and reactions, in which he is related to as if he were his progenitor.

Parallel research findings reviewed in the Namesake and Replacement Child Models buttress their conceptual convergence. For example, we have found that when a parent names a child for himself, the parent may be claiming the child as his more emphatically than by simply giving the child his last name. This finding is parallel to one discussed in the Parent–Child Resemblance Model, where parents prefer children to resemble them, especially if a parent is pathologically narcissistic or, more generally, has a deeply wounded self. Likewise, there is a parallel between research finding that a mother may name her child for the man she claims to be the child's father to assign a claim of paternity and research showing that a mother and her relatives tend to assert that the baby looks like the alleged father.

Both the Namesake and the Replacement Child Models clearly hint at developmental changes in the way a namesake feels about his name and himself and how that influences how he may relate to others. We have seen that a young boy's self-esteem and identification with his father appear to be enhanced by resembling him, whereas in adolescence the older boy may have more ambivalent feelings about being like his father. Likewise, being a paternal namesake seems to benefit young boys, if only by enhancing the likelihood that their fathers will feel responsible to provide financial and other support. But we also saw that namesakes with the "Jr." suffix are disproportionately represented in troubled and troubling populations and are more likely to have been physically abused by a parent. While problems in the area of identity are most clearly pointed to, other areas of difficulty are also likely.

We can see a parallel between the Parent–Child Resemblance and Namesake Models in that naming a child for oneself may be an example of wanting to live on by making one's child like oneself. The greater likelihood that parents will name an adopted child for one of them might, in part, reflect a compensatory "memic" attempt at parental immortality when genetic immortality through the child is not possible.

Although the Namesake Model is weak conceptually in that the identity between individuals with the same name is not genetic, it is very strong in other respects. For a parental namesake, the model reflects key features of self-cloning and the asymmetry of power and dependency that is part of the parent–child relationship. In fact, a parent naming a child for himself symbolically illustrates how a parent can impose a hoped-for identity onto an offspring from the moment of birth. Whether that is the result of namesaking one's child depends on the parental attitudes that go with it, in rearing the child. Although there is an overall historical trend away from naming a child for a parent, we must be cautious about surmising parental psychology when a parent names a child for himself. After all, the practice is still common and is strongly influenced by cultural and ethnic traditions. These and other sociocultural, ethnic, and religious traditions and values are bound to influence attitudes about cloning, too.

The Namesake Model also helps to conceptually link the Child of the Famous and Parent–Child Resemblance Models. Namesaking for a famous person would draw social attention to the fact that a child is a clone of a famous person while at the same time making it clear that the parents are the parents of this clone of a famous individual.

WHAT THE MODELS AND THEIR INTEGRATION IMPLY
FROM A VARIETY OF PERSPECTIVES AND LEVELS

Now let us apply the models separately and together to make predictions from the several perspectives of the clone and others. We will focus predominantly on the clone and how she may be affected at each successive stage of her life and then on the clone's "parents." We will conclude by briefly considering the perspective of a clone's siblings and individuals promoting or facilitating cloning.

The Clone

The Life Cycle of Clones

At various points, we have referred to certain stages in Erik Erikson's very accessible psychosocial theory of human development. His theory organizes our commonsense notions about the seasons of a person's life. It describes eight stages we all must negotiate as human beings. We have most often referred to stage five, "identity versus role confusion." Here I briefly describe all the stages and state what I think we can reasonably infer for clones from the models taken together.

But, before proceeding, let me emphasize that while certain developmental tasks, themes, or issues are central to a given developmental stage, they are not at all confined to that stage and may be revisited in different ways at various points in one's life. Other important issues fall even less neatly into a developmental stage. We will examine them separately after going through Erikson's stages.

Stage 1: Basic Trust versus Mistrust. In Erikson's first stage, corresponding to Freud's "oral" stage, the infant develops basic trust if the primary rearing parent meets her basic needs by responding empathically to her. Being able to do so requires the parent to discern when an infant is uncomfortable, along with a willingness and ability to figure out why and to do what is needed to soothe the child. Is the child hungry, wet or soiled, cold or sick, in need of stimulation or having too much of it, or just being held in a way that does not feel right?

Depression or other major mental disorder, substance abuse, or pathological narcissism interferes with a parent's ability to be able to read correctly and respond appropriately to the child, to help her reestablish a comfortable homeostasis. In the Stepchild Model, we see that a stepparent may be less motivated to do so. The nonprogenitor parent might feel "as if" a stepparent to the clone.

In the Parent–Child Resemblance Model, we see how parental narcissism can undermine a child's basic sense of trust in her first relationships to another human being. Self-cloning by a parent might be significantly motivated by a narcissistic desire for self-resemblance in an offspring, and so could carry this risk. And in the Replacement Child Model, we see how parental depression from an earlier loss may interfere with bonding to the child and relating empathically to him.

If a child is a "replacement," then his parents relate to him as if he were the child they lost. These issues would also be relevant if a clone were created from tissue from any dead individual, whether or not a child. To the extent that any clone would be created from another person, dead or alive, he or she would be a replacement. This would not only be true in a genetic sense. The clone would also be a living embodiment of transference toward the progenitor for his parents and other people who knew the progenitor, personally or in the "parasocial" way that people "know" a celebrity or other public person.

Of course, any person, whether or not a clone, is bound to be an object of transference for others, especially one's parents. In that sense, everyone is to some degree a replacement. However, a clone might be at greater risk in this regard, created as an embodied enactment of the parents' transference toward the person cloned.

Stage 2: Autonomy versus Shame and Doubt. The child in this stage struggles with shame resulting from failures in continence and doubt from difficulties with self-control and self-determination. This stage corresponds with Freud's "anal" stage. Also occurring within it is Mahler's separation-individuation subphase, a time when the child is taking her first steps in becoming a separate individual with her own independent will. It is important that her parents provide the child with approval and a sense of safety as she takes these early steps toward autonomy.

Parental expectations of a child clone at this stage might be particularly unrealistic, as the child's progenitor (be it a rearing parent or someone else) would have already negotiated this difficult phase. The parent, as everyone normally does, would have repressed any memory of it and may find it hard to imagine that the child clone of herself, another adult, or older child is not like her progenitor—continent and otherwise in control of herself.

The Replacement Child Model is particularly appropriate to this stage. If the child replaced had successfully negotiated this stage, then his clone may be perceived as not trying hard enough. To the extent that a clone's parents explicitly or even implicitly communicate this, he will be likely to suffer from shame and self-doubt. Parents for whom the clone is a replacement may have difficulty tolerating his earliest attempts at separating from them, steps necessary to develop an autonomous self.

The Parent–Child Resemblance and Child of the Famous Models also help us understand unrealistic parental expectations of a child clone in this stage. The child may begin to develop a "false self" to try to give her parents what they want. We see the same risks relating to unrealistic parental expectations described in the Adoption Model. In the Stepchild Model, the nonprogenitor parent may be less tolerant of the child's failures in continence and control, with increased risk for emotional or physical abuse for the child clone. The Namesake Model reviews evidence for greater parental abuse of those with the "Jr." name suffix.

Stage 3: Initiative versus Guilt. This stage corresponds to Freud's "oedipal" period, during which gender awareness and identity develop, along with conscience, and fantasy and play become important. The Adoption Model is probably most useful in anticipating what a child clone might go through in this stage. In particular, we would expect complications in resolving the family romance fantasy. If the child knows that he is a clone, then he would be deprived of this normal fantasy of being not the child of the rearing parents but, rather, the offspring of royalty or someone famous. This fantasy helps a child deal with the developmentally normal disappointment he feels with his parents and also helps him to deal with the oedipal conflict by making his parents "not really" his parents. If the clone's progenitor were someone other than a rearing parent, perhaps even a celebrity or famous person, then reality would greatly complicate any realistic resolution of the family romance fantasy. The Child of the Famous Model helps us anticipate difficulties that might be faced by such children.

If the progenitor of the child is a rearing parent, the nonrearing parent would usually be of the opposite sex, and complications may result in experiencing and resolving oedipal issues. The Adoption and Stepchild Models illustrate this. The latter model suggests that the unrelated nonprogenitor parent might be more likely to respond in humiliating, aggressive, or even seductive fashion to the child's oedipal overtures.

The Parent–Child Resemblance and Namesake Models both suggest that being a clone of a rearing progenitor might enhance a clone's identification with that same-sex parent and could be a force to facilitate resolution of the oedipal conflict. However, less tolerance of any identification with the opposite sex, suggested by the Namesake Model, might lead to other problems later on. Perhaps more important, and surely problematic in resolving the oedipal phase, would be the child being able to imagine himself as his same-sex parent grown up, when that parent, his genetic double, is in an intimate marital relationship with the child's other rearing parent.

Stage 4: Industry versus Inferiority. This stage corresponds to Freud's "latency" period of childhood, wherein the child is developing physically and intellectually and building social competencies, mainly with peers of the same sex. The Parent–Child Resemblance Model predicts a positive connection to the rearing parent at this stage, if the parent is the clone's progenitor. The Adoption Model also helps us think about this stage, not because it is like cloning but because it is so different, with parents and child being unlikely to resemble each other strongly in an unrelated adoption. The Stepchild and Parent–Child Resemblance Models converge in predicting a greater risk of emotional abuse by the nonprogenitor parent, leading to feelings of inferiority in the child.

The Parent–Child Resemblance Model predicts that the progenitor parent might expect a self-clone to be like her in her competencies. In addition, she might not be tolerant of the clone as she works at becoming competent. A clone would also risk evoking jealousy from a narcissistically motivated parent if she should perform better than expected.

The Child of the Famous Model predicts that the achievements and competencies of the progenitor might be used as a template, which the clone would be expected to match. If he tries to emulate it, then he might not find his own way and may feel that his accomplishments are not really his own, as they had been expected of him all along. If he fails to be like his progenitor, then he risks disappointing his parents and himself.

The Namesake Model raises concerns that the parent may have unrealistic expectations that the child be like his progenitor, whether that is one of the rearing parents or someone else. The Child of the Famous Model raises the same concerns as the Namesake and Parent–Child Resemblance Models. The child's difficulties in meeting such expectations might not manifest in psychological difficulties until the next stage.

Stage 5: Identity versus Role Confusion. In this stage, the adolescent begins to decide for himself what is important to him and what he wants for his life, exploring different options and ultimately defining a unique, relatively stable, autonomous self-identity. In the context of the Adoption Model, we have applied identity status theory and research to cloning and have concluded that clones would be at high risk for a "foreclosed" identity status. This identity status describes well what some philosophers and legal scholars claim cloning risks: the absence of "an open future."

The Namesake, Replacement Child, Parent–Child Resemblance, and Child of the Famous Models all raise concerns for clones in this stage, with risk of identity "foreclosure" and perhaps rebellious unsociability in reaction to the pressure to conform. We have seen a divergence between juniors and boys with Roman numeral suffixes in conformance, hinting that how a clone experiences his rearing could make a tremendous difference with regard to psychological and psychosocial outcome.

Stage 6: Intimacy versus Isolation. In this stage, people develop intimate relationships with peers, usually of the opposite sex, or fail in this regard and become isolated. The Identical Twin Model suggests that a clone could have a diminished capacity for intimacy with people other than his progenitor. This model also suggests that a clone might have a special sense of connection to other clones. The Adoption Model also predicts greater difficulties in establishing intimacy, and the Replacement Child and Child of the Famous Models may lead us to infer the same.

The decreased femininity of paternal namesakes might imply a diminished ability to identify with a member of the opposite sex. If that is correct, then it could suggest a decreased capacity for empathy with women—a detriment to intimate heterosexual relationships. If research were to confirm this conjecture, then it would have clear implications for clones.

Stage 7: Generativity versus Stagnation.   This stage refers to being able to foster a new generation and be creative in other ways. A clone may find it difficult to imagine that she can parent a sexually reproduced child, for she never was one. The Adoption Model is relevant here, in that adoptees may feel conflicted about sexual reproduction, given their origins, and handicapped in knowing how to be parents, having never known their "real" parents.

A clone may feel that she knows more about parenting a clone than a sexually reproduced child and so might prefer to rear a clone. Like anyone else, clones would be at risk to rear their own children as they had been reared or to react against their own personal history in a way that saves their children from those problems only to impose others on them.

A clone may find it difficult to be creative and productive in other ways, too, comparing himself to his progenitor and feeling that he can never equal his accomplishments. Particularly relevant here are the Parent–Child Resemblance, Child of the Famous, Replacement Child, and Namesake Models.

Stage 8: Integrity versus Despair.   This stage occurs at the end of the life cycle, when people reflect back on their lives. Is a person satisfied that he lived a meaningful life with integrity, or is he full of regret?

The Child of the Famous, Parent–Child Resemblance, Replacement Child, and Namesake Models all suggest that a clone might feel as though she lived her life under tremendous pressures to be like her progenitor and to satisfy the unfulfilled aspirations of her parents. We might expect a clone to reach this final stage with a sense of despair that the life she lived was never really her own. As a result, we might expect a greater incidence of late life depression and suicide.

Issues over Many Stages for Clones

Body Image.   The child's first sense of himself is of his body. The "body ego" develops as the child discovers his body.[2] Difficulties are most likely to arise in later childhood, preteen, and teenage years, the same ages when eating disorders develop. Children of these ages are strongly influenced by idols, including their notions of what an ideal body is. If a person were a clone, then she might be inclined to think of her progenitor's body as the model for her own—not only as what it is destined to be but as what it should be. As a result, she might well feel compelled either to strive for it or to reject it actively. Either way, a clone's sense of ownership of her body may be impaired. The problem is likely to be compounded when the progenitor was selected because of fame or celebrity, especially if that status was due, in large part, to bodily appearance. Seeing her genetic double age and perhaps become decrepit in advance of herself is likely to accentuate the clone's attention to her own body as she anticipates the same changes.

Mortality. It is worth examining mortality separately from the context of Erikson's stage eight, "integrity versus despair." Although facing one's own mortality is part of that stage, it does not address how the death of another may affect the lifelong development of the self. The Identical Twin Model would predict that, for a clone, losing the progenitor/rearing parent would be a more profound loss than losing a nonprogenitor genetic parent and certainly more than losing the nonprogenitor parent. In theory, we could predict the same from the Parent–Child Resemblance and Namesake Models. One part of this may be because the progenitor's death foreshadows the clone's own. Beyond reactive grief, life after the loss might be subject to despair over a sense of irreplaceable loss of self.

Dominance, Control, and Dependency. We first encountered these issues with identical twins, finding that one twin is almost always dominant. Such power is limited by the fact that the twins are of the same age and cognitive level, and their dependency is more mutual than it is consistently asymmetric. Such is not the relationship of parent and child, wherein the parent is consistently dominant and able to exert control and the child is dependent on the parent for his very survival. The Parent–Child Resemblance, Stepchild, Adoption, Child of the Famous, Replacement Child, and Namesake Models better illustrate how these issues might be played out in cloning. These models demonstrate the power of the parent–child relationship to impose parental expectations, including unconscious ones through the vehicle of transference. We have seen that the most consistently documented negative consequences are for Erikson's stage five, "identity versus role confusion," but also evidence from which to infer greater problems in every other stage.

The "Parents" of the Clone

Now let us see what we can conclude regarding the "parents" of a clone. To do so, the first question we need to address is: "Who are the clone's

parents?" Helping us with this is the Adoption Model, which has introduced us to the notion of "genealogical bewilderment." A clone's genetic
grandparents might be almost as confused as the clone as to who exactly his
parents are and in what way. Seeing a child closely resembling their own by
virtue of genetic identity is apt to be a bewildering experience for the parents of the progenitor of a self-clone—the clone's social "grandparents." As
they see the spitting image of their own little boy or girl grow up all over
again, they might feel more entitled than most grandparents to offer their
"two cents" of advice in the clone's rearing. Perhaps they would also feel
more entitled to want to intervene, perceiving their grandson as more of a
son. That said, psychologically, a clone's "parents" are his rearing parents,
one of whom may or may not be his progenitor. It is where progenitor and
parental status do or do not coincide that complications might ensue, not
only for the clone but also for the "parents," separately and together.

Helping us to understand possible motivations for becoming a parent
by self-cloning are the Assisted Reproductive Technologies and Arrangements, Adoption, Parent–Child Resemblance, Replacement Child, Child
of the Famous, and Namesake Models. The desire for self-resemblance in
one's child may be seen, in part, as biologically driven by evolutionary
forces. We have seen that the natural human wish for immortality and
pathological narcissism may intensify these. The latter can lead to
parental expectations based on narcissistic transferences, such as the
"twinship" transference, in which the child would be expected to be just
like the parent or, at least, her idealized self-image. A parent who felt deprived of empathic parenting by his own parents may also look to the
child to provide him with what his parents did not supply. These transferences would inevitably lead a parent to be extremely disappointed in
his or her real child. The Namesake Model suggests that in reaction, such
parents might even become physically abusive of the child.

The Stepchild Model may help us understand the nonprogenitor parent
of a couple, for only one parent would be the child's progenitor. A variety
of factors would probably mitigate the risk that the nonprogenitor parent
to the child clone would be less attached to him. We would expect better
outcomes if that parent rears the child from the outset and if that parent,
as the mother, carries the pregnancy. We might assume that the surrogate
mother's ability to think of the child as "not mine," discussed in the Assisted Reproductive Technologies and Arrangements Model, would not
apply in these situations. That is because the mother would know that although the child is not hers genetically, she has not agreed to relinquish
the child at birth. Nonetheless, the nonprogenitor mother might, at times,
feel excluded from the special relationship of progenitor father and clone.
In that relationship, a father might feel as much like a mother as a father
to his son, and like an older brother as well.

The nonprogenitor parent might have to deal with an intensified oedipal struggle from the clone, as the boy comes to realize that his father is having sex with his mother and that he is "as if" his father. Based on the Stepchild Model, it is conceivable that some nonprogenitor parents might respond to the child in an emotionally abusive way, and sexual molestation is not out of the question.

The Replacement Child Model helps us understand those parents who would choose to clone a loved one—dead, dying, relative, or friend. These parents, motivated by unresolved grief, would probably very much want and expect the child to closely resemble the beloved progenitor. The Replacement Child Model shows us that the child's birth is unlikely to resolve their sorrow. Instead, we would expect their grief to be perpetuated through disappointing reenactments of transference toward the lost loved one with the child clone.

Some couples might wish to adopt a clone of a person of allegedly superior health, talent, accomplishment, or fame. The Assisted Reproductive Technologies and Arrangements, Adoption, Child of the Famous, and Namesake Models help us understand the psychology of wanting to be the parent of such a child.

The Assisted Reproductive Technologies and Arrangements, Stepchild, Adoption, Parent–Child Resemblance, and Replacement Child Models are best at indicating the possible effects of cloning on the marital couple. Recall that the Assisted Reproductive Technologies and Arrangements Model suggests that the marital relationship may be enhanced when the child is genetically the father's, even if not the mother's. This finding seems complementary to one pointed out in the Namesake Model—the use of paternal namesaking by the mother to reassure the father of his paternity and assure his financial support and involvement with the child. By analogy to cloning, we might expect that some women would want to clone their husbands to try to enhance the fathers' attachment to their children, their clones. Perhaps this would apply more generally to individuals in homosexual or lesbian relationships, as well. Maybe the prototypically more "feminine" or less dominant member might advocate for cloning of the prototypically more "masculine" or more dominant member of the pair.

Siblings of the Clone

Nonclone siblings might feel, perhaps accurately, that the progenitor/rearing parent of the clone favors him over them. Studying the relationship of non-namesake siblings to a parental namesake could shed light on this.[3] We might expect a child nonclone with a sibling cloned from a parent to be jealous of the clone and to feel angry and alienated from the progenitor parent. On the other hand, as a child nonclone enters adolescence, she might be happy to have a sibling who is a parental clone or clone of some idealized other person. That clone would be a magnet for parental expectations that sexually reproduced siblings might get to escape.

In some cases, the clone may have a living sibling as a progenitor, and here the Replacement Child Model would apply. This would be the case whether the parents intended the clone as a "replacement" for a child who has disappointed them deeply or one they love so much that they wanted another just like him. The sibling/progenitor might feel replaced in either case. When children are relatively close in age, sibling rivalry can be intense; the first child typically feels replaced by the second. One may predict an even more intense rivalry and resentment of the clone by his progenitor sibling.[4]

Individuals Promoting or Facilitating Cloning

Cloning requires an egg donor to provide the ovum that would be enucleated and into which the progenitor's somatic nucleus would be inserted. Though perhaps in most cases, the egg donor would both carry the pregnancy and be the rearing mother, that trinity of roles could be dissociated for reasons of reproductive health, age, or even convenience. The Assisted Reproductive Technologies and Arrangements Model helps us understand the motivations of a woman who would not rear the child but may want to participate in her conception or gestation. We have seen that a history of prior loss and poor self-esteem resulting from early psychological trauma is common, with the woman contributing her body's reproductive capacity in the hope of emotional restitution. We might expect similar motives to apply to women offering to facilitate cloning, perhaps combined with a desire for fame.

The Assisted Reproductive Technologies and Arrangements Model also helps us understand possible motives of the progenitor who does not intend to rear the child but who offers his DNA for use in cloning. Not much is known about the psychology of sperm donors, but concern has been raised regarding excessive narcissism. We might raise the same concerns about somatic cell nucleus donors for cloning. The issue of disclosure of parentage, explored in the Adoption and Assisted Reproductive Technologies and Arrangements Models, becomes even more relevant with this kind of cloning, as this progenitor is the child's only genetic "parent."

Because human cloning cannot occur naturally, cooperating clinicians would be required. The full range of motives of clinicians involved with assisted reproductive technologies may be relevant and is a matter about which nothing is known.

We also do not know much about those advocating politically for reproductive cloning, but some homosexuals and those who have lost a loved one are among them. The Replacement Child Model might help us to understand the latter; and the Assisted Reproductive Technologies and Arrangements Model, and perhaps the Parent–Child Resemblance and Namesake Models, might help us to understand the former.

Profit is a primary, even if not exclusive, motive of entrepreneurs, and their involvement in cloning should be expected, even in the "not-for-profit" guise. The Adoption and Assisted Reproductive Technologies and Arrangements Models might best help us to understand what such involvement would look like.

CONCLUSION

Each of the eight models of cloning points separately to risks of particular psychological and psychosocial problems for clones, their rearing parents, and others in their lives. In many instances, we have also found good reason to believe that some of these risks would be greater with cloning than with any of the existing practices and phenomena on which the models were based. Integrating the models gives us greater confidence in some of those possibilities.

Although many have expressed misgivings that a clone would face problems in developing her own identity, we have grounded concerns about reproductive cloning in eight models and have looked at every stage of the life cycle. Each model has its own relevant psychological and psychosocial theoretical constructs, as well as supporting clinical data and research. *By integrating the models, we appear to have an adequate basis to infer that a clone would be likely to face extra difficulties in negotiating every stage of psychological and psychosocial development.*

Individuals falling into the eight models described here are clone-like with regard to various dimensions, themes, and issues, as human beings. *The key psychological and social consequence of being a clone is sure to derive from being perceived and related to as if he or she were someone else.* For the clone, that someone else would be their progenitor. For the rest of us, that other person is usually not so manifestly clear. If we happen to belong to one or more of the groups described by one or more of the eight models, then we might have a keener sense of it. *But in a more general sense, we are all clones because we were all objects of parental transference.*

Transference is a pervasive and universal characteristic of all human relationships. To a greater or lesser degree, we are all perceived and related to by others as if we were other people in their lives, present or past. As we have seen in the Parent–Child Resemblance Model, one of those others people may also be idealized or devalued aspects of a parent's self.

Not to discount the role of genetic heritage, as far as we can tell, how a child is reared appears to be the most important factor in how a child in any of the eight models turns out psychologically. Likewise, child rearing is bound to be the major determinant of psychological outcome for a clone. At one extreme, parents might raise a child clone with a highly detrimental insistence that he be like his progenitor or some idealized image of him. At the other end of the spectrum, the parents of a clone could be just as good and loving as the best of parents in providing encouraging support to facilitate the child's development as a unique and autonomous individual. However, the parents of a clone, well intended as they might be, could find it more difficult to positively parent their child and might easily slip into destructive patterns of parenting. *The reason for this prediction boils down to one thing: In addition to their own internal images of what they expect from the child, they also have a real, though idealized, template for the child in the person of the progenitor.*

As we have seen, a child needs a parent to be attentive and empathically attached to him as an infant. This fosters secure attachment in the child. Fostering the child's autonomy requires parents to help him separate and individuate in basic ways as an older baby and to respect his efforts to gain mastery and control over bodily functions as a toddler. The parent needs to be gentle but clear about generational boundaries as the child struggles with his desire for exclusive possession of the parent of the opposite sex, as he goes through the oedipal phase, and must allow him to resolve it by identifying with the same-sex parent. Parents need to encourage the child to develop his own interests and competencies, and allow the child to develop his own unique identity. If parents are to succeed in these essential psychological tasks of parenthood, they must be able to delight in the differences as much as in the similarities between the child and themselves.

That is a key reason why narcissistic parents cannot do any of these parenting tasks well. But rearing a clone, especially a clone of oneself or of a famous person, might exacerbate even the mildest of unhealthy narcissistic tendencies in a parent, as well as threaten to turn normal healthy narcissism pathological. Parents would need to make a conscious effort to avert that outcome and probably would need psychotherapy to help them become more aware of their unconscious transferential patterns. Some have pointed to the problem of parental expectations and have even suggested that parents of clones might be educated out of the likely strong propensity to impose such expectations on the clone. While such education could only be a plus, the unconscious is relatively impervious to didactic instruction. *We need to be most concerned about the unconscious expectations that derive from transference.*

In the main, the challenge for parents of a clone in rearing their child well is no different from the challenge of rearing a noncloned child. However, parents of a clone would be likely to face a variety of personal issues in transference feelings and reactions to their child, intensified by the fact that the clone is, by definition, a "replacement" of self or other. Parents rearing a clone might be more likely than other parents to feel compelled to reenact or struggle to more positively contend with unresolved grief and loss, as well as childhood memories of empathic failure by their own parents. This seems all the more likely because individuals might choose cloning as a route to try to avoid these very issues.

An additional challenge would arise for the parents of a clone of one of them. The exclusive genetic connection of the progenitor parent with the clone might well lead that parent to view the child as "mine" and not "ours," with the other parent feeling excluded and alienated as a parent and perhaps also as a marital partner.

Just as the eight models help us to understand that reproductive cloning probably incurs an increased risk of problematic psychological and social consequences, many and perhaps most people within the groups described by these models manage to be emotionally well adjusted. However, when a person highly resembles a parent, is a child of a famous person, is a replacement child, or is a parental namesake, he or she has to deal with something extra and beyond the inherent challenges of simply being human. Cloning would be another "something extra" and one potentially many times more potent than many of those described by the models. In large part, that is because any clone not only would be a clone but, at the same time, would belong to at least several of the groups around which we have constructed models of cloning. At a minimum, all clones would belong to the Assisted Reproductive Technologies and Arrangements, Parent–Child Resemblance, and Replacement Child groups. And some might belong to additional ones.

Should there eventually be clones among us, their parents should try not to add to the psychological and social burden a child is likely to incur simply by being a clone. As a good example of what I mean, let me share an anecdote involving my son. When Noah was seven, his lone pet fish Wilt was joined suddenly and inexplicably after several months by a tiny version of itself or, rather, herself. I was excited that we apparently had a clone by parthenogenesis living with us. When Noah asked for my thoughts about a name for Wilt's "son," I jokingly suggested Wilt Jr. Noah, the wise rearing "parent," rejected the idea of making Wilt's clone even more clone-like by naming her for her progenitor, an example that parents of human clones would do well to emulate.

In this chapter we have integrated the models focusing on the psychological and psychosocial consequences, primarily for the clone and her parents/progenitor. But we can also reasonably anticipate wider social and cultural consequences of cloning. That is the subject of the next chapter.

10

Wider Social and Cultural Implications of Cloning

We have noted some social and cultural implications of cloning in the context of each model. In this chapter we will review themes of social relevance suggested by the models and take a step or two beyond them. In the next chapter, we will continue our examination of social and cultural implications of cloning, focusing on the very personal areas of intimacy, sex, and sexuality. Let us begin by examining major social and cultural themes emerging from our models of cloning.

UNSETTLING IDENTITY

A clone created by somatic cell nuclear transfer would be younger than his or her progenitor and hence not identical in the same way identical or monozygotic twins are—clones by embryo splitting. This is only one reason that societal expectations of identical twins are not the best approximation of societal expectations of a clone. The Parent–Child Resemblance, Child of the Famous, and Namesake Models are superior in this regard. Even so, the Identical Twin Model still provides us with important insights into the unsettling effects of nearly indistinguishable resemblance. In that model, we speculated that part of maturing in our world involves accommodating or acculturating to the presence of identical twins. Cultures that do not foster accommodation to the uncanny existence of two people who so highly resemble each other often kill one of the pair, whereas others, perhaps in reacting against that impulse, practically deify them in their foundational myths.[1] Like identical twins, clones by somatic cell nuclear transfer represent a challenge to our assumptions about what it means to be a unique human being.

Perhaps part of the disquieted societal reaction to the prospect of cloning might be related to the socially naive reaction of the immature mind or of so-called primitive cultures to identical twins. The Namesake Model helps us to understand that how a young child typically responds to name clones or namesakes is a consequence of cognitive immaturity, one in which "paleological" conceptions of identity are normal. The same sort of immature or "primitive" thought process might help us understand children's reactions to identical twins. The results of this type of cognition might also contribute to some of society's childlike fascination with cloning, including its current discomfort with it. As cloning's novelty diminishes, society may eventually be less discomforted by expectations that progenitor and clone would be identical.

THE MEDICALIZATION AND
COMMODIFICATION OF REPRODUCTION

Nature and reproduction are being "commodified" as part of a broader impact of medicine and biotechnology on society. A human being resulting from medical procedures necessary for asexual reproduction might be viewed at once as less human and more animal- and even plantlike, and at the same time be perceived as a technological product.

We first encountered the medicalization and commodification of reproduction in the Assisted Reproductive Technologies and Arrangements Model, wherein medical intervention can help achieve pregnancy or prevent or terminate one. Asexual reproduction takes medicalization of reproduction to another level, not by helping to free the natural process of reproduction but, rather, by inducing a process of reproduction not naturally occurring in humans. In a free market society, commodification of asexual reproduction could be a socially natural concomitant to its necessary medicalization.

In our examination of every model, we consider the possibility that a clone could be adopted, quite likely at an embryonic stage, and given birth to by either the rearing mother or a surrogate. Both proponents and opponents of cloning people tend to agree that there should not be a commercial market in embryos, yet it would take money and expertise to clone people.[2]

In business jargon, "production on demand" of allegedly superior child clones for adoption might prove to be an efficient and profitable alternative to "market" embryonic clones to prospective adoptive parents. Such businesses might have a competitive edge over other adoption agencies offering sexually reproduced children. The availability of clones for adoption, most likely of people possessing sought-after physical or mental abilities, some having achieved fame or celebrity, is bound to impact adversely the adoption of sexually reproduced children. Adoption of embryonic clones would certainly engender resentment by large segments of the population. Probably most antagonistic to the practice would be those most directly hurt by it—those sexually reproduced children in need of adoption who are never placed or whose adopted parents later wish that they had adopted a made-to-order clone.

Discrimination against clones might add to social resentment of them. The discrimination would be of a new kind—one not based on race, color, creed, gender, or sexual orientation but, rather, knowledge of their clonal status, a matter we will examine shortly. Health and life insurance companies could apply to clones the same sort of justifications they would like to use to deny coverage for people they regard at higher genetic risk for certain diseases. Until a cloning technique is demonstrated to produce normal babies, without special medical problems in infancy or childhood, insurance companies would surely consider them "high risk" and might deny them coverage. Even clones supposedly "genetically enhanced" to have superior health profiles or selected from particularly healthy progenitors might not be excepted. But if cloning were to become more common, and the results were predictably good, it would be an insurance actuary's dream. Insurance companies might then engage in reverse discrimination against nonclones. We might then expect a societal backlash against the alleged clonal "elite."

CLONING FASHION

In the Assisted Reproductive Technologies and Arrangements Model, we have seen the crude and deadly results of what we might call "genetic fashion" in the use of sex-selective abortion to express societal gender bias. Cloning would be a subtler vehicle to assert genetic fashion.

Cloning could help to further an ideal standard for a person, however fickle that standard might prove to be. From this perspective, cloning might become, in the realm of reproduction, an instance of what has been called the "tyranny of fashion."[3] Cloning would not create an ideal. Ideal standards already exist in people's minds—certainly influenced by fashion—and these would be expressed in cloning choices in society. Once those choices have been expressed, they might further influence cloning fashion.

The choices of parents wishing to adopt a clone might follow the general rules of fashion in deciding what are desirable physical, mental, and social attributes, including the social and cultural standing of the progenitor. In this regard, we might benefit from studying the history of preferred body types and the psychology of fads.

If cloning is a means to try to reproduce, if not a specific person, a person with a good chance of having many of the desired characteristics of the progenitor, then cloning could become a eugenic tool of fashion and more obviously pernicious ideologies. This could occur in several ways. First of all, the selection of specific people with particular traits to try to replicate could become a form of eugenics. Second, cloning could enable the use of genetic technologies to change the genome.

We should hope that eugenic ideology is never again codified into law and that, if society decides to permit cloning, except for specific circumstances, it never regulates who may or may not be cloned. We should know, however, that culture exerts as much power as law and that cultural pressures, as invisible as the air we breathe, can be harder to detect and resist than any systematic policy. For example, cultural pressures for self-censorship frequently make censorship laws unnecessary. Likewise, eugenic legislation might not be necessary for cloning to become a de facto instrument of a kind of "stealth eugenics" resulting from cultural influences on individual choices. Cloning fashion would be uncodified eugenic fashion. At the very least, cloning would be a vehicle to express societal conceptions of the ideal person, varied as they might be within a culture and between cultures. Standards set by cloning fashion might even result in diminishing the socially perceived value of people who are the first of their genotype.[4]

In addition to whatever else it might mean, cloning a famous person or celebrity would be a fashion statement. The frequency with which children are named for famous people might indicate how often prospective parents might want to clone them, and their choices in these names would betoken who is currently in fashion. We get a glimpse at the likely social consequences of cloning a famous person in the social reactions and expectations of a child of a famous person, who is also a namesake and highly resembles the celebrity.

CHILD ABUSE AS A SOCIAL ISSUE

It is in a society's interest that its children be well treated, and so there are laws mandating physicians to report cases of suspected child abuse. Abused children not only suffer themselves but also are more likely to require societal intervention to treat mental illness and are more prone to engage in antisocial and criminal behavior. All this makes child abuse an important concern for society as a whole.

The propensity to abuse one's child resides within an individual parent, but it emerges in and may be facilitated by certain features of that parent's family, group, and culture.[5] Might we expect the acceptance of human cloning within a society to have any effect on the inclination of parents within that society to abuse their children, whether clone or nonclone?

The specter of parental abuse for clones is clearly suggested by the Stepchild Model, where we have seen that the single most powerful predictor of child abuse is the presence of a stepparent in the home. Although several mitigating factors would probably be present for most nonprogenitor parents, they would otherwise be akin to a stepparent. Perhaps the Parent–Child Resemblance and Namesake Models are more compelling in suggesting concerns about an increased risk of child abuse for clones. In those models, we have anticipated that a clone would inevitably disappoint his parents' heightened expectation that he match the idealized mental template that his parents hold of who the progenitor is, was, or might have been.

Child sexual abuse can be emotional, physical, or sexual. When it occurs within the family, sexual abuse is incestuous. We have reviewed reasons in the Stepchild Model to be concerned that psychological barriers to incest might be eroded for the nonprogenitor parent of a clone. In the next chapter we will explore this area further.

In earlier chapters, we have seen that parents of a clone might feel more entitled than parents of a sexually reproduced child to have their child comply with their wishes and to even more fundamentally be exactly as they wish her to be. But an increased risk of various forms of child abuse might not be confined only to clones. In a society where cloning is a fact, parents of nonclones might also come to feel more entitled to children who meet their expectations and may be more disappointed if they do not. Some of these parents might vent their heightened frustration and anger on their less-than-"clone-perfect" children, possibly resulting in a general increase in emotional and physical child abuse.

CLONAL STATUS AS A
NEW ASSIGNED COMPONENT OF SOCIAL IDENTITY

In contemporary parlance, to call someone a "clone" is an insult, implying that the person has no identity or will of his own. This slang is a metaphor for a social identity status awaiting the first cloned human beings. When actual clones appear, we should expect the social meaning of this status to evolve to reflect the more complex reality of their existence.

In the Adoption Model, we have looked at the concept of assigned social identity status as a component of self-identity. Gender and race are two important social identity statuses, ones with some genetic basis. In addition, the circumstances of a person's birth can dictate an assigned social status, hence the concept of "illegitimacy." Adoptive status is often a derivative of the "illegitimacy" of out-of-wedlock birth and is a way of trying to legitimize the illegitimate by legally conferring parenthood on people who cannot claim it on genetic grounds. Clonal status would become another assigned social status, one that is based both genetically and on the facts of the clone's reproductive origin.

Her adoptive status may leave the adoptee feeling socially marginal. In earlier chapters, we have seen that social attitudes toward adoption and assisted reproductive technologies play a role in the decisions parents make in whether and how to disclose a child's origins to her and in how the child herself feels about disclosing the facts of her origin. Social policy and law regarding these issues have been evolving in favor of greater openness and disclosure.

Given the currently predominant social attitudes toward reproductive cloning, a clone's parents might elect not to disclose his clonal status to him or others. Individuals aware of their clonal status might also prefer to keep it secret. While understandable, nondisclosure would leave such social attitudes unchallenged and unchanged. Society as a whole will surely wrestle with the issues of disclosure versus secrecy as cloned children mature and share their feelings about knowing too much or too little about the truth of their origins. Whatever their parents may have told them, eventually the truth will be hard to hide.

The ambivalent desire to know the truth of one's origins is surely connected with the desire for knowledge versus the wish to remain ignorant of conception and the sexual act that made it possible. Individual and societal attitudes about disclosure versus secrecy regarding parentage and progenitorship cannot be divorced completely from individual and societal attitudes about knowledge versus ignorance of sex. We will examine this in depth in the next chapter.

GENEALOGICAL BEWILDERMENT

We first encountered the concept of "genealogical bewilderment" in the Adoption Model and by analogy anticipated it for clones and how society thinks about them. Through "self-narratives," we all tell ourselves who we are.[6] A person's self-narrative usually begins with whence he came—when and where he was born and who his parents are or were. Children of consanguineous marriages, especially of incest, face the challenge of explaining to themselves, and, if they dare, to others, who their parents are. The clone's self-narrative must start with the fact that she has a progenitor—singular and genetically identical to her—as opposed to parents, two in number, to whom the offspring, while resembling both, is identical to neither.

Society faces a genealogical challenge in thinking about clones, too. Perhaps this would be most clearly evident in how society thinks about clones of famous people, but the more commonly encountered mundane but terribly important matter of child custody would more frequently demand thoughtful attention. To give it the proper attention, society will have to move beyond notions informed only by the genetic identity of clone and progenitor.

The self-narrative of a clone would require new ways of expressing kinship relationships. Randolfe Wicker, a cloning advocate, believes that multiple clones of the same progenitor would result in a "second tier of relationships that never existed before." He calls people related in this way members of the same "clan" and believes that between members of the same clan there would be "immediate understanding" and a sense of connection and even responsibility.[7]

There may be scientific support for this conjecture in Richard Dawkins's "selfish gene theory," to which we have referred several times already. By appealing to Dawkins's theory, we might well predict a greater sense of loyalty among those genetically identical to each other. In the selfish gene theory "a gene might be able to assist *replicas* of itself that are sitting in other bodies. If so, this would appear as individual altruism but it would be brought about by gene selfishness."[8] Of course, a clone would be an exact genetic replica of her progenitor. Altruism at the level of the gene and individual organism might be extended to the level of the clan of clones of a particular individual, with the progenitor and his clone or clones each genetically identical instances of what Dawkins calls an "extended phenotype."[9]

Progenitor and clone might develop new personal pronouns. Recall the twin pair who had a name for themselves together before either had a name for his own self or one for his twin. In some languages, certain personal pronouns denote an especially close relationship between people. For example, with the pronoun *tu* the French have a familiar form of *you*, used by people who are close to one another. Perhaps familiar forms of the personal pronouns *you* and *we* might emerge for clones, clonal "siblings," and progenitor to denote not emotional closeness but, indeed, genetic identity. Personal pronouns have "subject-creating properties" that help to "establish the capacity of the person to posit him- or herself as a subject."[10] Any new pronouns that cloning might engender would presumably serve the same function but in the context of tying the subject to a genetically identical counterpart.

SOCIETY AND THE FAMILY ROMANCE FANTASY

We first encountered the "family romance fantasy" in the Adoption Model, have revisited it several times since, and will do so again here and in the next chapter. In the family romance fantasy, a child imagines that she is adopted and that her "real" parents are much better than the ones with whom she seems to be stuck. For at least a brief period, as a way to cope with oedipal issues and developmentally normal disappointments, children almost universally entertain the fantasy that their parents are not their parents. A scrap of reality that helps encourage the fantasy is the child's knowledge that some children are, in fact, adopted. In the Adoption Model, we have seen that for adopted people this apparently universal childhood fantasy is complicated by the reality that their rearing parents are actually not their "real" biological parents. Complications produced by this fact may help to produce psychological difficulties for some adoptees.

We have already examined some possible psychological complications for the clone in the family romance fantasy. Problems resolving the oedipal conflict might occur if a child knows that he is the child of one parent (and an exact copy at that) and not the other. Clearly, there are social ramifications when a whole group of people might be at heightened risk to have psychological problems along certain lines.

Adoptive parents often promulgate the "myth" that their adopted child was "chosen" to avoid the messy sexual truth of his actual origins and the sad tale of him having been given up or even abandoned. The story that "the stork brought you" told to nonadopted children also shields them from sexual facts and relieves adults from the uncomfortable task of explaining "the facts of life" in an age-appropriate way.

With the stork story, the parents also plant an early seed of doubt in the child that they are her true parents. It would appear that the stork story might be the parents' companion to the child's family romance fantasy. For the parents, the stork story reflects their wish to avoid explaining the child's sexual origins but perhaps also their own romantic wish that their route to parenthood was pure and devoid of the fact that a sexual act was intrinsic to conceiving this child. Mixed in with this might be a bit of wishing that they were not really the parents of this sometimes difficult child. Being able to tell one's little boy truthfully that he is a clone of his father would be a sexually sanitized story a parent could feel good about, as well as one that might delight a little boy who wants to be "just like daddy."

A sexually reproduced child might imagine himself to be a clone, not of his same-sex parent but, rather, an adopted clone of a famous person. Imagining oneself a clone of the same-sex parent would be counter to what motivates the family romance fantasy, and knowing that one is, in fact, a clone would be a powerful bit of knowledge that would make holding the fantasy more difficult.

In the Child of the Famous Model, we have seen how the desire to have famous children may be, for the parents, a marriage of the family romance fantasy with the romantic fantasy of fame. In the Namesake Model, we have seen how the wish to have famous children could be realized by adopting a clone of a famous person. Were celebrities to be cloned for adoption, the family romance fantasy would cross over into the realm of possible reality. Any child in that era, clone or not, could realistically entertain the possibility that she is the adopted clone of a celebrity.

Perhaps deriving from the family romance fantasy is the wish not for a parent who perfectly understands the child but, rather, for an identical twin who would "have to understand" because he and the child are the "same." Such a "twinship fantasy" variation of the family romance fantasy might underlie the fact that some people believe they were born twins, despite a lack of medical proof.[11] By analogy, as an expression of the family romance fantasy and its variations, in a world where cloning exists a person might wish she were a clone or one of a group of clones. And some parents might want a clone of themselves to create a twin who would "have to understand" them the way their own parents failed to do perfectly.

Because the family romance fantasy is frequently found in delusional psychotic ideation, once the first humans are cloned we should expect that it will become more common for a psychotic individual to believe mistakenly that he is a clone or perhaps one of several clones of the same progenitor.[12] Some might believe that somatic cell samples had been taken from them without their knowledge or consent to produce a clone. Such people would be on the lookout for their alleged "brother" or "sister" clones, clonal children, or progenitors—either wanting to be or trying not to be discovered.

THE WISH FOR SELF-PERPETUATION AND THE DENIAL OF DEATH

At many points in this book we have examined the normal human refusal to accept mortality—either our own or that of those we love. With cloning, a person might hope to perpetuate himself more precisely than he might have been able to through sexually conceived children, children who are genetically only half his.

An individual could also further his understandable wish to deny the death of a loved one by cloning that dead or dying person. In the chapter on the replacement child, we have seen how cloning might be sought as a means for an individual or family to deny the death of a loved one. But cloning could also serve that role for society, as well. If a replacement clone were of a famous person, then that clone could serve a role as a socially shared living representative of the dead. The denial of loss of an important public figure or widely popular celebrity could be abetted by the possibility that she could be cloned, even if a clone or clones were not in fact created.

Reverence for particular deceased members of a society can be important in binding certain societies together. Ancestor worship in Japan is one example.[13] Clones might naturally enter into this system in such cultures as the undead representatives of the dead. Perhaps such cultures would be particularly attracted to the idea of cloning children as replacements not of children but, rather, of revered elders, to become living memorials to them after they die. Some Mexican and Russian Orthodox funereal rites place the deceased at the banquet table.[14] A clone of the "dearly departed" could substitute at the table for the deceased. More generally, clones of a dead progenitor could do the same in postcontemporary rituals for the dead—augmenting or replacing the open-casket good-byes of contemporary Western culture.

THE DESIRE TO CLAIM OWNERSHIP

As an expression of personal autonomy and industry, the desire for ownership can become selfish acquisitiveness. In the realm of child rearing, a parent who feels that he owns his child undermines the child's efforts in developing autonomy and his own identity. In the Namesake Model, we explored a symbolic act with practical consequences for child rearing, the assignment of a name. We have seen that there is a higher likelihood of negative mental health consequences when a father names his son for himself and also assigns the boy the suffix "Jr." We have also learned that naming a child for a parent can be a way for a mother to claim the child for his alleged father, probably hoping that he will accept the responsibility she has assigned. Unfortunately, a sense of ownership of the child by the father is sometimes concomitant.

We have found that though parental namesaking is still common in our society, there is a historical trend away from it. To the extent that cloning practices might parallel namesaking practices, we might expect less self-cloning now than might have been the case if cloning had been available when parental namesaking was more common. However, cloning provides an opportunity for counterhistorical forces to exert themselves, providing individuals for whom ownership is very important with something new and unique to possess—a clone.

Cloning would be a new and powerful instrument to assert parental dominance and control. Some parents of self-clones might believe that compared with any sexually reproduced child, a clone could be expected to more dutifully "honor" his father or her mother progenitor.

Sadly, some of my patients have had mothers who told them that they wished they had never given birth to them. Implicit in that explicit and deeply rejecting message is another one—that the child owes her existence to her mother. Even if being a clone is presented in a positive way, a parent/progenitor might still convey the implicit message that the clone owes her existence to her alone. Being genetically "owned" by a sole parent/progenitor, the child's risk of being emotionally "disowned" would be greater than ever before.

The wide range of models of cloning makes clear that many features intrinsic to cloning create powerful opportunities for the abuse of parental power. The result could be a heightened risk of destructive psychological consequences for the individual clone and negative social consequences for society as a whole.

SELF-RECOGNITION AND
THE PREFERENCE FOR SELF-RESEMBLANCE

The Parent–Child Resemblance Model has helped us to understand parents' strong wishes that their children resemble them, and the Namesake Model examines one technique used to enforce those wishes. But the preference for self-recognition and self-resemblance goes beyond the parent–child dyad. It is present at multiple levels.

At the molecular and cellular level, immune defenses against pathogenic invaders are essential. When this system is disabled, as in AIDS, severe recurrent infections occur. In cancer, the cells, having mutated, are not recognized as sufficiently different from one's own and so do not trigger an immune response. In contrast, it is now thought that autoimmune diseases, like rheumatoid arthritis and multiple sclerosis, might begin with the body responding too generally to an infection. As a result, the immune system mistakenly recognizes certain bodily tissues as alien and begins to attack them as well.

When the cerebral hemispheres of the brain are disconnected, the right hemisphere is better at recognizing others, whereas the left is superior at self-recognition.[15] But the neurology is more complicated than this because individuals with damage to the parietal lobe of the right cerebral hemisphere have problems in self-recognition. These people also frequently have more extensive damage to the right hemisphere, resulting in left-sided paralysis. As a medical intern, I helped care for one such patient. She complained that while she had expected a double occupancy room in the hospital, she had not expected to share a bed with another patient. Her unwelcome bedmate was, in fact, her own paralyzed left leg.

In dementing illnesses, such as Alzheimer's disease, the person may not recognize a previously familiar person and may claim that an unfamiliar individual is a familiar one. In Capgras syndrome, the affected person believes that a familiar person is not really that person but someone else—perhaps an inauthentic copy. Sometimes a manifestation of dementia, Capgras syndrome may also be seen in psychotic disorders without striking brain pathology.

In other realms of psychopathology, a patient with schizophrenia or other psychotic disorder may have auditory hallucinations, perhaps as a result of not recognizing her own internal speech, whereas in pathological narcissism, everyone is experienced as either "just like me" and good or dissimilar and bad.[16] The self and those seen as like the self are preferred, while at the same time, like the mythical Narcissus, the narcissistic person often does not recognize unwanted aspects of himself in others. The narcissist often reacts negatively to others based on unconsciously projecting unwanted aspects of himself onto them.

While disordered brain function, severe mental illness, and pathological narcissism can give us clues to mechanisms of self-recognition, psychological research with normal people also demonstrates its importance. People are inclined to feel more trusting toward a depicted "person" whose face has been digitally morphed to resemble their own more than toward one whose image was not experimentally altered to create self-resemblance.[17] Adolescence, a time of normally heightened self-involvement, is also a time of enhanced ability to recognize a mirror image of the self.[18]

Clinical experience teaches us that parenting that puts too great an emphasis on the child resembling a parent often has unfortunate results for the child, sometimes leading to mental illness. On purely genetic grounds, sexually reproduced offspring can only partially resemble a parent. Nevertheless, some parents feel disappointed at areas of major difference, perhaps most often, or at least most obviously, when the child is viewed as being of the "wrong" sex.

At the social level, kin selection is a manifestation of the preference for self-resemblance in others. At the social level, the preference for self-resemblance and its converse can be taken to intolerant extremes, manifesting in racism and genocide, as well as arrogant nationalism.

It should be clear that phenomena of self-recognition and preference for self-resemblance manifest on many levels of the hierarchy—from gene, to cell, to brain and mind, to social group and society. Because preference for self-resemblance is key to the "selfish gene" theory of evolution, we are able to close the circle from the social to the molecular and cellular level, and back again. Though the process may appear similar at each level, different mechanisms and rules apply, with differing results should things go awry. While "narcissism" at the level of the cell works to preserve the individual organism, it can also cause disease.[19] At the level of the gene, it may help to preserve the species but may not always be beneficial at other levels. For example, at the level of the individual, we have seen how interpersonally destructive pathological narcissism can be. At the level of the family, it can be tragic. And at the levels of social group and society, it can be catastrophic. As an embodied interface of self, family, and society, the clone is at great risk to feel the brunt of both the adverse effects of parental preference for self-resemblance and a complementary societal expectation that the child resemble his parent.

CLONING AS A "TECHNOLOGY OF SELF" FOR THE NARCISSIST

Michel Foucault defines "technologies of self" as technologies that, with or without the help of others, "permit individuals to effect . . . operations on their own bodies and souls, thoughts, conduct and way of being . . . to try to transform themselves" with the goal of attaining "happiness, purity, wisdom, perfection, or immortality."[20] Psychotherapy is a technology of self, one proven useful in helping people, including those with pathological narcissism, not to achieve perfection or immortality but, rather, to improve themselves and their relationships.

Cloning technology could be used as a new "technology of self"—not to cure but, in fact, to express pathological narcissism more fully. The narcissist seeks not improvement but perfection and would not be satisfied to live on through good works and meaningful relationships with others but, rather, would feel tempted to try to approximate, if not achieve, biological immortality. He might be inclined to try to achieve perfection and immortality through self-cloning, perhaps with genetic enhancement, or by choosing to parent a clone of a person who has achieved immortality through fame or accomplishment.

Such a person would probably never consider that parenting a child might incidentally help his own individual development, though the idea might occur to him if he could rear a self-clone. In this context, let us imagine a narcissist trying to justify self-cloning intellectually as a technology of self. He might reason like this:

> My life is too short to develop myself to the fullness of my potential. There are many great things I could have done, but I've been cheated by having

only one life to try to do just a few. A person is forced to choose one path and not another, and even then, idiots are likely to get in your way. By cloning myself, I can accelerate my personal development and that of my clone. He'll live on for me and in a way, be me when I'm gone.

Certainly, my clone would need to be reared properly to participate in my project, and this would require new principles of education and child rearing. Especially in the child's infancy and early childhood, I'll need a woman to help who appreciates just how special I am. I don't think I'd have much patience for changing diapers. But when the boy is a little older, I'll teach him to avoid the small mistakes I've made and encourage him to do things I know I could have done. In all of this I'll learn more about myself by seeing myself in him, and he'll learn from me. Although I could do this with a regular child of my own, that child wouldn't be me and mine the way a clone would be.

I'm thinking of both myself and my posterity. I can imagine multiple generations of clones of myself. Because they'll all be just like me, each parental generation could really learn from its own earlier experience and so perfect its own parenting with the next, or "n + 1," generation. In this way, cloning could be a real psychological benefit to both me and my clone (upon clone), especially as we accomplish great things. Cloning the self might even enable a biological version of the Buddhist notion of the progression of souls through reincarnation, with the advantage that the earlier form may persist to inform and advise the newer one. That's where I come in. At least at the beginning. . . .

Let us interrupt the rationalizations for self-cloning of this imagined narcissist, just as he appears to be edging close to delusional grandiosity. Most narcissists would not go that far but would probably sound the same themes and, up to a point, employ the same sort of reasoning. We are unlikely to be terribly effective in explaining that *each person is a separate self/soul.* That is because the radical individualism of the narcissist scarcely permits the individuality of another, especially anyone he explicitly views as an extension or reflection of himself.

What has been called "autonomy of self" is fine as an ideal, but it needs to be informed by fairness and empathy, which requires us to take the other's perspective.[21] Autonomy of one at the price of the subjugation of another is tyranny. In the case of progenitor and clone, the narcissistic progenitor would view the clone not only as an extension of herself but also as a new creation of her own self, interfering with the clone developing a new separate, autonomous self.

The "romantic" notion of the self "stresses the capacity of selves to create and re-create themselves."[22] Cloning would create an intergenerational conflict in the romantic notion of self—utopic from the perspective of the narcissistic progenitor viewing her clone as a re-creation of her own self and dystopic from the perspective of the clone who realizes that she can never be her own creation. This intergenerational conflict might exacerbate preexisting intergenerational tensions in society.

THE WISH TO BE
EMPATHICALLY UNDERSTOOD AND ADMIRED

We all wish to be understood and related to with empathy. It is essential to developing the basic trust of Erikson's first stage, but it is also a basic human need that is with us all our lives. When met, it enriches us, and when unfulfilled, it contributes to depression, though it is never as essential for our survival as it was when we were infants. It might be a major unconscious motive to self-clone for those who suffered from narcissistic parenting. Cloning oneself could be a way for such individuals to try to fulfill the wish to not be alone, to not be the only one, and to be understood and accepted.

Cloning might also come to appeal to society for the same reasons. People in our diminishingly communal society might be drawn to cloning as a way to try to avoid loneliness and abandonment by having someone to own and control as an individually customized possession. The concepts of alienation and anomie are relevant to this discussion. When social and cultural structures have eroded or broken down under various stresses, people often feel cut off from others and disconnected from society. In this state, the individual feels powerless and sees little possibility of self-determination within his society and culture. It is a state of mind associated with socially deviant behavior and suicide.[23]

Many individuals in a state of anomie are poor and would be without the financial resources to involve themselves in cloning in any capacity but as egg donors or surrogate mothers. But poverty is not the only determinant of this state, and some individuals with the financial resources might turn to self-cloning as a way to try to alleviate their sense of disconnection and powerlessness in relating to others. They might see self-cloning as a way to connect to another person and, at the same time, feel a sense of power and control in a relationship, the relationship of parent to child clone.

But cloning itself might also further cultural anomie—as people relate less personally and intimately to one another and more as consumers and patients to entrepreneurs and medical and auxiliary personnel in a medicalized and commodified reproductive endeavor. The family of a clone, especially one created out of a parent's anomie, is also likely to be a system relatively closed to the local community and wider society. If the family feels or is treated as deviant, one might expect a further sense of alienation and anomie to develop, encompassing the clone as well.

Regardless of whether it was utilized by any particular individual or family, reproductive cloning may also further a general sense of anomie in society. An already fragmented society could split further into those who clone and those who do not or cannot—the children of Adam and Eve versus the children of Adam *or* Eve. Would this simply be more diversity in a free society or, rather, serve to fragment society?

SOME POLITICAL ASPECTS AND IMPLICATIONS OF CLONING

Although the most vocal prominent public and political figures in our society currently condemn cloning, society seems to have an ambivalent fascination with and attraction to the clone-like. Its attitudes toward identical twins, parent–child resemblance, children of the famous, and parental namesakes exemplify this. We seem to be drawn to family dynasties in politics, where the names Kennedy and Bush guarantee interest and political advantage. In politics, as in the rest of life, the wish for the familiar and predictable is in conflict with the wish for the novel and exciting.

Society may not realize it, but it longs for clones, at least politically. Even now there is hope that the offspring will be like the famous political father or at least how the father is remembered in the ideal. When the son inevitably turns out not to be a "clone" of the father but only clone-like in name, appearance, or mannerisms, people ask in disappointment: "Why couldn't he be more like his father?" But that proves not to be a surprise. Often, the person society hopes will be his father's clone has spent much of his life trying to prove that he is his own person and not simply his father's son.

Royal families emphasize the clone-like, with names followed by Roman numerals running into high digits, and the Roman Catholic Church has done the same, with the numerical suffixes of popes going still higher. The stability, predictability, and safety of the known, along with good feelings and idealized memories about the earlier version, are all intrinsic to what society expects of the clone-like. In this context, the idea that some may wish to "immortalize" beloved political leaders by cloning them seems less like an improbable science fiction scenario than it might have otherwise.

A society that embraces cloning would probably become more conservative and traditional and less adaptable and tolerant of differences among its members. We might expect it to be more parent dominated, though not necessarily patriarchal. Although even the author of selfish gene theory asserts that culture is the major determinant of humanity's way of life, culture would be shaped, some might say warped, by the reality of human cloning within that culture.[24]

We will explore the concept of human nature in the final chapter, but it is important to note it here. Francis Fukuyama points out that our ideas about human rights are based on our ideas about human nature and that politics would be affected by any change in human nature that biotechnology and, in particular, genetic engineering might produce.[25] Gregory Stock believes that we may want to change certain aspects of human nature and should not be timid in trying to improve it through genetic means.[26]

Communist idealists used to say that a "new man" would evolve in their society, a man with a new and improved nature. I do not see a change in human nature based on cloning as an ideology or even through the psychological and social consequences it might produce for those most directly affected by it. Instead, if reproductive cloning changes human nature, it would mainly be through indirect but manifold effects on society as a whole.

The politics of abortion complicates discussions of cloning. We will examine this interface in the final chapter, "Implications for Cloning Ethics and Policy."

THE "AS IF REAL" SOCIETY

The theme of "as if" emerges from several of the cloning models. It is a theme relevant to the clone, her parents, and society. But cloning would only continue a trend in an "as if" direction for society.

The biblical directive against making "any manner of likeness" is part of the Second Commandment, following directly after the statement: "Thou shalt have no other gods before me." The injunction against any "likeness" discourages idolatry and polytheism, which threatened the integrity of the then-new concept of monotheism. The likenesses prohibited are of "anything on the earth, in the skies, or under the water" (Exodus 20:4). Abstract geometric images, letters, numbers, and other symbols are not included. On a practical level, what is prohibited by that commandment is any likeness of a living thing.

We are surrounded by such likenesses in our modern society, and it would be the rare person who would suggest that we dispense with them. They are not only a staple of marketing and commerce but important to science and medicine as well. The making of likenesses is intrinsic to the foundation of modern society, but it is also indispensable to its "as if" characteristics.

Likenesses need not only be visual. A likeness can be fashioned in any sensory modality. The telephone was a critical step in the modern making of likenesses, making it seem "as if" the person one is talking with is "there." Of course, a person's voice is heard "there," but the person is not really "there."

Technological advances in computers and telecommunication have greatly accelerated the making and transmission of likenesses. Artificial intelligence simulating human interaction is a bigger step in that direction. No person is either "there" or somewhere else in these artificially intelligent "beings," and the sense that one is communicating with a human being is illusory. Some interactive computer programs can fool many but usually not for long. In the not-too-distant future, increasingly realistic virtual reality possibilities incorporating visual and other sensory perceptual inputs, including tactile feedback, will be able to more and more realistically simulate sexual encounters.[27] The literally and figuratively seductive possibilities of an "as if" life might come to be more than many will be able to resist. Pornography and "virtual sex" technologies might facilitate the dissociation between intimacy and sex and interfere with the full maturation of a person's sexuality as an integral component of an intimate relationship.

A likeness need not only be appreciable by one or more of our senses. A likeness to a naturally produced emotional state is also possible. Drugs of abuse—alcohol, amphetamines, cocaine, and opiates—give the user a chance to try to replicate the sense of being loved or feeling powerful or simply to attempt to banish for a while the psychic pain a person may feel about her real life.

The individual fraud asks himself how to seem real and authentic without being either, but the quest to create the "as if real" is everywhere in our society. How does one get something to seem real and authentic without being either? It is an appropriate question for interactive sports video games, where the question "How real can they get?" is answered by yoking artificial intelligence to real athletes to produce "digital clones."[28] The question is also relevant to the corporate accounting scandals of 2002, in which "certified" financial statements were, in fact, "as if real" certifiable fakes.

"As if real" assessments can also lead to real death and destruction, though death viewed from a distance can be perceived with the same sort of "as if real" detachment of inflicting "virtual death" on the "bad guys" in a violent video game. The "as if real" has nowhere more dangerously insinuated itself than in warfare.

The "screen" has not only brought a wider world to us but also separated us from the world. During the American invasion of Iraq, national news anchor Tom Brokaw said that "television cannot ever adequately convey the sheer brute force of war, the noise and utter violence. . . . It somehow gets filtered through the TV screen, and that's probably just as well."[29]

People have been filtering reality naturally long before there ever was an external screen. Sometimes reality can be so overwhelming that a person naturally takes distance from it by screening it with memory images that psychoanalysts call "screen memories." A screen memory conceals others, representing a compromise between denial and the memory of a painful experience that is covered or screened by something less significant. Even as a screen memory conceals, it also hints at the presence of the underlying secret.[30] Screens outside our minds add another layer of protection from reality, concretely screening it for us prior to our own representation of an image of it in memory.

Separating oneself from reality may sometimes be a matter of psychological survival. Some realities may intrude so violently that individuals may develop Posttraumatic Stress Disorder or a dissociative disorder in which images from the event intrude unbidden and the person feels as if the traumatic event is happening all over again. Psychotherapists sometimes help these people gain some distance and control over these experiences by suggesting, occasionally with the help of hypnosis, that they put the disturbing post-traumatic memory images onto an imagined video screen in the mind. The person is then instructed to move the image out of sight or far away, make it less distinct, and finally "change the channel" to a soothing image of her choice.[31] Clearly, there is an important difference between utilizing an imagined internal screen to distance oneself from memories of a traumatic reality and experiencing reality in an "as if" way because of the literal or figurative screens society interposes between us and the world.

The reality of blood—that is, life and death—is compromised in our "as if real" society. This may also be relevant to the idea of "blood relations." Western society has evolved from a society where blood relations and bloodlines of descent were very important in alliances and where blood literally was often shed in war, epidemics, and famine. In those days, Michel Foucault explains, "blood was a reality with a symbolic function."[32]

Cloning, by preserving the "pure blood" or bloodline of the progenitor, returns us to the "symbolics of blood" of that earlier time, but the reality of cloning would be one that has been washed clean—clean of blood, including menstrual blood, and semen as well. The "surreality" of the clone viewed socially "as if" he were his progenitor threatens the "reality of blood."

Sexual reproduction has been an intrinsic part of the reality of life and death. Asexual reproduction might further "as if real" social perceptions and attitudes toward life and death. Cloning would allow us to avoid not only the wonder but also the fear that accompanies creating completely new human life and the linked terrors of sex, the body, and death.

RELIGION

I am not a theologian and so can only take a psychological and social perspective on religious views regarding cloning.[33] We have already ventured into religion in examining the biblical prohibition found in the Second Commandment against making likenesses. Any religion embracing all of the Ten Commandments as literal and absolute injunctions might object to reproductive cloning on the grounds that it would produce a likeness of a living or once living being. The possibility of a living likeness was not anticipated 4,000 years ago.

There are some scientists who would dismiss objections to cloning coming from religious figures or deriving from religious dogma. This strikes me as an unscientific mistake. A clergyman could not be expected to approach the topic of cloning scientifically but, rather, to do so with a perspective influenced by the framework of the particular beliefs and dogma of his religion. Whether or not one agrees with their religious views, at the very least, some ideas voiced by clergy about human reproductive cloning might also be understandable as metaphors. As such, they would be relevant, if only as clues, to the psychological and social impact of cloning on society.

Perhaps some objections coming from religious quarters are related to their conceptions of the self. A "theistic" notion of the self "allocates human souls a special place in the universe." In contrast, a "naturalistic" notion "regards the self as amenable to scientific reason, explicable in terms of biology, heredity, psychology, socialization, and the like."[34] Human cloning puts into conflict naturalistic notions of self against theistic ones of the soul.

Some laymen and theologians have asked the question: "Would clones have souls?" While the consensus is that they would, it is worth pondering what it is about cloning that leads some to pose the question in the first place. The clone's lack of genetic uniqueness is generally put forward as the reason for asking the question, but it might have more to do with the motherlessness of clones. Recall that on a biological level, a male clone would have as a mother only the donor of the enucleated egg, as well as, and only indirectly, the mother of his progenitor.

In the Adoption Model, we have examined the likelihood that a clone might feel motherless. From a social and religious perspective, society might view a clone's motherlessness as profoundly disturbing. Otto Rank tells us that in matrilineal societies, the mother is seen as the soul bearer.[35] But even in patrilineal cultures, this may also be seen. For example, in the Jewish religion, a child is considered to be born Jewish if his mother is Jewish. That is because it is believed that he receives his Jewish soul from her.

There is a danger that from some religious perspectives, a clone might be at risk to be viewed as the personification of cloning as a process and could be condemned, along with the process, as an "abomination." This could be a threat to a clone's psychological well-being, if not her personal safety. And if cloning became acceptable in one country, culture, or religion but not in others, ideologues disapproving of cloning and clones might generalize their negative views and prejudicially perceive any group that permits it.[36]

It is claimed that Islam and Judaism are more receptive than Christianity to the idea of cloning.[37] Perhaps Christianity is more threatened by it. After all, for the individual in the secular/scientific society, cloning might seem to be the next best thing to the resurrection promised to Christian believers. Perhaps another source of Christian opposition to reproductive cloning arises from its eerie reminiscence, if not heretic resonance, to Mary's virginal pregnancy with Jesus—a central mystery of Christian faith.

There is a "religion," or, rather, a belief system, which is said to claim 60,000 adherents worldwide, that not only enthusiastically promotes cloning but also is largely built around it—the Raelians. I hesitate to call it a religion because its "prophet," Rael, says, "If you can clone a human being, there is no God, and no soul." Rael believes that mankind arose from clones created by space aliens.[38]

It would be inappropriate to render any psychiatric opinion of Rael or pass judgment on the validity of his beliefs, but we can speculate about the appeal of his belief system to certain others in terms of the issues and concerns we have discussed here. It would appear that the Raelian belief system is an "ideology of self," promising both immortality and perfection through genetic engineering and cloning. The role of space aliens in this ideology might appeal as a disguised version of the family romance fantasy, with a powerful and omniscient alien parent substituting in the fantasy for a famous human one.

In the relationship of "God the Son" to "God the Father," Raelian beliefs appear strangely parallel to Christian and, specifically, Roman Catholic ones, but for Raelians, a space alien serves as a kind of "God the Father" to Rael, the "prophet," his son, born to a human mother. Raelian beliefs might appeal to some fertile females who, by donating eggs to help cloning efforts or serving as gestational surrogates for clonal pregnancies, would get to be "as if" "the Blessed Virgin."

Religious attitudes toward sex and sexuality are undoubtedly relevant to religious attitudes toward cloning. Perhaps I simply do not understand their full meaning, but it seems to me that the archaic terms *beget*, *begat*, and *begotten* as used by clergy lend a welcome fog of mystery and sanctification to reproduction and the sexual reproductive activities of our ancestors. We tend not to want to think that our forebears, especially our parents, ever had sex. If we do admit that reality, then we tend to want to doubt that they ever had it again after conceiving us, despite evidence to the contrary, including our younger siblings. If we are forced to acknowledge that our parents had sex to conceive us, and continued to have sex afterward, then the final defense is to rationalize it as sanctified. Their sanctified acts of "begetting" put any sexual activity of our own in question, unless it is also sanctified under the institution of marriage.

If we condemn asexual reproduction, then even nonsanctified "normal" sexual activity, in contrast, becomes morally more acceptable. *A surely unintended consequence of religious condemnation of cloning as immoral is that it conversely moves sexual reproduction and, specifically, the sexual aspect of it to a higher moral plane.* We will delve further into the wish of individuals and society to deny that sex has any role in reproduction in the next chapter.

CONCLUSION

In this chapter, we have seen that cloning itself is likely to have profound social and cultural consequences. To bring as many important features of the panorama into focus as possible, we reviewed key social themes anticipated in the cloning models, took some of these in new directions, and raised a few others.

It would appear that the initial societal resistance to human reproductive cloning might give way to powerful social and cultural forces consonant with it. Of course, cloning can only meet in a medicalized, commodified, and "as if" way man's desire for immortality for himself and his wish to deny the mortality of those he loves. But some might be tempted to grasp at it, if only because society as a whole is increasingly becoming medicalized, commodified, and "as if." Normal mechanisms of self-recognition and preference for self-resemblance will encounter a new "technology of self" in cloning, and the pull might be irresistible to some. The universal wish to be empathically understood and admired powerfully persists as the soft underbelly of the narcissist, and he might hope that his clone will meet these needs.

But we need not focus solely on the narcissistic individual or even a culture of narcissism to see that cloning might eventually be welcomed. Our society appears to have an affinity for the clone-like, especially in politics. It might also be happy to have asexual reproduction obviate an issue about which it is squeamish—the sexual origins of human beings. In addition, a society that values wealth and ownership could easily come to see having a self-clone as the ultimate "possession." A society that needs to be reminded that it should consider "the best interest of the child" in decisions affecting him might come late and give insufficient thought to such a development. For the most part, at present, society does not now embrace the idea of reproductively cloning people. This chapter shows why there might come to be a shift in momentum toward it.

The next chapter focuses on other social and cultural aspects of reproductive cloning that are also highly personal, those dealing exclusively with intimacy, sex, and sexuality. On these matters, we will explore what reproductive cloning may imply for the individuals directly involved—the parent/progenitor, his or her spouse, and the clone himself, as well as for society as a whole.

11

Intimacy, Sex, and Sexuality

We have already touched on the issues of intimacy, sex, and sexuality in the context of various models of cloning. Now let us integrate what we have concluded to try to understand how cloning might affect intimacy, sex, and sexuality from a wider societal perspective. As in the last chapter, we will transcend the models. In doing so, we shall rely heavily on psychoanalysis and related psychological theory and, to a lesser extent, sociobiology and evolutionary psychology.

INTIMACY

The achievement of intimacy is a central human issue. If all goes well, the capacity for it is usually achieved in young adulthood. Erik Erikson focuses specifically on intimacy in his sixth stage of human psychological development, "intimacy versus isolation." Although the term is often used more generally, let us define *intimacy* as a closeness of one autonomous person to another.

One of the several paradoxes of intimacy is the "need to be separate to be close."[1] When one's own personal identity is sufficiently established, a person can relax boundaries of self in certain areas while maintaining them in others without fear of feeling submerged or engulfed. All this requires deep mutual trust between two people and a degree of independence and maturity that allows both individuals to feel that they can survive apart from one another.

The mutual bonding and attachment of infant and mother are the earliest basis of intimacy. If a mother responds empathically to her infant's physical and emotional needs, then her infant becomes securely attached and develops a basic sense of trust. Having been able to implicitly trust one's mother is at the foundation for all later trusting relationships.

But let us be clear that the close relationship of child and parent that is an important basis of later intimacy is not, itself, an intimate relationship. That closeness has its foundation in the necessary dependency of the child and her desire and need to trust and rely on the parent. When parents are attentive and empathically responsive, the child can feel good about being close to them, but a close parent–child relationship and an intimate one are two different things. An intimate relationship is one of two people on roughly equal footing with each other in terms of power, dominance, and dependency. Each person in an intimate relationship loves and appreciates the other as a separate self.

When people speak of intimacy, it is usually in the context of a couple, usually a heterosexual marital couple, and it is typically assumed that a sexual component is an important expression of their intimacy. However, we may think of close friendships as intimate relationships, including so-called platonic relationships between men and women. Close friendships between people of the same sex typically do not become sexually intimate unless the individuals are homosexual. However, a person may discover that he or she has homosexual feelings and might possibly be homosexual in the context of close friendship. Even when sex is not part of an intimate relationship between members of the opposite or the same sex, sexual tension may be in the background.

We all know that intimate friendships between individuals of compatible sex and sexual orientation frequently progress to sexual intimacy. Sexual intimacy can reinforce the general intimacy of the relationship but sometimes undoes it by taking the relationship to a level of intimacy for which the couple may not be ready. Mutual infatuation is not intimacy, and neither is a sexual relationship that may come out of it, though it is sometimes possible for both to mature into true intimacy.

WHAT DO THE MODELS
OF CLONING SUGGEST ABOUT INTIMACY?

We have examined the issue of intimacy directly or indirectly in each of the models relevant by analogy to cloning. We have found that psychological theory, research, and clinical data converge to suggest that clones might be prone to difficulties in establishing and maintaining intimate relationships. The capacity of intimacy is built upon successful negotiation of Eriksonian developmental stages prior to the one in which it is the central theme. Lingering residues of unresolved difficulties from those earlier stages can interfere with it. For this reason, difficulties anticipated for clones in each of those stages would likely adversely affect their capacity for intimacy. We reviewed these in chapter 9 and discussed other relevant factors in chapter 10.

RELATIONSHIPS AMONG INTIMACY, SEX, AND SEXUALITY

A person's sexuality has several components, two of which are gender identification and sexual orientation. As children, gender quickly becomes a basic component of our identity, and early on we typically come to identify most strongly with the parent of the same gender. But we should understand gender identity as different from sexuality, including sexual orientation.

According to classic psychoanalytic theory, to understand immature sexuality, we must realize that the earliest manifestations of sexuality relate to basically nonsexual bodily functions, such as feeding and the development of bowel and bladder control. Freud described it as "polymorphous perverse." To the child during Freud's oral phase, which extends into the second year of life, sucking, biting, and chewing are erotic activities. During the anal phase, ages two to four, the main erotic activity shifts to anal and rectal regions, and during the phallic phase, beginning during the third year until about the end of the fifth, erotic activity becomes linked physiologically and psychologically to urination.

The very young child explores his or her body and tries to elicit pleasure from it even before masturbating with sexual fantasies at a later stage and certainly before orgasm is possible. Sexual orientation as a component of sexuality starts out fluid, becoming less so with maturity. Some people may not settle on their sexual orientation as heterosexual, homosexual, or bisexual until adulthood, and experts now agree that people are not simply heterosexual or homosexual, there being a variety of ways to be heterosexual and homosexual.[2]

From a developmental point of view, intimacy, sex, and sexual orientation are separate dimensions. In general, a person's earliest sexual experiences are autoerotic. Young children of both sexes touch themselves where it feels good, much to the chagrin of many parents. Aside from these early experiences, males and females appear to diverge. Much of the development of a male's sexuality occurs outside of a personally intimate relationship with another person. Some males never progress beyond the developmentally normal adolescent perceptions of females that Stephen Levine calls "default heterosexuality." In effect, it is a heterosexual least common denominator. Layered on top of this, if all goes well, a man develops the capacity for intimately relating to a woman as an individual whom he can uniquely appreciate for qualities other than her sexually exciting physical attributes.[3] Although biological differences may account for some of this, it is thought that at least some of it arises from the fact that a boy must stop identifying with his mother to identify himself as male, whereas a girl does not need to push away from her so hard.[4]

In contrast, females typically develop the capacity for intimacy early. Because a girl never needs to stop identifying with her mother, girls and women feel their identity as females enhanced by an intimate connection with a male, whereas boys and men typically do not feel their identity as men enhanced by an intimate connection with a female: "Thus, women are more likely to experience feelings of love, attachment, and connection even as young girls, with linking of full sexuality to romantic love taking a longer time to develop over adulthood." In contrast, "males start out with strong sexual feelings in adolescence and over the course of adult development (if it occurs normally) become progressively able to experience attachment and love with the same partner with whom they feel sexual excitement, passion, and release."[5]

In mature sexuality, people of both sexes, secure in their sexual orientation and other aspects of their personal identities, are capable of and prefer a sexual relationship with someone with whom they feel personally intimate. For such adults, intimate personal connections often incorporate sex in a very natural way.

FROM SEX AND SEXUALITY TO REPRODUCTION

As sex is often dissociated from intimacy, as we saw in chapter 2, there are many ways in which sex may be dissociated from reproduction. Cloning, while the newest of these, would be fundamentally different. It unlinks reproduction not only from the sexual act but also from sex at the biological level. At that level, sex is the incorporation of genetic information from two individuals of opposite sex to constitute a unique genome of an embryo that has the potential to develop into a unique new person. The result and presumed evolutionary purpose of sex is to promote genetic and biological diversity and, hence, enhanced adaptability to a changing environment. Sexual organisms derive an adaptive evolutionary advantage from their diversity. This is thought to occur either by bringing beneficial mutations together into a particularly favorable combination or by bringing deleterious mutations together in such an unfavorable way that it purges them from the genome.[6]

The physical pleasure of sexual intercourse serves the purpose of reproduction, but it also helps to reinforce other kinds of closeness and intimacy in the parental relationship.[7] That helps the adult couple serve as a secure nucleus for the rearing of their children.

The intimate biological merger of genes from two people does not occur with cloning. For this reason, cloning also makes it possible to dissociate reproduction from personal intimacy between people of the opposite sex. About these facts, let us now consider the feminist perspective on cloning, pornography and "virtual sex," and the implications of cloning for homosexuality.

A VIEWPOINT ON THE FEMINIST PERSPECTIVE ON CLONING

We could think of cloning as another step in the sexual liberation movement. By allowing reproduction without sex, it would indirectly further sex without reproduction. Although it might be liberating for women, if they wished, to have children without the involvement of men, some feminists view it differently. To them, cloning as an assisted reproductive technology would be a medicalized expression of male domination of women in society.[8]

Not wishing to discount these concerns, we might want to reframe them. While many feminists assume that males are inherently sexually aggressive and wish to dominate women, research indicates that this is not the case for the majority of men. Men preoccupied with sexual domination appear to be compensating for fearful fantasies regarding the size and potency of their penis, damage to it during sex, fear of rejection by women, and fear of homosexuality. The most common sexual fantasies in men are not those of sexual dominance but, rather, those of the "omni-available woman" and of lesbian sex. Both serve to compensate for those same fears.[9]

Fantasies of the omni-available woman reveal the man's wish for not only a sexually available and appreciative woman but a woman without individual importance.[10] At a societal level, cloning could also potentially diminish the individual importance of women by turning them into actual or potential egg donors and living incubators for the gestating clone. In the world of men's sexual fantasies, it would follow that women who advertise themselves as available to donate eggs for cloning might also be sexually available. On a broad social level, knowing that there are women available to donate eggs for cloning might encourage the fantasy of sexual availability without individual identity of women. An alluring photo of several cleavage-revealing women from the pro-cloning Raelian sect lying shoulder to shoulder seemed, at least to this adult male, to blur the difference.[11]

Knowledge of breeding systems might also inform the feminist perspective on cloning. Richard Dawkins points out that conflicting interests
between males and females are evident in the breeding systems in animals.[12] In most animal species, females are more invested than males in
their offspring, a fact that sociobiologists understand as a natural consequence of pregnancy and early maternal care for the infant, as well as the
much smaller number of reproductive cells they produce. Humans are no
exception. A woman produces only hundreds of eggs in a lifetime,
whereas a man generates countless sperm over many more years.[13] The
use of cloning as a reproductive option would create a new alternative
breeding system that could exacerbate the potential for conflict between
the sexes. We should consider how such consequences might affect not
only child clones but also any member of society and society as a whole.

PORNOGRAPHY, SEX, AND SEXUALITY

In the previous chapter we saw how both asexual reproduction and
pornography, especially its most technologically advanced form, virtual
sex, could be thought of as instances of the "as if real" in our society.[14] Reinforcing this understanding is the fact that pornography and asexual reproduction have literally converged in a pornographic website that also
auctions models' services as egg donors.[15] From there, it is a short diabol-
ical step to selling tissue samples from models for reproductive cloning
efforts. The fantasy of actually cloning women as sexual objects will hopefully never be realized, not if anyone entertaining it realizes that the clone
would start life like anyone else—as an infant.

We might think of heterosexual pornography as the photographic
"cloning" of women as objects of male sexual fantasy. As mentioned, the
receptive woman of fantasy is typically without individual identity, a characteristic imagined for clones. Just as the pornographically depicted omni-
available virtual "clones" of heterosexual pornography lack individual
identity except as sexual objects, so too might egg donors for cloning lack
individual identity except as sexual objects of another sort. As mentioned
previously, the line between sexual object for egg donation and sexual object to the heterosexual male's imagination might seem to blur.

CLONING AND HOMOSEXUALITY—
SOME PRACTICAL IMPLICATIONS

It might eventually become possible for a homosexual couple to create an
asexually reproduced offspring with a combination of their genomes

through gene splicing or through the fusing of very early embryos to create chimeras.[16] This would make it possible for a girl baby to have two mothers, one of them most probably also the woman who gives birth to her. The same technology might also make it possible for a boy baby to have two fathers, though a gay couple would need the help of a surrogate mother. Except for the more limited roles of egg donor and gestational mother, men would no longer need women to have boy babies, and women would not need men at all to make girl babies.

While these would be new biological "facts of life" to add to the age-old ones, aren't two parents, one a man and one a woman, necessary for successful child rearing? The preliminary answer seems to be no. Although still an unsettled issue, children reared by gay and lesbian couples seem to fare as well as children reared by heterosexual ones.[17]

Though those research findings are not well known, we might suspect that even if they were, it would only partially mitigate the negative societal response to homosexual families. For a society unsettled by homosexuality and nontraditional family forms, cloning threatens to make male and female homosexuality itself seem more "natural" by providing an option for genetic children. Their families would then become more comparable to those of heterosexuals.[18]

Perhaps society might become more comfortable with the idea of homosexuals rearing a child if that child's sexuality were not overwhelmingly influenced by the fact that her parents are homosexual—something the limited research in adoption by gays and lesbians seems to indicate.[19] However, in the Parent–Child Resemblance Model we have seen that there are strong genetic determinants to homosexuality, so it is reasonable to presume that a clone of a homosexual would have a great chance of also being homosexual. The idea of homosexual couples rearing homosexual children might be more discomforting to many in society who would wish to "protect" the potential child, not only from the family environment created by homosexual parents but, prior to that, from the "homosexual genes" that would help to predetermine his fate.

DISCREDITED PSYCHOANALYTIC IDEAS ABOUT HOMOSEXUALITY: HOW THEY MIGHT APPLY TO CLONING

Although Sigmund Freud never viewed homosexuality as pathological, many other psychoanalysts thought that same-sex erotic preference was a manifestation of narcissistic choice and a perversion.[20] In this, psychoanalysis came to reflect the ideas of the larger science and culture of the day as to what was normal and what was perverse.[21] The view that homosexuality is invariably pathological is now politically and scientifically suspect, given the powerful evidence for a strong biological predisposition toward it.

This is not to say that homosexuality is completely predetermined biologically or that social influences are unimportant. Nor is it to say that all homosexual individuals are sexually normal and healthy, but the same might be said about heterosexuals. On the other hand, confused sexuality, fetishism, and other perversions are indications of psychological problems that may be found in both heterosexual and homosexual individuals. We'll examine sexual perversion in more detail later in this chapter.

Although the idea that homosexuality is a matter of choice has been largely discredited scientifically, we ought not to completely discard those earlier psychoanalytic thoughts about it. They might apply to those who would prefer cloning as a mode of reproduction. Those psychoanalytic misunderstandings of homosexual preference as perversion might help us understand those who would take normal asexual preference for self-resemblance in offspring to the extreme of self-cloning.[22] For example, substitute the word *self-cloner* for *homosexual* in this statement made by the psychoanalyst F. H. Allen over sixty years ago: "The homosexual is often one who chooses a body like his own because of his terror of the differences of the woman" and "his lack of strength to support such a difference."[23]

But what if a clone could be genetically modified when it is only a fertilized egg to make its gender opposite that of its progenitor?[24] While we cannot apply those older psychoanalytic ideas to such a situation, we can use others. The fascination with the hermaphroditic image has been said to "represent a striving for wholeness, a striving that is not sexual but ontological."[25] Cross-dressing provides a transvestite with a kind of "direct sexual relationship [to himself]" and can be understood as the "logical consequence of [the fantasy of] hermaphroditic completeness."[26] Self-cloning of an opposite-sex clone could permit the progenitor a kind of hermaphroditic completeness. Such a progenitor might wish to move beyond the symbolic value of the new hermaphroditic image of one person existing as two people of opposite sexes to the reality of progenitor and clone having sex with each other. Perhaps such a progenitor might try to rationalize incest with his opposite gendered clone by equating such incest with masturbation.

THE APPARENT ASEXUALITY OF CLONING

Our exploration of cloning requires us to see how it relates to sex. This is because society's fascination with asexual reproduction might derive in part from its long-standing fascination with sex and what Michel Foucault has called "peripheral sexualities," including transsexuality and sexual ambiguity—topics of increased interest recently.[27] Although it may not seem it, some of society's fascination with cloning might be because it deeply resonates as a "peripheral sexuality."

Cloning and Erotic Desire

First let us look at the components of erotic desire and see how they might relate to cloning. The psychoanalyst Otto Kernberg characterizes erotic desire as the search for pleasure through "closeness, fusion, and intermingling that has a quality of a forceful crossing of a barrier and becoming one with another person."[28] It may be a stretch, but we could think of the fusion or penetration of an enucleated egg as symbolically possessing an "erotic" quality. It is possible that there might be a few people for whom obliterating an egg's identity through enucleation would also be a reproductive and symbolic fulfillment of sadistic rape fantasies.

The longing for oneness with another might, in the case of self-cloning, be with the clone as a representative of the self, while a progenitor other than the self might be selected, in part, because the person wishes she were close to them. Identification with one's sexual partner is another characteristic of erotic desire, in particular with one's partner's excitement and orgasm.[29] Cloning is completely nonerotic in this sense.

One important characteristic of erotic desire that cloning would share is the sense of transgressing prohibitions implied in all sexual encounters. Sex may be said to be symbolic of defiant transgression against the prohibition against sexual intimacy with the opposite-sex parent and victory over the same-sex parent or oedipal rival. However, "the simplest and most universal [symbolism of sex] is the transgression against conventional morality."[30] Self-cloning could be used as an instrument to triumph over the progenitor's oedipal rival by fathering oneself. By achieving the result asexually, a person could still maintain his repression of forbidden oedipal sexual desires.

Cloning, Narcissism, and Perversion

Some human concerns are universal. Erikson captured many of them in his model of psychological and psychosocial development through the life cycle. The way individuals negotiate these concerns varies from person to person, and some do so more successfully than others. Much abnormal behavior can be understood as maladaptive coping with universal human concerns, with aspects of the behavior entirely appropriate to an earlier stage of psychological development.

This is not to say that one's genetic heritage and individual biology do not play an important role in psychopathology, but there is an interaction among them, one's immediate and broader social and physical environment, and the process of individual physical and psychological development. For example, males with low activity of monoamine oxidase A, an enzyme crucial to the metabolism of certain neurotransmitters, are more likely to engage in antisocial criminal behavior like rape, robbery, and assault—*but only if they had been abused as children.*[31] The normal adult relates to others in a way that acknowledges their personhood. Realizing that other people are sentient beings, he is able to consider their perspectives and feelings, and doing so helps him to govern his behavior. An individual with low monoamine oxidase, after having been abused, becomes more likely to engage in the sorts of criminal behavior in which the perpetrator is typically deficient in appreciating the personhood of others.[32]

If we conceive of reproductive cloning psychologically as a new but *asexually disguised form of sex,* then we should also consider how specific coping patterns evident in neurotic sexual disorders and sexual perversion might be expressed in cloning. In neurotic sexual disorders, sexual function and pleasure are inhibited and the consequences are, for the most part, private. Sexual perversions, however, are of great social relevance, given the damage that a sexual pervert can do to others. For this reason, we will focus on perversion.

Pathological narcissism, as reviewed in the parent–child resemblance chapter, is another coping style that can also be quite damaging to others. Both pathological narcissism and sexual perversion do not admit to the personhood of other people, except to use it as a tool of manipulation. Individuals with sexual perversions are inherently narcissistic, at least in the area of their perversion. Both the sexual pervert and the pathological narcissist are deficient as mature adults. As with the pathological narcissist, we can condemn the damage inflicted by the sexual pervert and hold him accountable yet still try to understand him.

Perversion Defined

Sexually perverted behavior deviates from accepted adult norms in choice or aim. Classic psychoanalysis saw the norm as heterosexual genital intercourse. More modern views incorporate homosexuality within a broader norm, but some types of homosexuality do meet the general definition of perversion.[33] Because the term *perversion* has been used pejoratively, some authorities in the area prefer the term *sexual aberration,* but even they still use the term *perversion.*[34] Because the term has been hurled as an epithet or slur, I use the word reluctantly but will use it nonetheless—hopefully with care, precision, and accuracy.

"Extreme" sexual behavior that is "fixed and exclusive" is perverted, while the same activities when part of an otherwise normal session of sexual intercourse may be considered normal. In the various perversions, one finds particular sexual fantasies, masturbatory practices, sexual props, or special requirements for a sexual partner. Examples are fetishism, transvestitism, voyeurism, sadomasochism, and pedophilia.[35]

Briefly defined, a fetishist must focus on an inanimate object or a part of the human body to attain sexual arousal and orgasm, whereas a transvestite must fantasize himself or actually dress as a woman. A voyeur feels pleasure, as well as anxiety, in looking at intimate, mostly sexual activities of others. A sadomasochist requires physical pain in a sexual relationship, either inflicting it or experiencing it. A pedophile prefers or exclusively fantasizes about and pursues sexual activity with children.[36]

As mentioned earlier, sexual perversions can be contrasted with neurotic sexual disorders, such as erectile failure, premature or delayed ejaculation, and avoidance of sex. The common denominator to the neurotic sexual disorders is the inhibition of pleasure. In contrast, the common denominator in the perversions is the facilitation of pleasure, and the thought of having sex without indulging in those behaviors arouses so much anxiety that they typically become the precondition for sexual pleasure.[37]

Sexual perversion is much more common in men than in women, and some perversions do not occur at all in women.[38] This may be because, between the two sexes, the male needs to feel more secure to perform his role in sexual intercourse and because castration anxiety fears are also greater in men.[39] Another factor may be related to the fact that because a boy's first identification is with his mother, he needs to "disidentify" with her to identify himself as male. Hence, problems negotiating the separation-individuation phase of development may interfere with the consolidation of his self-identity as a male.[40] Although confusion about gender identity is not itself a perversion, to the extent that being able to identify oneself as male is very important for a boy to feel separate and secure, it may be a factor in the development of perversions.

Characteristics and Psychological Functions of Sexual Perversion That Might Also Be Found in Cloning

We may understand sexual perversions psychologically as serving certain defensive functions. They all relate to universal human concerns, and while some are gender related or gender specific, most are not sexual. Most of us have coped well with these much earlier in our development, so some might seem far-fetched or strange to us as adults. Others may still be with us, but we manage to keep them at bay successfully without psychological symptoms. Some concerns are generic and nonspecific, and others are more specific to sexual perversion. We shall consider the major issues here and see how some individuals might employ sexual perversion to cope with them. If reproductive cloning were to serve these same functions for certain individuals and for society as a whole, then we could think of it as an *asexual perversion*. Now, let us take a look at these concerns, how sexual perversion may be an attempt to answer them, and how cloning might also be an attempt.

Feeling Whole, Alive, and Special: Cloning and Narcissistic Dimensions of Perversion

We all want to feel securely alive and whole and, beyond that, special as individuals. Normally, an individual adequately establishes all this in childhood and adolescence. Unfortunately, emotionally abusive environments can severely undermine the basic trust, autonomy, and self-confidence of a child and the later establishment of secure individual identity by an adolescent. Sometimes a young child in such an environment recruits his immature sexuality to try to cope with problems in these basic areas. As mentioned earlier, immature sexuality is polymorphous perverse and relates to basically nonsexual bodily functions.

Polymorphous perversity persists in sexual perversion. Rather than understanding it for what it is—a disguised expression of immature and infantile sexuality used to defend against anxieties and fears—the sexual pervert may feel his perversion to be important to his uniqueness as an individual and may take pride in it. Without it, he may not be able to sustain a personal identity.[41]

Self-pride is also a central concern of the narcissist. As young children, we naturally feel self-satisfied and perfect until we repeatedly run up against our limitations in the real world. The narcissist refuses to recognize her limitations and is deeply wounded when forced to do so. More than the rest of us, she seeks to recover her sense of perfection in her "ego ideal." Likewise, the pedophilic sexual pervert may also seek to attain his ego ideal, but he does so by romanticizing a child as representing it.[42]

The pedophile, having engineered a relationship with a child, feels better because he may honestly feel that he is "in love" with his sexual prey. Freud wrote of the "cure by love" whereby the individual falls in love with "what he once was and no longer is, or else what possesses the excellences which he never had at all."[43] The pedophile's attempt at self-cure is not a cure but, rather, the interpersonal expression of his psychopathology. The child he "loves" is left confused at best and may be traumatized for life. Their relationship is not mutual, as the pedophile is in love with his own ego ideal projected onto the child.

Self-cloning might also be an attempt to personify and possess one's own ego ideal in a child. But in serving this function for the progenitor, the clone would be depersonified, though hopefully never as absolutely as the victims of childhood sexual abuse are. If he were trying to create his ideal self in the person of his clone, the progenitor might be enacting an asexual perversion. Just as sexual perversion tends not to be mainly about sex, the wish to parent through asexual reproduction might not mainly be about wanting to parent a child lovingly, helping him develop into a capable and independent adult.

Cloning as a romanticization of the clone, whether of a self-clone or a clone of an admired other, might be undertaken to enhance the rearing parent's self-esteem, in a kind of *narcissistic asexual pedophilia*. There might be a risk that the romanticization could eventually become sexualized, but the greater risk would be that the parent will "fall out of love" with the child with less dramatic but also detrimental consequences.

Because our society itself manifests many narcissistic features, we need not limit ourselves to thinking about the narcissistic individual as a rearing parent of a clone. The bioethicist Leon Kass argues that in our modern society, ideas of individualism unlink us to our ancestors and traditions, making us "projects for our own self creation, not only as self-made men but also as man-made selves," with "self-cloning" an "extension" of such "narcissistic self-creation."[44]

In our "culture of narcissism," cloning might become the ultimate consolation prize in dreams of self-fulfillment.[45] The progenitor might expect her clone to be the living embodiment of her "ego ideal"—that is, everything she had wished herself to be. Alternatively, she might wish to parent a clone of a living or dead representative of her ego ideal—in other words, a clone of a famous person.

As a culturally accepted perversion, the ancient Greek practice of homosexual relationships between men and mid- to late adolescent boys might shed light on possible motivations for self-cloning by narcissistic individuals.[46] But it is even more relevant to helping us understand self-cloning in a modern narcissistic culture. In this regard, the gender and sexual orientation of the self-cloning progenitor are irrelevant to the parallels.

It is believed that through a homosexual relationship with a boy, the ancient Greek man sought to produce a son in whom his soul would survive. Through that relationship, the man fertilized, both spiritually and otherwise, "the living image" of his own soul, materialized in the boy, as representing the man's ideal self. In this, it helped for the boy to resemble the man physically as much as possible.[47] This kind of homosexuality is not primarily sexual but, indeed, an attempt to be reborn "as a spiritual, intellectual, and *physically similar replica of oneself*" through a process of *"perfectly individualized self-perpetuation."*[48] The parallel of this ancient form of pedophilic homosexuality to self-cloning is obvious.

It is also parallel to a phenomenon that Ernest Becker has described some modern men, "true geniuses," as wishing to "bypass the family as the instrument of . . . procreation" as well as "the woman and the species role of his own body." Such a man may "try to procreate himself through a linkage with young gifted men, to create them in his own image. . . . It is as though he were to try to duplicate himself exactly [in] spirit and body."[49] Self-cloning would make this possible for any aspiring genius of means, regardless of gender or sexual orientation.

For some, not necessarily pathological narcissists but people on a search for themselves, self-cloning could be an attempt to find what we might call the "homuncular self." The homunculus is the little person frequently envisioned to be inside the mind.[50] It is a metaphor for the self, of course. Metaphors symbolically give physicality to concepts, sometimes embodying them.[51] Indeed, it might be said that metaphors are cognitive evidence for an "embodied mind."[52] As a homuncular self, the self-clone would be an externalized embodied self-image.

So, self-cloning might be a kind of misplaced search for the unknown and unexplored self or the true self from whom an individual may have been alienated during a difficult childhood. For a person with Dissociative Identity Disorder, self-cloning could even be an expression of the search for an undiscovered child alter personality.

Cloning as an Asexual Perversion Offers a False Solution to Unresolved Oedipal Problems

The oedipal situation forces on the child the reality of both gender and generational differences. These differences are important to understanding the oedipal situation and how it can go awry.[53]

According to classic psychoanalytic theory, the young boy wants to possess his mother completely, including sexually, but feels relatively inadequate to the task given the small size of his penis compared with that of his rival, his father. But the boy feels absolutely inadequate when he compares his small body to that of his mother and what he might imagine of her genitals. The boy fears castration by his father but may also feel inadequate, rejected, and even denigrated by his mother. If the frustration with his mother and anger toward her are great, then the boy may also fear that she will castrate him.[54] One way to understand perversion is to see it as an attempt to deny castration anxiety through an aspect of sexual behavior that seems safer.[55] For some progenitors, this might also be a perverse motivation for cloning.

Another sexually perverse response would be to deny that gender and generational differences mean anything at all. There is consensus that "the perverse act often represents an attempt to obliterate differences, whether between the sexes . . . between the generations . . . or by exploring the differences in a controlling way."[56] An even more general intolerance of difference, of the separate and complete self of the other, is a hallmark of perversion. Disrespect of generational difference is central to pedophilia; intolerance of sexual difference is central to perverted forms of homosexuality; and control and exploitation of sexual differences may be found in sadistic heterosexuality.

In this context, we should recall that the normal evolutionarily driven preference for self-resemblance becomes an intensified imperative for the pathological narcissist, an individual also unable to tolerate differences. When manifested in sexual and generational dimensions, intolerance of difference is a hallmark of sexual perversion.

The pervert despises genital intercourse, as it evokes in him feelings of inadequacy. Its origin is thought to be in his sense that he would be unable to satisfy a given woman the way his father could his mother, along with fear of the inside of a woman's body.[57] Cloning would afford a means to avoid these feared places, while medicalized control, if only through the proxy of the cloning clinic, would permit a kind of perverted mastery of it, making these spaces less threatening and mysterious.

The progenitor of a clone could declare oedipal victory by disposing of his need to compete with his father sexually. By impregnating a woman through cloning, the progenitor would achieve asexually the same "conceptual" result (literally and figuratively) as his father, the sexually potent and preferred male in the oedipal triangle. As a perverted solution to the oedipal situation, there would be no danger of castration because no sex is involved, and other insecurities originating in generational difference would not matter. Genital size and potency would be irrelevant to providing somatic cells for nuclear extraction and procedural insertion into an enucleated ovum. By re-creating himself as a clone, the progenitor would do so through an act that, though biologically asexual, might unconsciously symbolize incest, in defiance of his father. As the literal father of himself, the self-cloner would eliminate his father altogether as an oedipal competitor.

Although cloning does not offer a solution to having been excluded from his parents' special relationship, the progenitor could purchase some comfort by getting to feel like a "primo-progenitor"—the biggest and original daddy of himself. Thus, cloning oneself could be a perverted solution to persistent and unresolved oedipal fears.

Whereas for the progenitor, cloning could be an asexually perverted attempt to reverse his earlier oedipal defeat, we would expect the child's clonal origin to present the child with magnified difficulties in exactly the same areas—the acknowledgment of generational and sexual differences. Such difficulties might result from being reared by a progenitor/parent likely to feel that he has an enhanced generational prerogative in knowledge and authority over his clone. Others might arise from the clone's own, and not unexpected, genealogical bewilderment, an issue explored in the Adoption Model.

Incest

The wish for and the avoidance of incest are at the center of the Oedipal complex, so let us turn our attention specifically to incest. While the young child may fantasize about incest, he represses those wishes and does not, in any case, have the power to act on them. This is far from the case for adults with incestuous desires.

Seeing incest in self-cloning, the bioethicist Leon Kass states his case against cloning by arguing for the "wisdom of repugnance." Disrespect of generational differences is inherent in one repugnant act he likens to cloning—father–daughter incest and its "dreaded result . . . —to be parent to one's sibling."[58] He fails to mention mother–son incest and incest by the same-sex parent, but cloning as an asexual perversion more clearly points to them. Let us focus on mother–son incest to see what it can teach us.

A child's mother generally tends to be the parent who provides most necessary care to the body of the infant and young child to keep the child clean, healthy, and warm. Biologically, mothers are the only parent equipped to provide one type of care, nursing. It is thought that the infant may experience early maternal care as erotically seductive. Most mothers exercise strong psychological defenses against also experiencing elements of the experience as erotic.[59] Providing body care to the infant is thought, in a sense, to immunize against incest, presumably because of those defenses. Perhaps this is why mother–son incest is the least common form of incest.[60] An incestuous mother violates a most basic taboo with psychological, cultural, and even biological bases.[61]

Perhaps the very fact that Kass omits mother–son incest among a list of acts he finds most repugnant might indicate that it is even more repugnant than the others. It may be hard to imagine the idea of a mother exploiting the normal sensual experience of her child in receiving maternal care to his body, shaping and encouraging it to premature development of sexuality with herself as the focus, but it happens. The betrayal of trust that what the caretaker does is in the child's best interest compounds the violation of the boundary between generations. When the person is the primary caretaker, the violation is most extreme. This does not mean that clones would be more subject to incest.[62] The perversion would probably only be expressed in the asexual realm.

Cloning as an Asexual Perversion Permits Avoidance of Mutuality

The dominant motifs in the psychoanalytic literature on perversion have been oedipal guilt and the need to repress the Oedipal complex, but in more recent years, the focus has shifted to conceptualizing perversion as a defense against genuine relatedness to others.[63] Reflecting this change in emphasis is a definition of *perversity* as "the sexualization of the avoidance of mutuality."[64] By analogy, if cloning were used to avoid mutuality, then we could conceive of it as an asexual perversion.

The pervert not only does not relate genuinely to others; he is also fundamentally alienated from himself. His perversion can be understood as an attempt to find "personalization" through sex.[65] For some, self-cloning could have the same goal, perhaps as an attempt to repair the results of having been "depersonified" as a child.[66]

Perversions require the presence of another person to be enacted, usually in reality, sometimes only in fantasy.[67] But the pervert is unable to relate to the other as a whole person or even as a person at all.[68] We might understand the pervert's lack of relatedness as a result of his not registering his father as a "significant presence" and of his having had a mother who "lavished intensive-body care" on him as an infant and young child "but in a rather impersonal way." As a child, she treated by him as her "'thing creation' rather than as an emergent growing person in his . . . own right."[69] With the likelihood that the progenitor as *asexual* pervert would treat his clone as his "thing creation," he would also put the child at risk of becoming perverted himself—but *sexually* perverted.

Typically, a child who later becomes a sexual pervert internalizes his mother's idolized view of him as an "idolized self," usually around the oedipal phase when his mother suddenly withdraws from him, having become aware of their overly intense mutual attachment. Feeling abandoned and overwhelmed by the separation, he wards off panic by becoming attached to and invested in idolized internal images of himself.[70]

Later, he relates to the object of his perverted interest in the same idolizing, idealizing way, using the other person to mirror himself defensively in an attempt to hide feelings of inferiority or worthlessness.[71] The object of a sexual pervert's perverted interest is actually perceived as a vital piece of himself or a "selfobject," which he feels he needs to feel whole and alive through a process Arnold Goldberg calls "sexualization."[72] The clone might serve the same purposes for the progenitor/parent.

Just as the sexual pervert does not acknowledge the personhood of the object of his perversion, cloning as an asexual perversion does not acknowledge the personhood of the clone and other participants in the cloning process. The woman would be little more than a uterus or a donor of an egg to be enucleated, and the clone would be an object for idealization and narcissistic gratification. Thus, *the apparent asexuality of self-cloning may be parallel conceptually to the sexualization of perversion.*

Adam Phillips writes of a sixteen-year-old boy's wish to find a girlfriend who would be his clone. About this Phillips says, "The fantasy of cloning a girlfriend is the fantasy of not needing a girlfriend."[73] I once treated a young psychotic man for whom that fantasy became a delusion.

Ron told me at length about his girlfriend Ronnie. I realized that Ronnie did not exist in external reality when he told me that he could see her whenever he wished by looking at himself in the mirror and asking her to be there. His face would then appear to him to morph within a few seconds to become hers!

The wish to not need a girlfriend is like that of the pervert, but most boys move beyond it. For those who do not, relating to women fetishistically, that is, as body parts, is the answer. For others, women are replaced by pornography, prostitution, and perhaps in the future, virtual sex. These boys as adults might wish to reproduce without the involvement of real women, except as paid surrogates. Such a man might be relieved to perpetuate his genes, not "contaminated by that of his sexual partner."[74]

Cloning as an Asexual Perversion Could Be a Form of Revenge to Try to Master Psychological Trauma

A person may develop a perversion as a response to childhood trauma but with the central feature of taking revenge, doing to someone else what had been done to him.[75] It is a sad fact that children who have been sexually abused are more likely to grow up to abuse children themselves. Psychoanalysts explain it as an example of "identification with the aggressor," in which an individual tries to master the experience of having been a powerless victim by victimizing those less powerful than himself.

An individual abused as a child might wish to self-clone to try to give herself a childhood free of the trauma she suffered. Of course, the clone would not be her but, rather, her genetic double. Tragically for both clone and progenitor, the clone might even be at risk for abuse. That could happen if the progenitor felt compelled to reenact the abuse she underwent to try to gain a sense of mastery over experiences she had been powerless to stop or prevent.

An individual abused as a child might even wish to clone an abusive parent, perhaps not being conscious that she had been abused. Unaware of her motives to reverse her position from powerless child to powerful parent with respect to a clonal representative of her abuser, the progenitor might be inclined to abuse such a clone.

Cloning as an Asexual Perversion Might Be a Way to Deny the Reality of Death

We have explored the denial of death several times already, but here is another take on it. Sexual perversion may also be an expression of denial that the person is, before anything else, an animal and consequently fated to die.

Central to the concept of sexual perversion is the idea of "fetish," which you will recall is "an inanimate object or a part of the human body that a fetishist needs in order to attain sexual arousal or orgasm."[76] The classic psychoanalytic view is that by focusing on the fetish, the fetishist alleviates his fears of castration. However, we might also understand his use of the fetish to represent a grandiose means to cope with existential fears by liberating himself from "earthbound flesh."[77]

Other sexual perversions include sadism and masochism. The former appears to arise from the wish to deny separation, and the latter, in the fear of separation. In sadomasochistic relationships, the mode of relating is the attempt at repair. The sadist denies loss by punishing the other person. The masochist denies loss by identifying with the lost other and finding substitutes.[78] We might understand cloning of the dead or dying or of those whose loss is feared regardless of health and age as an asexual form of sadomasochism to retain or to deny the loss of the other person.

The rite of swallowing sperm found in Gnostic sects and medieval devil worship may be seen as a form of self-fertilization. These cults also advocated incest as an avenue to rebirth.[79] Asexual perversion, expressed through asexual reproduction, might also be symbolic of self-fertilization and incest motivated by the desire for immortality. As such, cloning might carry with it a risk of actual homosexual or heterosexual incest.

Ernest Becker eloquently elaborated on the psychological link between sex and death. He explained that "sex and death are twins." After all, "animals who procreate die": "Nature conquers death not by creating eternal organisms but by making it possible for really complex organisms to emerge in the place of simple—and almost literally eternal—self dividing ones." Thus, nature conquers death through sexual reproduction but only through "the defeat of individuality" that is intrinsic to it.[80]

Becker understood sexual taboos as an expression of man's need to defeat death by triumphing over sex, to try to perpetuate himself as a spiritual and eternal being.[81] The sexual pervert tries to triumph over sex, and hence death, by violating those same taboos. As an asexual perversion, asexual reproduction—and self-cloning, in particular—by triumphing over sex also attempts to vanquish its twin, death. The egoism inherent in the wish for immortality is *"antisexual,"* and asexual reproduction could become a vehicle for its expression.[82] Just as the sexual pervert may feel almost worshipful of his sexual fetish, and spiritually as well as sexually lifted by it, there may be those who view cloning the same way. The perverted "spirituality" of the fetishist may help us understand those pursuing reproductive cloning with a similar fervor, especially those who misunderstand it as a path to personal immortality.

Sexual perversion not only can help us anticipate possible individual motivations for cloning but also can help us to understand some possible meanings of reproductive cloning for society. In this regard, it seems likely that some of society's combined discomfort and fascination with the stark sexual facts of life has an oedipal basis, but some of it may also be deeply existential. More than sixty years ago, Otto Rank pointed out that "the biological self of natural procreation is denied from the beginning since it implies the acceptance of death."[83]

EFFECT OF ASEXUAL REPRODUCTION ON SOCIAL ROLES PLAYED BY SEX, SEXUAL REPRODUCTION, AND THE FAMILY

Sociobiology recognizes that "all cultures are heavily influenced by, bound to, limited by and incorporate our animal natures." Sex is part of our animal nature, and because we are the products of sexual reproduction, a basic biological characteristic, "any cultural characteristic takes as a given . . . our biological characteristics."[84]

Despite current problems with reproductive cloning, clonal origin by itself eventually might not constitute significant risk of pathology or change any of the basic biology of an individual. However, from social and psychological perspectives, *cloning would definitely change a basic biological characteristic of humanity by adding an asexual alternative to sexual reproduction.*

As such, cloning starkly confronts our society with some of its internal contradictions regarding sex, the family, and individuality. The "familial society . . . sanctions procreation, not sexuality."[85] Cloning presents our society with a paradox. Reproductive cloning would be consistent with antisexual cultural biases in our society, yet by removing the sexual glue heretofore necessary for human reproduction, it may fundamentally undermine the family as the most fundamental basis of society.

Even when a child resembles one parent more than the other, every sexually reproduced child, because he possesses genetic material from both parents, necessarily bears some resemblance to both. We have repeatedly seen the importance of self-resemblance to the narcissist, but we must understand that his psychopathology is layered on top of an even deeper preference for self-resemblance arising from inborn mechanisms favoring kin selection in evolution. When a child is not linked genetically to a parent, we have seen that he or she appears to be at risk for not being as well treated as a child who is genetically linked. By its merger of half the nuclear genetic material from each parent, it may be that sexual reproduction provides a biological substrate to produce offspring who *resemble each parent enough to appeal to kin selection mechanisms in both parents.*

Over thirty years ago, in examining the possibility of cloning humans, Joshua Lederberg wondered why asexual reproduction does not naturally occur in vertebrates—animals with backbones.[86] For many such animals, parental care and rearing of offspring are critical to their survival, and when it comes to the mode of reproduction, this feature of parental behavior is surely more important than any anatomical characteristic. Reframing Lederberg's question, we may ask why asexual reproduction does not naturally occur among organisms whose offspring require parental care and rearing to survive. To this question, we may answer that perhaps it is because *asexual reproduction might not be conducive to altruistic rearing of a cloned offspring by anyone but his or her progenitor.*

It might be argued that it would be advantageous for a child to resemble his father more than his mother, providing clear evidence of his likely paternity. That might compensate for the father's relative disadvantage in bonding to the fetus because the mother carries the pregnancy and undergoes hormonal changes that he does not. In the Parent–Child Resemblance Model, we saw that unwed mothers and their families spontaneously comment that babies resemble their fathers more than their mothers. Research that morphed photographs of the faces of adult research subjects with those of children's faces showed that it was more important to a man that a child look like him than it was to a woman that a child look like her.[87]

Although it might be advantageous for a child to get more than half of his genes from his father to be treated altruistically by him, this cannot be accomplished by sexual reproduction. Cloning, however, makes it possible for a child to get all his nuclear DNA from one parent, guaranteeing genetic identity to that parent and, if only genes mattered, very strong phenotypic resemblance. Although this might enhance the parent's investment in and altruism toward the child, throughout this book we have noted the *psychological and social dangers of too close a resemblance of one person to another, particularly when one is rearing the other.* In this regard, asexual reproduction might produce too much parent–child resemblance, whereas sexual reproduction appears usually to produce enough.

But how should we think about "enough" in this context? Perhaps the psychological concept of "good enough mothering" can give us some direction.[88] It describes the fact that no mother is perfect and that in her lack of perfection, she should not be judged to be an abject failure. Most mothers are able to provide mothering that is simply "good enough." In fact, much better than "good enough" turns out to be not so good, as it can deprive the child of the opportunity of facing necessary challenges he must learn to meet in dealing with the unavoidable frustrations of life. Similarly, we might also think of sexual reproduction as "good enough" in its psychological and social ramifications.

The merger of one-half of one person's genes with half of another's sets the biological stage for the child to resemble each parent sufficiently to reinforce that parent's altruistic inclinations toward his or her child. Evolutionary psychology, specifically informed by Dawkins's selfish gene theory, would lead us to conclude that evolutionarily derived parental investment and kin-selection mechanisms underpin those inclinations. Because parent–offspring resemblance resulting from sexual reproduction is "good enough" but not perfect, the altruism of both parents is adequately engaged.

The fifty–fifty genetic merger embodied in the child is also "good enough"—if not better—at providing a biological basis for the joint continued participation of both parents in the rearing of their shared creation. The joint effort of child rearing may strain a couple's relationship, but it also affords its individual members with important shared experiences that can deepen their connection with one another. It also provides opportunities for parents to mature further as individuals and as social beings in their community and larger society.

The fact that the child is not genetically identical to either parent may be as important as her resembling both partially. The absence of genetic identity is a reality that competes with the each parent's evolutionarily natural hope that the child will closely resemble them. When excessive parental narcissism does not complicate the picture, the genetic reality resulting from sexual reproduction may help each parent let go of the child as it increasingly makes sense to do so for her psychological and social development. Earlier we discussed how parents must carefully let go yet still be there for the child as she goes through the process of "separation-individuation" for the process to go well.

While separation-individuation describes a stage of early childhood development, in a more generic sense, it operates as a process in many subsequent stages of development, including into adulthood. For example, all parents remember having to let go the first time they put their child on a school bus. But good parents know that as hard as it may be, they must let go as their child develops greater competencies. They realize that it is only natural that their son or daughter becomes a member of the community and the larger society as an independent self, with his or her own identity.

Perhaps the "selfish gene" is at work in this letting go. After all, letting go is essential to the child becoming capable of intimate personal and socially cooperative relationships, and these enhance the likelihood of grandchildren to carry some parental genes one generation further into the future.

Helping parents to let go are societal rites of initiation for teenagers. Religious ones include Confirmation and bar and bat mitzvahs, and "sweet sixteen" parties and commencement exercises might be thought of as secular ones. Such rites not only help integrate the child into society but also serve to separate him from his nuclear family. By participating in them with him, parents recognize "that the child is no longer a 'private' person, but a social one."[89] Thus, the fact that children are sexually reproduced seems almost designed biologically to establish conditions that are conducive to the formation of people with a chance to become integral members of society.[90]

As discussed earlier, the child self-clone, in contrast, might be tied too tightly to her rearing family—especially so to her progenitor parent. This might impair her entry into the community and society as an independent and a socially competent person. Perhaps less evident is the fact that becoming parents through cloning might also deprive parents of a clone, both emotionally and socially. We might expect that this would be an indirect result of the absolutely asymmetric genetic link between the parents and child. As discussed in the Stepchild Model, parenting experiences that optimally are shared are less likely to be when a child is genetically linked to only one parent. This might stunt the growth that parents of a clone might have experienced individually and together in rearing a child who is genetically of both of them—but only in part—and who resembles each of them—but only "enough." Looked at this way, it seems fair to surmise that rearing an adopted child, genetically linked to neither parent, might demand more maturity from both parents and also foster even greater emotional and social growth for the parents as individuals and as a couple. I am unaware of any research on this issue.

Even if there were only a small number of cloned children, the option of parenting a clone might unleash social forces that could undermine the very same issues for all parents relating to their sexually reproduced children, each other, their communities, and society as a whole. This is one reason to think that asexual reproduction might well become a culture-transforming practice.

The anthropologist Claude Lévi-Strauss saw the basis of culture in practices of exchange between one person or group and another. He understood exogamy, or marriage outside of the group, as an "exchange" of women among family groups and the taboo against incest as both universal and fundamental to it.[91] Even though marriage is not needed for sexual reproduction, it still can create what is arguably the best social context within which to rear children.

Although Lévi-Strauss's ideas may be valid, our analysis suggests that the evolutionary task performed by the merger of half the genetic material of one individual with that of another may be the most fundamental basis for the social nexus on which society is built. Only sexual reproduction, but not cloning, naturally accomplishes this. Civil society helps to assure the well-being and survival of our species, especially by supporting individuals' and families' efforts in rearing children. *The partial merger of the genomes of two individuals to constitute a new and unique one may well be the basic biological fact underpinning the rest.*[92]

Cloning puts into conflict the "selfish" genes of the individual with the need for the survival of community and society absolutely necessary for the survival of humans as a social species. Reproductive cloning might turn out to be the Achilles' heel of shortsighted selfish genes, which up to now have managed to enhance their own survival despite their inability to anticipate wider consequences. Never before has there been the possibility for a human being to, in a sense, reproduce him- or herself like a hermaphroditic organism. The attraction of our shortsighted selfish genes to reproductive cloning might be quite strong, with the clone the hermaphrodite's spawn.

The asexual nature of cloning might also adversely affect society and culture by indirectly promoting perverse immature aspects of sexuality for individuals in society. Psychoanalysis discovered the composite nature of the sexual instinct, the universal disposition to bisexuality, and the polymorphous perversity of the normal immature sexuality of children. These discoveries led Freud to believe that achieving normal sexuality is difficult and to suggest that more people might be sexually perverse without the repressive influence of culture. While he advocated some sexual reform, he was firmly convinced that some degree of sexual repression is socially essential.[93] Cloning might increase the incidence of sexual perversion by decreasing the influence of certain sexually repressive forces in society.

Sexual modernism removes sexuality from reproduction and from the institutional context of marriage by furthering the "liberation" of reproduction from sex and, indirectly, sex from reproduction.[94] Cloning would be an expression of sexual modernism, but as "asexual postmodernism" it would also be generally socially detrimental at a most basic level by interfering with the necessity for the family, the most fundamental civilizing and regulative unit of society.

CONCLUSION

In this chapter we have examined the complex and interrelated topics of intimacy, sex, and sexuality. We have defined *intimacy* as a relationship that emphasizes the mutual appreciation of the separate personhood of the other in a close adult relationship, and we have reviewed its essential developmental building blocks in the parent–child relationship, clearly distinguishing that relationship from an intimate one.

We then moved from intimacy to examine sex and sexuality, beginning by examining three specific topics. In our analyses of pornography and the feminist perspective on cloning, we have found a common thread in heterosexual male perceptions and fantasies of women as creatures without individual importance.

We have also outlined some practical implications of cloning for gays and lesbians. The most important is that it would make it possible for them to have genetic offspring of their own without the involvement of individuals of the opposite sex—except that gay men would need to involve an egg donor and gestational mother. We also reexamined discredited psychoanalytic ideas about homosexuality and found them quite relevant to those who would clone themselves.

In the bulk of this chapter, we have examined the premise that, *psychologically and socially, cloning is only apparently asexual*. That required us to perceive the many shared features of cloning and erotic desire. In the process, we have discerned an important common denominator—the lack of appreciation of another's personhood.

After defining sexual perversion and distinguishing it from neurotic sexual disorder, we then focused on the characteristics and psychological functions of sexual perversions that might also be found in cloning. *We might think of cloning as an asexual perversion as our final model of cloning*, different in kind from the other models. Clearly, it is more speculative than the others—based less on research and more exclusively on psychoanalytic theory and clinical experience, but these are buttressed with ideas derived from evolutionary psychology and sociobiology.

We have seen how cloning might be an asexual perversion in how it parallels the ways sexual perversion can be used to try to cope with unresolved narcissistic and oedipal concerns. We also saw how cloning might be used similarly to avoid mutuality in relationships and as a means to try to master early psychological trauma. Then we focused on a topic that we have visited and revisited throughout this book—the denial of death. This time, we used sexual perversion as a new lens and, through it, saw how cloning might be used as a tool in the service of an *asexual perversion*.

In a prior chapter we found useful Adam Phillips's idea that people engage in "psychic cloning" all the time, making people, "including ourselves, in our own image of them." In this sense, cloning would be "a denial of difference and dependence and therefore a refusal of need."[95] It is an idea not inconsistent with conceiving of cloning as an asexual perversion. In fact, Phillips says, "Cloning seems to be a final solution to the problem of otherness. . . . In one fell swoop, cloning is a cure for sexuality and difference."[96]

Here, we should note that various forms of psychopathology manifest differently in different cultures and that there are historical trends in the incidence of its forms.[97] If it is true that "neurotic symptoms are attuned to the temporal sociocultural climate," then why shouldn't this also be the case for perversion?[98] If we accept this about perversion, and also the validity of conceptualizing certain societal phenomena as manifestations of psychopathology on a cultural level, *then we could consider reproductive cloning as a sign of asexual perversion at a cultural level.*[99]

The ancient Chinese practice of binding the feet of girls and women to make them small provides us with an example of a cultural perversion. It blurs the line between the creativity of the fetishist and cultural creativity.[100] If the tiny feet resulting from foot binding succeeded as a sexual fetish for men in Chinese society, then the practice was a cultural failure, producing women who could barely walk. *Cloning as a process, while scientifically creative, is not at all creative in its product, a genetic copy.* Cloning is also apt to be a failure as a cultural creation, perhaps at least as much of one as the binding of feet was for the Chinese. Thus, we may understand foot binding and reproductive cloning, respectively, as sexual and asexual perversions at a cultural level.

We need not diagnose any given individual wishing to clone as an asexual pervert to apply this concept of asexual perversion on a social and cultural level.[101] That is because cloning could serve asexually perverted social and cultural functions without doing so primarily for an individual. In addition, to the extent that there are times when everyone may focus on a sexual partner as less than a full human being, attending only to specific aspects of his or her anatomy, "perversions and so-called normality *intersect*."[102] By analogy, asexual perversion and normality could as well.

Several of the formal models of cloning presented earlier in the book give us reason to believe that more men than women would want to clone themselves. Given the male preponderance in sexual perversion, conceptualizing reproductive cloning as an asexual perversion leads us to the same conclusion. Recent research is not inconsistent with these predictions. Although it did not specifically ask about self-cloning, a telephone survey revealed that more than twice as many men than women approve of reproductive cloning of people, 26 percent compared with 11 percent.[103]

In this chapter, we see how cloning might critically affect some crucial roles played by sex, sexual reproduction, and the family. We have concluded that the partial merger of the genomes of two people to form a new individual through sexual reproduction might be the most fundamental biological basis for civil society. The "selfish gene" in evolution helps each parent of a sexually reproduced child to perceive adequate self-resemblance in that child because of the substantial but partial endowment of genes from both parents. The general result is that the parents behave altruistically toward their child, enhancing her chances of survival. At the same time, the lack of complete gene sharing between a sexually reproduced child and each of her parents, and the incomplete phenotypic self-resemblance that goes with it, may help each parent to appropriately "let go," as they must for the child to become her own person.

It is clear that as a child matures, she becomes progressively ready to be released from parental bonds to develop as an individual able to relate to others, eventually intimately, and also to be able to contribute to society as a unique individual.

Though reproductive cloning might seem to further the sexual liberation and personal freedom of those who would clone, to the extent that it might be an asexual perversion, it is fundamentally antisexual and not truly free. Emerging, as it is likely to, at the beginning of the twenty-first century, *reproductive cloning would not be a liberating climax to the twentieth or "the sexual century" but, rather, a socially antisexual anticlimax.*[104]

12

Implications for
Cloning Ethics and Policy

If we have decided that a democratic, free society is what we want, it seems to follow that people's wishes should be obstructed only with good reason . . . the onus is on those who would ban it [cloning] to spell out the harm it would do, and to whom.

—Richard Dawkins (1998), p. 66.

W e began this book with two key premises. The first was that analyzing models relevant by analogy to psychological and social aspects of asexual reproduction should help us to better anticipate and understand possible motivations for and consequences of cloning people. In this final chapter we will proceed from our second premise—that what we have concluded from our analysis should help us ground our thinking regarding cloning ethics and policy.

THE PSYCHOLOGY OF MORAL REASONING

Though science may ultimately discover a "biology of ethics," such an event would not obviate psychology or lessen moral responsibility.[1] Psychology can help us understand moral reasoning. That is important to us here because *how* we reason about the ethics of cloning may not be unimportant to *what* we specifically conclude about it. Sigmund Freud, Jean Piaget, and Lawrence Kohlberg each developed a way of thinking about the developmental psychology of moral reasoning, so let us review their ideas briefly.

SIGMUND FREUD'S CONTRIBUTION

For Freud, the "superego" is one part of his three-part structure of the mind and is largely unconscious.[2] The superego establishes and maintains a set of ideals and values, prohibitions and commands, that we commonly call "conscience." In addition to being a set of rules, the superego is also an agent that attempts to enforce compliance with these rules. It is the source of internal criticism and painful feelings of guilt and shame associated with noncompliance. Evolving out of a child's relationship to his parents, specific parental injunctions may become part of the superego, along with what the child thinks his parents would want him to do or not do. Fear of criticism and loss of his parents' love in not conforming to superego demands evolve into a wish to avoid self-criticism and loss of self-love.[3]

Just as Freud saw successful resolution of the oedipal conflict as critical to so much of psychological development, he also saw it as central to the development of mature moral sensibility.[4] We have seen, particularly in the last chapter, how aspects of an unresolved oedipal conflict might play out in the wish to clone. If resolution of the oedipal conflict is important to the development of conscience, then the moral reasoning of individuals with unresolved oedipal issues might be influenced by this fact. Whatever one might think of the role of the oedipal conflict in moral reasoning, the central point of psychoanalytic theory as it relates to morality is that much of a person's moral sensibility is unconsciously based and, as such, may not always be rational.

JEAN PIAGET'S CONTRIBUTION

Jean Piaget explained the development of moral reasoning in children as part and parcel of their cognitive development. Strict adherence to the rules and obedience to adult authority characterize a young child's earliest moral reasoning. Having not yet learned to be able to take the perspective of another, the young child assumes that others think and feel just as she does. This kind of cognition forces her to rely on the "letter of the law." Intent does not count for her; consequences do.

Out of this stage of moral development, which Piaget labels "heteronomous," the child's moral reasoning becomes "autonomous." This involves being able to critically consider rules and their rationale. The child comes to be able to selectively and appropriately apply these rules while being able to consider the perspectives of others as separate people with their own needs and wishes. From this, she develops mutual respect and fair reciprocity in relating to others.[5]

LAWRENCE KOHLBERG'S CONTRIBUTION

Lawrence Kohlberg took Piaget's theory of moral development and modified and elaborated on it. He described how an individual's moral orientation evolves out of unquestioning obedience to authority versus defiant entitlement, to simply conforming to convention and one's assigned role to avoid punishment or win approval, to grasping and adhering to certain rules with the feeling that rules must always prevail over individual wishes. Finally, the person may decide, after considering the rules, that principled individual belief must sometimes be given greater weight in deciding what is right.[6]

THE ETHICS OF ABSOLUTES

We have just reviewed immature developmental precursors of rule-driven ethics. Absolutist ethics is an ethical stance in which an action in question is judged to be either right or wrong based on whether it conforms with absolute moral rules. The rules exist independent of time, place and circumstance, and the consequences of following or deviating from them.

The sense of the absolute continues throughout all the levels and stages of a person's moral development. As adults, we may add qualifiers and conditions to our own personal ethical beliefs, but we still end up with rules to which we try to adhere absolutely. We do this even though it sometimes means adding a new qualifier to a rule so that we can deviate from it. In this very general sense, we all hold absolutist ethical views, even if we multiply and sometimes hypocritically amend them. An absolutist view is itself neither ethically primitive nor advanced, though how it is held and argued may be.

Individuals may have different opinions about what they believe are immutable moral rules. The only supporting argument some people consider necessary for any particular moral position is to claim that the Ultimate Authority supports it. But to base a moral position on what one understands to be a "God-given" rule only takes us so far. However, critical theological thinking can flesh out the context and implications of a particular absolute moral position, taking to a mature level what might have appeared to be morally primitive obedience to authority.[7] Being able to see the issue from the perspective of those with whom one does not agree is important to any mature ethical perspective, whether or not absolute in nature.

Abortion, Reproductive Choice, and Cloning as Absolutes

Reading this book is unlikely to change the fundamental moral judgment of anyone with absolutist views regarding cloning. Many people with such views base them on religious beliefs and oppose cloning of any type. The central issue for them is that whether cloning attempts to create embryos for reproduction or stem cells for medical therapies, it also destroys embryos in the process.[8]

In the debate and political struggle over abortion, those against it in any circumstance define themselves as "pro life." As a rule, with the occasional exception, they also oppose cloning of any type for the same reason and so may not consider other bases for moral judgment specifically regarding reproductive cloning.[9] They may need to soon, as *embryonic stem cells in both female and male adults can give rise to eggs,* as well as structures resembling blastocysts, probably early parthenotes.[10]

As a group, those who believe that abortion should be available define themselves as "pro choice" but are not of one mind about it. Almost all would endorse abortion if the pregnancy endangers the mother's health, and most would if there is a serious and uncorrectable fetal abnormality. Fewer might find abortion acceptable for a pregnancy that is simply unwanted—yet all still believe that every woman should be able to make her own personal decision in such a dilemma. Regardless of where they fall on this spectrum, "pro-choice" individuals bristle at the implication that they are at the "wrong" end of the dichotomy defined by "pro-life" forces. In the semantic wrestling match, they respond by emphasizing that they are on the "right" end of a dichotomy they define along the dimension of "choice."

By extension, some people who are pro choice on abortion might be reactively inclined to favor cloning as a "reproductive right," fearing that regulating or banning any kind of reproduction could set a precedent that might infringe on others. They might also sense that their pro-life opponents could attempt to construe their condemnation of reproductive cloning to imply that they are conceding that embryos are "unborn" people with rights that should outweigh the right to choose.

People opposed to abortion and people favoring reproductive choice both frequently fail to note other ethically important dimensions of cloning specific to reproductive cloning. This book describes a wide range of possible adverse psychological and social consequences of reproductive cloning—consequences unrelated to the destruction of embryos or to reproductive choice.

Many of the adverse consequences we might predict for reproductive cloning derive from abuses of parental power resulting from an enhanced sense of ownership and an accompanying sense of entitlement to control the clone. When and to what degree should a person be entitled to control what she identifies as hers? That dimension of cloning is more fundamental than those captured by the descriptors *pro choice* and *pro life* and is relevant to both reproductive and therapeutic cloning. Those advocating for the "rights of the unborn" recognize this issue as it relates to abortion, whereas those advocating for "reproductive choice" recognize the issue from another perspective.

When pregnant, a woman naturally refers to the embryo and later the fetus as belonging to her. Genetically, it is only 50 percent "hers," but it does reside within her body. A clone would be 100 percent genetically identical to its progenitor, and various models converge in suggesting that progenitors might feel an increased sense of ownership of clonal offspring well after birth. But identical genes should not entitle the first person to possess those genes or to, in any sense, *own* any subsequent copy in the person of a clone. Precedence is a powerful political factor and can be an overwhelming one in the parent–child relationship.

We might all agree that, ethically, the abuse of the power of precedence must be resisted, but people often differ on what constitutes abuse. Opponents of abortion see a woman who would terminate her pregnancy as abusing her power over the embryo or fetus. Advocates of reproductive choice see abuse of power in laws that interfere with a woman's right to choose what she does with her body and her life, even to the point of endangering it.

THE ETHICS OF ANTICIPATED CONSEQUENCES

Just as there is a developmental psychology underlying a mature ethics of the absolute, there is also one underpinning the ethics of anticipated consequences. A mature, rational adult and responsible citizen in a democracy wants to inform himself about matters that might help him assess possible consequences of any particular practice or policy.

Utilitarianism is a consequentialist ethical stance that aims to maximize positive consequences and minimize negative ones for as many people as possible. It sounds good in theory, but in practice, utilitarian ethics can be used to justify actions that most people would consider immoral. For example, if an action causes a few people to suffer greatly, then utilitarian ethics might not object if many others benefited from the same action.

One need not embrace utilitarianism to believe that trying to anticipate consequences is important to ethical decision making. We do this all the time without thinking about it. For the young child in Piaget's heteronomous phase of moral development, parental rules matter, not the rationales parents may offer. However, the prospect of parental disapproval and punishment as a consequence for the child is what makes their rules matter to him. The older child, in Piaget's autonomous phase of moral development, learns to be concerned about the possibility of harmful consequences, not only for himself but also for others.

As adults, we become capable of considering possible consequences more generally and broadly. We move beyond thinking about any given act in isolation and come to consider more than short-term consequences for both others and ourselves. As responsible adult members of a civil society, we also think of any act as one instance of a *practice*. Ideally, we try to consider fully how a given practice might affect not only the individuals directly involved but also all those who might be impacted indirectly by it, all the way up to the larger society. Science and logic alone cannot tell us what is right and wrong ethically. But to the extent to which science and logic can help us assess likely consequences, they can provide a basis for ethical judgment.

THE ETHICS OF "NATURAL LAW"

What Is Natural Law?

The term *natural law* describes an ethical concept defined as "a set of principles, based on what are assumed to be the permanent characteristics of human nature that can serve as a standard for evaluating conduct and civil laws. It is considered fundamentally unchanging and universally applicable. Because of the ambiguity of the word 'nature,' the meaning of 'natural' varies. Thus, natural law may be considered an ideal to which humanity aspires or a general fact, the way human beings usually act."[11] This ambiguity can create confusion, but it also affords an opportunity. Let us try to integrate the absolutist perspective of human nature *as it ought to be* with scientific theory and knowledge of human nature *as it is* and as we have applied it directly and by analogy to anticipate some of the possible consequences of reproductive cloning.[12]

Because the idea of "natural law" derives from one's conception of human nature, prescientific or more primitively scientific notions of human nature were shaped more by ideals about how people should be, rather than by how we actually are. In modern times, science informs, and sometimes misinforms, our conception of human nature as it is. At the same time, certain ideas about human nature "as it ought to be" have evolved and differ between people based on politics, culture, and religion. Sometimes the "natural rights" that devolve from natural law conflict, and society must sort out what and whose right should take precedence.

Too often, natural law is understood mainly in an abstract philosophical sense, and sometimes people conflate scientific understandings of human nature with nature itself. The claim that something is a "violation of natural law" has been used throughout history as an argument to condemn many scientific and cultural advances. The sociologist Michel Foucault has shown how that indictment has also been used to marginalize people who are different, especially sexually.[13]

A Psychology of Natural Law as Applied to Cloning

Our understanding of natural law as it relates to human reproductive cloning is informed by scientific and clinical attempts to understand human psychological and social nature. That understanding helps to ground our reasoning about possible psychological and social consequences of cloning.

There is a difference between something being natural or unnatural and it being consonant with natural law or a violation of it. For example, although asexual reproduction is not natural for humans, that fact by itself does not make it a violation of natural law. Many things are not natural for us, including flying in airplanes and taking antibiotics to treat bacterial infections, yet few would say that those are immoral acts. In fact, it might be argued that although it is not natural for us to fly, the invention of the airplane was a natural consequence of exercising our mental faculties. Likewise, although asexual reproduction is not natural for humans, it, too, would be a natural consequence of applied human intelligence.

The most basic thing psychological science teaches us about human nature applies to the animal kingdom as a whole. The simplest of organisms attempt to avoid pain and normally are drawn to pleasurable activities. Sigmund Freud recognized this most fundamental feature of behavior and labeled it the "pleasure principle." Of course, things get more complicated for people, who in the process of maturation quickly move "beyond the pleasure principle."[14] We might say that the biological, cognitive, and emotional basis for all ethics consists in avoiding pain and seeking pleasure. Without doubt, pleasure and pain serve evolution, and so there is also an evolutionary basis to these building blocks of the most primitive of "ethics."

As the infant develops, she develops a growing awareness that her wishes are not going to be immediately gratified, and she must learn to tolerate delay. With the socializing influence of parents, siblings, and others, she also develops a sense that she must take into account the wishes, needs, and feelings of others. The fundamental underpinnings of a civil society derive from those early socializing influences. More sophisticated conceptions of evolutionary biology and psychology demonstrate adaptive advantages for the child who moves beyond her developmentally primitive "ethics."

In this book, we have used Erik Erikson's ideas about the human life cycle enriched by others' insights about human psychological development as a framework to think about the big themes in human existence as a clone might face them. For people everywhere, attachment, trust, separation and individuation, autonomy, independence, intimacy, and feeling productive and generative are core concerns. If we can agree that these themes are essential to human psychological nature, then anything that furthers an individual's chances to achieve them would be congenial with natural law, and anything that undermines them would not be.[15] We might well consider as natural human rights the positive pole of each developmental and maturational issue embodied in every one of Erikson's eight stages.

Socializing a child, without unduly thwarting his striving for pleasure and wish to avoid pain, is consonant with the most basic of psychologically based natural law. Rearing a child means helping him to conform to certain standards of behavior necessary to survive and thrive in the social reality of his family and beyond. A family is more likely to raise an emotionally healthy child if it socializes him without unduly thwarting his quest for pleasure and avoidance of pain.

We find naturally complex cooperative hierarchical societies in nature among animals termed "eusocial."[16] But unlike the mundane ant or the exotic naked mole rat, humans have no similar universal "natural" society or culture. However, we might judge how "natural" a human society is by how well it facilitates or to what extent it interferes with an individual's ability to develop within it. In negotiating the eight Eriksonian stages of the human life cycle, individuals ought to be able to experience pleasure and no more pain than necessary in the process of trying to master the requisite developmental and maturational tasks. A society is great to the extent that its policies and their implementation respect humans' natural rights and facilitate people's pursuit of them.[17]

As we have seen in the previous two chapters, one reason to believe that reproductive cloning might cause wider social harm is that it would undermine the natural basic biological underpinning of society—the merger of genetic material from two individuals to create a new one. Thus, cloning would violate human nature at a deep social level, including society's biological basis in sexual reproduction. In general, when philosophers debating the ethics of cloning concern themselves with possible psychological and social consequences, they confine themselves mainly to the issue of the clone's individual identity.[18] This book demonstrates that there are good reasons to think that cloning might adversely impact a clone at every Eriksonian developmental and maturational stage. Our models clearly suggest that being of clonal origin is likely to be a congenital psychological and social handicap for a clone, interfering with his ability to live a happy and fulfilled life. For this reason, being of clonal origin is a violation of a clone's natural human rights.[19]

The Natural Rights of the "Unconceived"

Increasingly, people agree that every born human being has natural rights. However, pro-life and pro-choice forces disagree about whether the status of "person" should apply prior to birth. The pro-life position is that not only human life but personhood begins at conception. From this perspective, even a blastocyst or early embryo is an "unborn" person with natural human rights. In contrast, the pro-choice position typically assigns personhood only after birth.

Regardless of when one considers a developing human to be a person with natural human rights, consider this proposition: Every potential child has natural human rights. As I am using the term, a *potential child* is not developing or "unborn" but, in fact, *a child prior to conception*. With the possible exception of a child or two who might possibly be cloned by the time this book is published, as a class, cloned children only potentially exist. Even so, thinking about what is best for these *unconceived* children can help us think through what is best for any and all children.

Every unconceived child should have a right to be reared by adults who, in addition to providing for her basic physical needs, will be unselfishly loving and try their best to help her to negotiate the critical early stages of her personal development. That means that they will not intentionally or unwittingly try to thwart the development of her own unique self so that she has a chance of living a happy and fulfilling life that is *hers*.[20]

If the idea that people who only potentially exist have natural rights seems strange, then consider this: People without grandchildren sometimes ask themselves and each other whether a particular practice or policy might adversely affect the planet we will leave to our grandchildren. The idea that the *unconceived* have natural rights can help us move beyond the vague abstraction of a future generation to the more concrete and imaginable idea of particular unconceived individuals constituting that future generation. Cloning would make an unconceived child vividly imaginable as a specific person, for we would know his exact genetic forebear.

If we have good reason to believe that particular individuals or couples wishing to become parents through cloning are likely to be poor parents, we would be likely to feel more of an ethical imperative to try to prevent them from fulfilling their wish. With sexual reproduction, we could have a more difficult time imagining a particular unconceived child who might be at high risk to become a victim of parental abuse or neglect. In contrast, we would be able to "see" clearly the face of an unconceived child clone. This might make it harder for us to ignore his or her likely fate.

We may not be able to or want to try to stop people likely to be poor or abusive parents from having sexually reproduced children, but as the bioethicist Arthur Kaplan argues, "A liberty right to reproduce" does not entitle a person "to technological aid in having children." Neither does it eliminate our obligation to think about "whether it makes sense to limit that right if the mode used for the creation of children is not in the child's best interest."[21] The issue comes into sharper focus than ever with cloning, hence the need to recognize *the rights of the unconceived.*[22]

Because a clone would have been *preconceived genetically* with the intention that he be like his progenitor, a child clone would be more likely than other children to have been *preconceived psychologically* by his parents. Whether unconceived or already born, every child should have the right to be free enough of parental preconceptions that he or she has a chance to develop his or her own individual identity. In this, we may understand the "rights of the unconceived" as subsuming the idea that every child has a "right to an open future."[23]

Of course, parental preconceptions are a central feature of many of the practices we used to model cloning. Does that give those practices the same status as cloning, ethically? To a degree, yes, but genetic preconception can take parental preconception to the ultimate. Occurring in the context of the extreme asymmetry in power and dependency between parent and child, the psychological risks inherent in reproductive cloning would be even greater. In addition, the preconception of cloning is deliberate and involves others in society. For these reasons, the decision to reproductively clone should not be left only to individuals. It is proper that society protect the rights of the unconceived before reproductive cloning genetically preconceives them.

The Special Issue of Consent

Inextricably tied to the rights of the unconceived is the issue of consent. We need to be concerned about it from a number of perspectives. Most of the time, the prospective nuclear DNA donor would be in a position to give or deny consent to be cloned. However, if the intended progenitor were a young child, she might not be able to understand sufficiently to consent legally. A clear conflict of interest would exist if her parents wanted to clone her. And although it would be impossible for a deceased person to consent to be a somatic cell donor for the purpose of cloning, he might choose to do so in his will.

In contrast to a potential progenitor, it is clearly impossible for a potential clone to give consent to actually be conceived. It is up to society to represent the rights of this unconceived person. His inability to consent is no different from that of any person ever born. What would be different when it comes to consent is that the clone not only would lack the ability to consent to be born but also would lack the ability to consent to be born a clone, with all that this would mean. Another important difference is the heightened contrast between the absolute inability of the intended clone to consent and the enhanced ability of his or her rearing parents to consent to have a child—exactly this child. Reproductive cloning would highlight this asymmetry and so contribute to the already magnified disparity in power and dominance inherent in the parent–child relationship.[24]

We should note that whatever view we may hold on abortion, we can embrace the notion of rights of the unconceived without contradiction. Agreeing that clones should not be brought into existence to protect their rights as unconceived human beings advances neither "abortion rights" nor the "right of the unborn." Cristina Page and Amanda Peterman, women on opposite sides of the abortion debate, recognize that the two sides often miss opportunities to agree about important issues, such as preventing unwanted pregnancies and encouraging adoption.[25] Another issue about which both sides could agree without compromising their other positions is to support the rights of the unconceived by opposing human reproductive cloning.

The best interests of the child might be best protected if society embraces the notion of the rights of the unconceived. Respecting those rights would also help to protect the best interests of a civil democratic society.

EMPIRICALLY BASED NATURAL LAW—A CONVERGENCE OF ABSOLUTIST AND CONSEQUENTIALIST ETHICS

Nothing in the full range of human desires is unnatural. That includes wishes to do things that are universally considered immoral, destructive to others, and even undermining of civil society as a whole. Remember the sentiment of the ancient writer Terence: "Nothing that is human is alien to me." Stated another way, human nature encompasses the full range of emotions and includes wishes most of us would consider repugnant. It also includes the performance of actions ranging from the most loving and altruistic to the most cruel and selfish. Unchecked by laws acknowledging natural human rights, human nature has the potential to make for a society where life is "nasty, brutal, and short."[26]

Human nature alone cannot tell us whether a given practice is right or wrong, but a scientific knowledge of human nature can help us think systematically about possible consequences of a given practice. We can then decide whether we want to risk those consequences and, on this basis, ethically judge the practice.

Perhaps the Golden Rule is the best and simplest consequentialist guide to making an ethical judgment. Expounded ages ago by cultures and religions around the world, it asks us to consider the other person's perspective as our own and to treat him as we wish to be treated ourselves.[27] It is an excellent procedure to restrain natural, though uncivil, human propensities. From the Golden Rule we can derive most of the explicit rules of that ancient absolutist guide, the Ten Commandments.

PONDERING POLICY GUIDELINES

We might be able to derive our own "commandments" regarding reproductive cloning by *applying the empathic perspective of the Golden Rule to the "unconceived" other of the clone.* In doing so, we must bear in mind the considerable risk of psychological and social harm we have predicted for the clone. The challenge then becomes one of formulating secular "commandments" as guidelines for public policy. Let us use our knowledge of human nature as we have applied it to cloning to inform our consideration of cloning policy. Doing so will help to assure that our policy guidelines are consistent with our understanding of natural law.

When it comes to policy regarding reproductive human cloning, it seems to me, society has a range of choices:

1. We can encourage it without legal restriction.
2. We can permit it without legal restriction.
3. We can sanction it within certain legal guidelines.
4. We can disapprove but tolerate it legally.
5. We can ban it and punish those who violate the ban.

Let us examine each option in turn.

1. Encourage Reproductive Cloning without Restriction

The foregoing analysis and discussion make it obvious that this policy option is ethically unsupportable. It is clear that we ought not to encourage reproductive cloning.

2. Permit Reproductive Cloning without Restriction

This policy might seem to be the most libertarian—but only for those wishing to clone. We could only endorse permitting cloning without restriction if we give short shrift to the natural human rights of the unconceived. Even so, that which society disapproves and even condemns, it does not always restrict or outlaw. It is a matter of political orientation rather than ethical judgment, natural law, or the science of human nature that might lead us to permit a practice, unrestricted, that we disapprove of as ethically wrong.

3. Sanction Reproductive Cloning under Specific Circumstances or for Specific Purposes

If we are to sanction reproductive cloning within certain guidelines, then we must consider whether there are any circumstances in which reproductive cloning might possibly be justified. Several of the following circumstances and purposes have been discussed in the context of the various cloning models, but with the benefit of our prior analysis it is worth reconsidering them here.

To Give a Deceased or Dying Loved One a Chance to Live on Genetically in a New Life

This is one of the primary motivations some people have for reproductive cloning, and it is one that we have seen carries great risk for the clone. Although the wish is an understandable human one, it would not be justifiable ethically, as it would likely infringe the natural human rights of the unconceived. Let us be clear that *a new life identical genetically to an old one is still a separate life with a separate self.*

To Allow Infertile Heterosexual Couples to Become Genetic Parents

This is, perhaps, the area where there is the greatest interest in reproductive cloning. Multiple models point to problems for the clone in having one and only one genetic "parent" or progenitor. In the last chapter, we reason that simulating the genetic results of sexual reproduction with cloning might eliminate or greatly ameliorate the anticipated risk of adverse psychological and social consequences. If so, that particular form of cloning might become ethically acceptable.

Although that form of cloning would likely be preferred by infertile couples interested in cloning they might not want to wait until genetic engineering makes it possible. However, the host of psychological and social problems we have predicted to result from cloning not engineered to incorporate nuclear DNA from two people makes reproductive cloning as we now know it ethically unacceptable.

To Eliminate or Better Treat Heritable Diseases

Although therapeutic cloning does not reproduce an entire organism to develop in utero and live life outside the womb, one motivation for reproductive cloning might be therapeutic. Reproductive cloning could allow genetic engineering interventions to correct defective genes before they have a chance to exert detrimental effects. Correction at the earliest stage would also free germ or reproductive cells and hence subsequent generations from carrying the defective gene.

Certain genetic disorders may enhance certain universal human vulnerabilities, such as those to infection, bleeding, and aging. Beyond increasing these, everyone has inherited vulnerability to some disease or diseases. We would all like to be free from the threat of heart disease, cancer, diabetes, hypertension, and Alzheimer's disease.

Among the genetic disorders, some are so highly heritable and horrific that we might wish to employ reproductive cloning to enable the use of genetic engineering to correct the defective gene. That would free the clone and all subsequent generations from their ravaging impact.

We should distinguish reproductive cloning with a therapeutic intent from therapeutic cloning to produce stem cells.[28] Most of us are familiar with the nightmare scenario of reproductively cloning a person in order to use him as a possession for "spare parts." Parents who conceive children in the hope that the new child would be a good match immunologically to donate an organ needed by an existing child contribute to this image. With a child cloned from the original, there would be no doubt that the needed organ would perfectly match the recipient immunologically. Creating a person to be a source of spare parts is not what therapeutic cloning is about.

Therapeutic cloning provides hope for cures or better medical treatment for people with many diseases. These include many of the genetic disorders for which reproductive cloning with a therapeutic intent might be entertained, but they also include diseases that are not necessarily genetic. As explained early in the book, in therapeutic cloning, cells are extracted from an embryo to clone specific bodily tissues for medical use, particularly transplantation. Type I or juvenile diabetes is one example of a disease that might be cured by therapeutic cloning to produce stem cells.[29] Without contradiction, we can condemn reproductive cloning and at the same time, if we choose, support research with embryonic stem cells.

To Create Longer-Lived and More Capable Individuals than Ever Before, Possibly with Capabilities that No Human Has Ever Possessed

Clones might be genetically engineered for the purpose of genetic enhancement.[30] The entire genome of the natural world could be tapped. There is nothing fundamentally different about the genes of our species and those of another, except the number of chromosomes. The nucleic acids of DNA are the same in all species: guanine, cytosine, adenine, and thymidine.

We had better think long and hard about genetic enhancement and transgenic splicing of DNA as a form of eugenics. This goes beyond cloning with the aim to create a child with a particular genetic makeup. That has the potential of decreasing the degree to which other individuals are valued and respected in society. Creation of multiple clones from a single source would be even more likely to foster this attitude.[31] But genetic enhancement, including transgenic splicing, might in addition change human nature, with adverse consequences for the family, society, and culture. Aside from possible individual psychological harm to the clone, the practice of genetic enhancement through cloning might also be indirectly destructive to society.

Although medical and public health advances in the past several hundred years have greatly decreased infant mortality and reduced other causes of premature death, any use of cloning to extend longevity well beyond the natural range would constitute a change in human nature. We had better think through the possible psychological and social consequences of attempting to fundamentally alter life expectancy.[32]

Human stem cell researchers realize that they must exercise care in selecting which human stem cells should and should not be inserted into animal blastocysts for study. No one wants a mouse born with human brain or reproductive cells.[33] In such studies, and in the genetic engineering of any animal clone, care must be taken to not "humanize" the animal in appearance or behavior or have it produce human sperm or eggs. Altering an animal's nature so fundamentally might by comparison severely impact our view of our own.

To Permit Gays, Lesbians, and Single Heterosexuals to Become Genetic Parents

Though perhaps not by choice, it is a fact that many heterosexuals, mainly women, are single parents. But some individuals decide to become parents despite being single. If she does not opt to adopt, a childless woman, late in her reproductive years, may still wish to have a child who is linked genetically to her. Sexual reproduction, generally by donor insemination, can allow such a woman to bear a child who is genetically hers. Lesbian couples are also finding their way to parenthood in this manner. Cloning would also enable gay men to become genetic parents without the child having any nuclear DNA from a woman, though, of course, one would need to serve as an egg donor for somatic cell transplantation and one would have to carry the pregnancy.[34]

Our models abundantly illustrate the psychological and social risks of self-cloning. We must reject it as a route to parenthood, regardless of the sexual orientation of the parents and most especially for individuals single by choice or circumstance. However, as mentioned previously, we may want to consider accepting cloning if the nuclear DNA of a clone could be constituted from both members of a couple or from a single person and another individual of his or her choice in order to approximate the genetic results of sexual reproduction.[35]

The legal scholar William Eskridge Jr. makes the argument for gay and lesbian families, claiming that "a thriving society is one that accommodates the needs of its productive citizens," helping them to personally flourish. He believes that social policy should help gays and lesbians feel safe in forming lasting intimate relationships that minimize the risks of interpersonal exploitation while at the same time helping others in society to accept these persons previously excluded because of their sexual preferences.[36] If one accepts that it would be good to help gays and lesbians to exist securely as couples and be able to constitute families, rearing children, then why wouldn't it also be good for them to be able to have children linked genetically to both rearing parents?

It may be hard to separate that question from whether a minimal increase in the gay and lesbian population would be bad, good, or of no concern for society.[37] There may be those who will be alarmed at the prospect that such clones would be likely to carry the "homosexual gene" in a double dose. Unless specific adverse effects might be anticipated, we may want to support research that would make joint genetic parenthood possible for members of a gay or lesbian couple.

In the meantime, we need to contemplate possible evolutionary, psychological, and social ramifications of what Joshua Lederberg has called "tempered clonality." The term describes his assessment that mankind, and any particular society, might not suffer adverse consequences from cloning if the practice accounted for only a small percentage of births.[38] Though clonal reproduction engineered genetically to approximate sexual reproduction would still be cloning, it would lack the critical feature that is at the root of many of the psychological and social risks we anticipate for cloning—creation of an offspring who is an exact genetic copy of one progenitor. However, the practice of even this sort of cloning in society would still be a special instance of "tempered clonality."

4. Disapprove of Reproductive Cloning but Legally Tolerate It

Another policy option would be to disapprove of cloning while tolerating it. We should be able to apply the same reasoning to this option as to the option of permitting reproductive cloning without restriction. Although the law would not restrict cloning in either case, the key difference between them is one of societal attitude—permissive versus disapproving.

Societal disapproval might be a relatively effective deterrent to cloning—but only for some people.

5. Ban Reproductive Cloning and Punish Those Who Would Violate the Ban

If we ethically disapprove of reproductive cloning, then why not clearly and unambiguously ban it legally and punish those who would violate the ban? For one, perhaps doing so would force reproductive cloning "underground." This would make it harder to identify and offer help to individual clones. Another reason not to make cloning illegal is that doing so might stigmatize the offspring of law-breaking parents, turning them into pariahs. Recall the problems faced by children of the infamous, discussed in chapter 6.

Perhaps the most important reason not to legislate against human cloning is that doing so could seriously hamper promising medical research. That is because pending legislation in the United States, disproportionately driven by the politics of abortion, does not distinguish between reproductive cloning and therapeutic cloning. Whether to make such a distinction is currently also a bone of contention at the United Nations.

ALTERNATIVES TO REPRODUCTIVE CLONING AS SOURCES OF HOPE AND COMFORT

If our society decides not only to discourage reproductive cloning but also to make it illegal, then we might wish to leave open the option to pursue genetic engineering endeavors to try to have it approximate the results of sexual reproduction. Of course, such research should proceed only with animals, until it is perfected. As discussed earlier, we would not anticipate this sort of cloning to be nearly as problematic psychologically and socially as cloning that produces a genetic copy of a preexisting person.

Of course, many individuals interested in reproductive cloning might not be at all interested in cloning that would not result in an exact genetic copy. This would signal underlying unmet emotional needs—in their minds best met by the idea of a self-copy. Society must address these needs and provide alternative ways of meeting them. *A total ethics of reproductive cloning must move beyond the right and wrong of cloning itself to explore avenues of meaning and hope for those contemplating it.*

Such an ethical perspective requires us to acknowledge that reproductive cloning offers hope in overcoming limitation, loss, and a sense of personal lack to infertile couples and intrinsically infertile homosexual pairings, as well as to those who otherwise may feel unfulfilled. In a naive and misleading way, it also offers hope to everyone who lives and dies and whose loved ones will do the same. If we take reproductive cloning away as an option, then we must offer alternative sources of hope.

We should make these people aware that nonreproductive or therapeutic cloning for stem cell research offers hopes of curing previously incurable diseases or mitigating their effects. It is conceivable that it might even enable production of gametes (sperm and eggs) for the infertile—so that they might reproduce sexually. They should also consider that their pursuit of reproductive cloning might ultimately be self-defeating if it contributes to societal opposition to all cloning, including therapeutic cloning.[39]

We must be sympathetic to those who would wish to utilize reproductive cloning in a process to forever eliminate genes coding for heritable diseases in a family. However, if we were to approve of reproductive cloning to enable germ line modification, then it would be harder for us to object to the cloning that others might attempt to justify with rationales that feel equally compelling but which to us might seem inadequate or worse. *Whatever benefits reproductive cloning might promise, we need to weigh them against the substantial risk of psychological and social harm this analysis strongly suggests it would produce.* Fortunately, however, our increasing knowledge of the genome is making it possible to know who is carrying what genes, and that knowledge promises to make germ line modification unnecessary. Genetic counselors will do what they have always done, only much better—provide the information to help individuals and couples decide what risks they are willing to take in conceiving a sexually reproduced child. In addition, treatments resulting from therapeutic cloning for stem cell research should increasingly make the diseases for which defective genes code more controllable, if not fully curable.

As a society we can do more to encourage adoption and try to help those who might wish to clone to view adoption as a better way to become a parent. Granted, adoption carries its own risks. We have reviewed many in the Adoption Model. Based on the many conceptual parallels and oppositions between clones and adoptees, we have raised concerns about possible adverse psychological and social consequences for clones. However, we have good reason to think from the rest of our analysis that the problems an adoptee may encounter would overall be less significant than those a clone might be likely to face. Furthermore, clones cannot exist without active efforts to create them, whereas millions of children in need of adoption already exist worldwide. Adoption may be better than, or as good as, any assisted reproductive technology in satisfying most motives for having children—except that an adopted child is not genetically "part" of his or her parents.[40]

Just as clinicians offering currently available assisted reproductive technologies should engage couples in a discussion of their motives and assess whether they might be satisfied in ways other than having a child, society should be asking the same of those who would consider reproductive cloning.[41] Recall from Erikson's "generativity versus despair" stage that having children is but one way to be generative.

In addition, religion can offer a vision of a good life as one based on good and kind acts. Most religions teach that human life is always less than perfect. Limitation, loss and lack, and death are inevitable and must be accepted. Religion also frequently emphasizes the idea of an afterlife to help people to accept these regrettable facts of human existence.

Most religions teach that a person may be able to live on as a spirit or in good works and kind acts. Most religions teach that these paths to immortality are superior to those of the "flesh"—that is, personal or genetic immortality. Religions might do more to explain why.

In emphasizing spiritual immortality, however, religion must be careful lest it inadvertently help to further interest in the illusory physical immortality that cloning might seem to offer. Value systems, religious or secular, that de-emphasize spiritual immortality, and instead foster the idea of living a fulfilling and socially responsible life through good deeds and works, might do more to undermine reproductive cloning.

Individuals who would spend large sums of money on reproductive cloning, hoping to replicate either themselves or someone they love or admire, might consider the more lasting "immortality" that may be achieved by donating those funds to a worthy cause. In chapter 8 we discuss names as the simplest of "memes" and namesaking as the simplest sort of attempt at "memic" immortality. Although there is no name more immortal than "anonymous," if an individual wished, he could contribute charitably in his own or a loved one's name. While being a namesake might be harmful for a person, it is apparently not detrimental to a foundation, scholarship, or other endeavor to carry its founding donor's name. By charitably vitalizing an existing endeavor or bringing one newly to life, the donor is apt to memically "outlive" any cloned offspring who might carry his or a loved one's nuclear DNA one generation further into the future. In addition, he might be more likely to have a wider societal impact as a charitable donor than as a cloner.

A ROLE FOR PSYCHOLOGICAL ASSESSMENT AND TREATMENT

Despite my efforts to try to be sensitive to those who might wish to clone, the reader might still have incorrectly formed the impression that because we have concluded that cloning risks great psychological and social harm, anyone who would wish to clone is a bad person. That would be under-standable, given that our primary focus has been on likely psychologi-cally pathogenic behaviors of such parents and their resulting adverse psychological consequences for the cloned offspring. But let us remember to compassionately consider the psychological needs of those who would clone if they could.

In each of the models relevant by analogy to reproductive cloning, we have explored a wide range of possible human motivations for it. Most fall into or derive from four not mutually exclusive areas: (1) reaction to loss, (2) a sense of personal limitation, (3) denial of death, and (4) a sense of not being loved—either by oneself or by another.

Though we cannot avoid making ethical judgments about acts of re-productive cloning, condemning the people involved would be antitheti-cal to trying to help them to fully understand their motives and, as a re-sult, choose to abandon their efforts. Even when the desire to clone is an expression of psychopathology, let us not lose sight of the fact that all of the many possible motivations a person may have to clone would still be fully human.

The first clinicians to describe the "replacement child syndrome" hoped that psychological treatment for the bereaved parents might "prevent the senseless arithmetic of adding a pathetically warped new life to the one already tragically ended."[42] It truly does seem foolish to utilize the so-phisticated science and technology of cloning to create more replacement children. We must try to understand what motivates those adults wishing to become parents by cloning and try to help them understand, too. Through the Replacement Child Model and all the other models relevant by analogy to various psychological and social dimensions of cloning, we have come to understand the risks of harm that might well accompany re-productive cloning, as well as many possible motives to clone. Let us hope that this book helps some people contemplating cloning to recon-sider—perhaps through science and logic or mainly through empathy with the individuals in the cases that illustrate the models, people all clone-like in some way.

Throughout this book we have seen how various unconscious motiva-tions might be expressed in the wish to clone. For this reason, in addition to doing a good clinical interview of anyone wanting to clone who wishes to discuss the matter, the mental health professional may want to formally assess personality and also arrange for what is called "projective testing" to assess unconscious motives and conflicts specifically.[43] If a major psy-chiatric disorder is diagnosed, then appropriate medication may help to alleviate the person's suffering. By helping a person to feel better, suc-cessful treatment with medication by itself might also result indirectly in him to think differently about cloning.

Psychotherapy would surely be indispensable for a person whose suffering has led her to hope that cloning would alleviate her emotional pain. Within the context of a trusting relationship with an empathic, compassionate, and competent psychotherapist, the individual may be able to develop healing insight into the underlying sources of her distress, while medications alleviate the biologically based psychic and physical symptoms.

The psychoanalyst Adam Phillips believes that "psychoanalysis . . . was a cure for cloning before cloning existed, the cure as the precursor to the problem."[44] We should encourage individuals considering cloning to explore their wish to clone in a psychotherapy that acknowledges the importance of unconscious motivations, not necessarily psychoanalysis.

When a couple wants to clone, they might initially be assessed together. However, a person is more likely to discover and acknowledge deeply unconscious motives in individual psychotherapy. Even with experience in therapy, it is likely that some individuals and couples will persist in their decision to clone. Hopefully, whatever insight they gain from the process will help them to rear the child in a way that minimizes the risk of emotional damage that our models suggest as likely. These parents might benefit from ongoing treatment to help them parent the child clone and also to help them identify if and when the child might also need psychological help.

It should be a natural right of every unconceived child—clone or not—to be born to parents not so burdened by unresolved personal conflicts as to be unable to accurately perceive and genuinely relate to the child as a unique individual. If a parent is so burdened, then the already born child ought to be able to expect that the parent will try to address his problems so as to parent that child as best he can.

Of course, people are born into and encounter all sorts of hardships in life, but this does not justify deliberately creating people more likely to face such difficulties. Although a cloned person should be as resilient as any other human being, he would not be any less vulnerable. Given our assessment that he would run a substantial risk of psychological harm from being a clone, he would need society's compassionate understanding and might also need psychological help. Just as psychotherapy has helped countless others to rise above their difficulties and live satisfying lives, it has the same potential to help a person who also happens to be a clone. Issues and approaches in the psychotherapy of people representative of the various models of cloning should help suggest useful ways in which a psychotherapist might conceptualize a clone's problems and conduct the psychotherapy. Just as each individual case in the eight separate models in this book helped us to understand cloning by analogy, a psychotherapist could utilize the models separately and together to help an individual clone in psychological distress.

CONCLUSION

In this chapter we have overviewed the developmental psychology of moral reasoning. We have seen that some ways of thinking ethically about reproductive human cloning might possibly reflect less mature stages of moral reasoning.

Then we examined three general ethical approaches—the ethics of the absolute, the ethics of consequences, and the ethics of natural law. In the process, we have tried to disentangle reproductive rights and the rights of the unborn from reproductive cloning so that we could clearly focus on the unique ethical essence of reproductive cloning. That essence might be best captured in the concept of the "rights of the unconceived."

Our understanding of natural law enabled us to integrate our understanding of the ethics of the absolute with the ethical implications of consequences. Psychological and social science theory and research, as well as clinical experience, should inform our understanding of human nature—that is, natural law as it applies to humans. Because the various models relevant by analogy to various psychological and social dimensions of reproductive cloning overwhelmingly predict adverse consequences from it, we have concluded that human reproductive cloning would be ethically wrong.

We have considered a variety of scenarios to help us think through whether there ought to be some exceptions to this conclusion. Although we might have sympathy with one or several scenarios, in every case, *the rights of the unconceived* clone outweigh every argument and rationale to permit cloning.[45]

In considering possible exceptions to the absolute ethical judgment that reproductive cloning is wrong, we did not so much amend that conclusion as arrive at a different one. *Cloning per se would not be wrong, but many of the psychological and social consequences likely to flow from the genetic fact that cloning gives the clone the nuclear DNA of only one person would be.* However, if cloning could incorporate nuclear DNA from two people, perhaps regardless of gender, then it might be ethically acceptable.

If society is to strongly discourage reproductive cloning and perhaps also make it illegal, it must address the motivations of those who would wish to clone. If cloning should take hold, society might look to how harmful social practices have been expunged from society, not so much by the force of law but, rather, by social forces. For example, it was not through the force of law that the Chinese eliminated the widespread practice of binding the feet of females in a relatively short period of time.[46]

Like so many other things we might want to prevent, it will be impossible to prevent cloning. If we accommodate to it or criminalize it, then we ought to consider the repercussions. Making it illegal might force it completely underground, and this might make it harder to identify and help the clones. The illegality of their origin might contribute to a sense of illegitimacy and shame for clones. On the other hand, although accommodating cloning might ameliorate that harm, if cloning were to be more frequent as a result, then the overall harm could be greater.

Among his many creative ideas, Otto Rank proposed the idea that birth is a psychological trauma to the infant. There is no evidence of lasting adverse psychological sequelae to being born, but we have good reasons to think that *being born as a clone might prove to be a continuing trauma throughout the person's life.*

Perhaps some technologies are too potentially dangerous to be pursued. This argument has been made by none other than Bill Joy, chief scientist at Sun Microsystems. He argues for "relinquishment" of our right to pursue knowledge in such cases.[47] If a technology had only a malevolent purpose, then I would agree, but generally it is not the technology itself that is good or bad but, rather, the purpose to which people might put it.

The renowned sociobiologist E. O. Wilson says that human "biological constraints exist that define zones of improbable or forbidden access" in the future trajectory of human history.[48] While he is not referring specifically to cloning, clearly it lies in such a zone. Also, without reference to cloning, Wilson warns of dire social and psychological consequences if humans were to try to adopt the social system of a nonprimate species.[49] As we have seen in our examination of reproductive cloning, we might run the same sorts of risks by adopting this primitive mode of reproduction, even though only a scientifically advanced human social system could make it possible.

Whether one accepts it literally or understands it mythically, the biblical story of the tree bearing the forbidden fruit in the Garden of Eden is relevant to this issue. That tree was not, as is often misunderstood, the tree of knowledge but, in fact, *the tree of knowledge of good and evil.* Eating its fruit conferred on man a moral capacity that made him more fundamentally like God than simply being created in his image. We can still explore knowledge as long as we remember that, because the first woman and man ate from that tree, we have the terrible knowledge of good and evil and the responsibility to behave accordingly. These facts are complicated by the reality that there will always be those interested in utilizing knowledge with evil intent and others who will mistake the bad for the good and vice versa.

But there is more to the story that is also of great relevance to us here. After man ate from the tree of knowledge of good and evil, God expelled Adam and Eve from the Garden of Eden, not only to punish them but also to prevent them from eating of the *tree of life* and becoming Godlike in one more way—by becoming immortal (Genesis 3:22). Neither reproductive nor therapeutic cloning will make anyone immortal, though some might be seduced by the illusion of immortality that attaches to reproductive cloning.

We need to be circumspect, about not only cloning but also other biotechnologies, which might bring adverse consequences.[50] Governmental regulation of biotechnology, cloning included, might be desirable, even necessary, but let us base it on clear thinking, drawing distinctions between reproductive and therapeutic cloning, and utilizing our knowledge of human nature as it is relevant by analogy to cloning.

Even as we resist pursuing reproductive cloning of humans, let us see what medical researchers can achieve with therapeutic cloning and support their efforts. At the same time, it seems reasonable and ethically justifiable to continue to cautiously explore reproductive cloning in animals. With the idea that it might be a permissible form of reproductive cloning for humans, we might also want to consider exploring whether, through genetic engineering, cloning could be made to approximate the results of sexual reproduction. If it could, then we may want to consider this form of cloning to be an ethically acceptable reproductive option for humans, helping childless couples regardless of sexual orientation to have children with whom they are linked genetically. Whatever we decide in that regard, we must not forget the millions of children in need of adoption.

Over seventy years ago, Otto Rank wrote that "a correct statement of the problem is more important than any attempt at solution."[51] I have tried to be clear and unbiased in stating the possible problems and promise of human cloning. Cloning policy should respect our complex human nature.

Notes

PREFACE

All names of individuals used in case studies or for illustrative purposes are pseudonyms.

1. "The Rights and Wrongs of Human Cloning," session at the American Association for the Advancement of Science Annual Meeting, Philadelphia, March 1998.

2. National Bioethics Advisory Commission (1997). *Cloning human beings: Report and recommendations of the National Bioethics Advisory Commission* (Rockville, Md.: National Bioethics Advisory Commission).

3. President's Council on Bioethics (2002). *Human cloning and human dignity: An ethical inquiry* (Rockville, Md.: President's Council on Bioethics).

4. "Homo sum; humani nihil a me alienum puto."

5. Bottum, J (2001). Against human cloning (editorial), *The Weekly Standard*, 7 May: pp. 9–10.

6. McGee, G (2000). Cloning and new kinds of families, *The Journal of Sex Research* 37: pp. 266–72.

INTRODUCTION

1. Alexander, B (2001). You again, *Wired*, February: pp. 120–35; Pickrell, J (2000). Experts assail plan to help childless couple (news), *Science* 291: pp. 2061–63.

2. National Academy of Sciences (2003). *Scientific and medical aspects of human reproductive cloning* (Washington, D.C.: National Academy Press), pp. 41–42, 114–23 (table 1), 124–35 (table 2).

3. Feinberg, J (1980). The child's right to an open future, in *Whose child? Children's rights, parental authority, and state power*, ed. Aiken, W, and LaFollette, H (Towata, N.J.: Rowman and Littlefield); Pizzuli, F C (1975). Asexual reproduction and genetic engineering: A constitutional assessment of the technology of cloning, *The Southern California Law Review* 47: pp. 476–584.

4. See, e.g., Pence, G E (1997). *Who's afraid of human cloning?* (Lanham, Md.: Rowman and Littlefield); Robertson, J (1998). Cloning as a reproductive right, in *The human cloning debate*, ed. McGee, G (Berkeley: Berkeley Hills Books). See also Jaenisch, R, and Wilmut, J (2001). Don't clone humans, *Science* 291: p. 2552.

5. The notable exceptions are Nancy Segal, an experimental psychologist and authority on twins, and Adam Phillips, a child psychoanalyst. Even in the National Bioethics Advisory Commission report of 1997, no section addresses possible psychological and social consequences of cloning, and the President's Council on Bioethics report in 2002 contains only a small discussion.

6. Pence (1997), p. 135.

7. Lederberg, J (1966). Experimental genetics and human evolution, *Bulletin of the Atomic Scientists* 22: pp. 4–11.

8. Spemann, H (1938). *Embryonic development and induction* (New York: Hafner [reprint]).

9. Schilder, P (1933). Psyhoanalyse und Biologie, *Imago* 19: pp. 168–97.

10. Classic examples of models in modern physics illustrating this are the particle and wave theories of light. The diffraction of light is best explained by modeling it as a wave, whereas modeling light as a particle best explains its energy-transmitting properties. Physics, as "queen of the sciences," lends itself best to building models that can be tested and proven to be true or false. Compared with the "hard" science of physics, psychology and the social sciences are "soft." I try to make up for this by employing multiple models. We can have the most confidence in those predictions that converge and complement one another.

11. This is in the spirit of the unifying "bio-psychosocial" diagnostic schema in contemporary psychiatry incorporated into the contemporary guide for psychiatric diagnosis. See American Psychiatric Association (2000). *Diagnostic and statistical manual IV-TR* (Washington, D.C.: American Psychiatric Press).

CHAPTER 1

1. The standard diagnostic guide, the "DSM," provides a framework for assessing the biological, psychological, and social components to every case, but connecting meaningfully with a patient requires more than that. Psychoanalytic theory informs my listening, along with knowledge of individual and interpersonal psychology, cognition and behavior, and family therapy. See American Psychiatric Association (2000). *Diagnostic and statistical manual IV-TR* (2000) (Washington, D.C.: American Psychiatric Press).

2. Happily, once I developed the models, I was also able to use them to help my patients.

3. The original serotonin reuptake inhibitor is fluoxetine or Prozac. Now there are others.

4. Right-handers almost always are left-brain dominant for motor and speech functions, whereas the cerebral dominance of left-handers is often less asymmetric and sometimes is reversed in identical twins.

5. Left-handedness occurs more often in the first- than in the second-born twin, with the twin's left-handedness determined not genetically but, rather, by prenatal environmental factors. See Orlebeke, J F, Knol, D L, Koopmans, J R, Bommsma, D I, and Bleker, O P (1996). Left-handedness in twins: Genes or environment? *Cortex* 32: pp. 479–90.

6. Segal, N L (2000). *Entwined lives: Twins and what they tell us about human behavior* (New York: Plume [Penguin Putnam]), p. 205.

7. Rarely, the split is incomplete, and "conjoined" twins result. Previously, such twins were popularly known as "Siamese twins" after the pair of brothers, Eng and Chang, made famous by P. T. Barnum. Very early embryos can also be intentionally split to create clones, an intervention that raises a few of the same questions as cloning by somatic cell nuclear transfer. See Macklin, R (1994). Splitting embryos on the slippery slope: Ethics and Public Policy, *Kennedy Institute of Ethics Journal* S 94: pp. 209–25.

8. A bit later, and apparently even in adults, some cells are "pluripotent." These "stem" cells are able to form more than one kind of tissue under certain influences.

9. Temporally different environments for monozygotic twins have also become possible. Embryos frozen for in vitro fertilization could be implanted at different times. If some were identical twins, what has been called a "delayed twin" could result. Conceivably, the delay could be lengthy, though, given the limits in reproductive years for women, probably less than a generation in length. However, adoption of a frozen embryo could result in the birth of a "twin" a generation or more later than his or her firstborn "twin" to a woman other than the biological mother.

10. Asexual reproduction may occur naturally for a variety of nonmammalian species of animals through a process called "parthenogenesis." In it, an egg essentially fertilizes itself and begins the process of embryological development. It occurs infrequently in nature in certain fish species. Typically, "parthenotes" do not live as long as their sexually reproduced counterparts. By necessity, all parthenotes are female.

11. Segal, N L (1997). Behavioral aspects of intergenerational cloning: What twins tell us, *Jurimetrics* 38: pp. 57–61.

12. Parthenogenesis might become another avenue to cloning humans.

13. Wilmut, W I (1996). Viable offspring derived from fetal and adult mammalian cells, *Nature* 385: pp. 810–13.

14. National Academy of Sciences (2002). *Scientific and medical aspects of human reproductive cloning* (Washington, D.C.: National Academy Press), pp. 41–42. National Academy of Sciences (2002), pp. 42–51, describes possible molecular mechanisms for these defects. Most of these are "epigenetic," which means that although the DNA sequence has not changed, the way the gene is expressed has.

15. Segal (1997), p. 63.

16. Each of these cloning scenarios is covered in future chapters.

17. Zazzo, R (1976). The twin competition and the couple effects on personality development, *Acta Geneticae Medicae et Gemellologiae* 25: pp. 343–52.

18. Segal (2000), pp. 60–61.

19. Kendler, K S, Martin, N G, Heath, A C, and Eaves, L J (1995). Self-report psychiatric symptoms in twins and their nontwin relatives: Are twins different? *American Journal of Medical Genetics* 60: pp. 588–91.

20. Zazzo (1976), p. 350.

21. Lewontin, R (1998). The confusion over cloning, in *The human cloning debate*, McGee, G ed. (Berkeley: Berkeley Hills Books), p. 130.

22. Lewontin (1998), p. 132.

23. Fischbein, S, Hollencrewtz, I, and Wiklund, I (1990). What is it like to be the parent of a twin? *Acta Geneticae Medicae et Gemellologiae: Twin Research* 39: pp. 271–76. Segal (2000) recounts anecdotes of similar preferences and choices made by identical twins reared apart, pointing to the importance of a genetic component.

24. Segal (2000).

25. Fischbein, Hollencrewtz, and Wiklund (1990).

26. Dibble, E D, and Cohen, D J (1981). Personality development in identical twins, *Psychoanalytic Study of the Child* 36: pp. 45–70.

27. Fischbein, Hollencrewtz, and Wiklund (1990).

28. Winnicott, D W (1971 [1967]). Mirror-role of mother and family in child development, in *Playing and reality* (New York: Basic Books), pp. 111–18; Bowlby, J (1969). *Attachment and loss, vol. 1: Attachment* (New York: Basic Books); Ainsworth, M S (1993). Attachment as related to mother–infant interaction, *Advances in Infancy Research* 8: pp. 1–50.

29. Segal, N L (2003). Meeting one's twin: Perceived social closeness and familiarity, *Evolutionary Psychology* 1: pp. 70–95.

30. This would be an example of "twinning transference," discussed in chapter 5.

31. Moore, B E, and Fine, B D (eds.) (1990). *Psychoanalytic terms and concepts* (New Haven: Yale University Press), pp. 180–81; Mahler, M S (1968). *On human symbiosis and the vicissitudes of individuation* (New York: International Universities Press); Mahler, M S, Pine, F, and Bergman, A (1975). *The psychological birth of the human infant* (New York: Basic Books).

32. Moore and Fine (1990), p. 180.

33. Burland, J A (1998). The developmental consequences of giftedness: Can there be too much of a good thing? presented at Psychiatry Grand Rounds, Albert Einstein Medical Center, Philadelphia, December 4.

34. Ackerman, P H (1975). Narcissistic personality disorder in identical twins, *International Journal of Psychotherapy* 4: pp. 389–404.

35. Moore and Fine (1990), pp. 149–50.

36. Segal (1990), pp. 102–3.

37. Joseph, E, and Tabor, J (1961). Simultaneous analysis of identical twins: Twinning reaction, *Psychoanalytic Study of the Child* 16: pp. 279–99.

38. Waterman, P, and Shatz, M (1982). The acquisition of personal pronouns and proper names by an identical twin pair, *Journal of Speech, Language, and Hearing Research* 25: pp. 149–54.

39. Zazzo (1976), p. 349.

40. Zazzo (1976), p. 350.

41. Segal, N L, and Ream, S L (1998). Decrease in parental grief intensity for deceased twin and non-twin relatives: An evolutionary perspective, *Personality and Individual Differences* 25: pp. 317–25; Segal (2000).

42. Engel, G (1975). The death of a twin: Mourning and anniversary reactions. Fragments of ten years of self-analysis, *International Journal of Psycho-analysis* 56: pp. 23–40.

43. Woodward, J (1998). The bereaved twin, *Acta Geneticae Medicae et Gemellologiae* 37: pp. 173–80.

44. Woodward (1988), p. 177.

45. Woodward (1988), p. 178.

46. Segal (2000).

47. Segal (2000), p. 185.

48. Segal (2000), p. 183.

49. We will explore the issue of survivorship further in the Replacement Child Model.

50. Segal (2000), p. 178.

51. The studies on grief reduction look at identical and fraternal twins but not at "singletons," that is, people born singly. To further explore the possibility that identical twins may have less capacity to attach to people other than their co-twins, they should also be compared with singletons on the intensity of grief and its reduction over time after losing a spouse, parent, other relative, or friend.

52. Zazzo (1976), p. 346; Pearlman, E M (1990). Separation-individuation, self-concept, and object relations in fraternal twins, identical twins, and singletons, *Journal of Psychology* 124: pp. 619–28.

53. Pearlman (1990).

54. Zazzo (1976), pp. 349–51.

55. Zazzo (1976), p. 349.

56. Tienari, P (1966). On intrapair differences in male twins with special reference to dominance-submissiveness, *Acta Psychiatica Scandinavica, Suppl.* 188, 42: pp. 1–166.

57. Sulloway, P J (1996). *Born to rebel: Birth order, family dynamics, and creative lives* (New York: Putnam) discusses sibling birth order.

58. Ramsey, P (1970). *Fabricated man: The ethics of genetic control* (New Haven: Yale University Press), p. 71, warns that "to mix the parental and twin relationship would be psychologically disastrous for the young."

59. Jamieson, J W (1998). The case for cloning, *Mankind Quarterly* 39: pp. 95–107.

60. Smith, J D (1988). *Psychological profiles of conjoined twins: Heredity, environments, and identity* (New York: Praeger); Clay, M M (1989). *Quadruplets and higher order multiple births* (Philadelphia: J. B. Lippincott), p. 5.

61. It is so new, and as yet uncommon, that we cannot use it as a model, but embryo adoption might eventually model cloning by making it possible for one member of a monozygotic twin pair to be born years later than another.

62. Segal (2000), p. 207.

63. Segal (1997), p. 67.

CHAPTER 2

1. Van Balen, F, Verdurmen, J, and Ketting, E (1997). Choices and motivations of infertile couples, *Patient Education and Counseling* 31: pp. 19–27.

2. Lesbian women and celibate heterosexual women can have sexually reproduced offspring without sexual intercourse with the help of assisted reproductive technology. Celibate or homosexual men can become fathers by arranging for a gestational surrogate.

3. Stotland, N L (ed.) (1990). *Psychiatric aspects of reproductive technology* (Washington, D.C.: American Psychiatric Press), pp. 1–12.

4. Notman, M T (1990). Reproduction and pregnancy: A psychodynamic perspective, in Stotland, pp. 13–24.

5. Rowland, R (1981). Women as living laboratories: The new reproductive technologies, in *The trapped woman: Catch-22 in deviance and control*, ed. Figueira-McDonough, J, and Sarri, R (Newbury Park, Calif.: Sage Publications), pp. 87–88.

6. Downey, J, and McKinney, M (1990). Psychiatric research and the new reproductive technologies, in Stotland, pp. 155–68.

7. Downey and McKinney (1990).

8. Berger, D M (1980). Impotence following the discovery of azospermia, *Fertility and Sterility* 34: pp. 154–56; Berger, D M (1980). Couples' reactions to male infertility and donor insemination, *American Journal of Psychiatry* 137: pp. 1047–49.

9. Czyba, J C, and Cheveret, M (1979). Psychological reactions to artificial insemination with donor sperm, *International Journal of Fertility* 24: pp. 240–45.

10. Landridge, D, Connolly, K, and Sheeran, P (2000). Reasons for wanting a child: A network analytic study, *Journal of Reproductive and Infant Psychology* 4: p. 333.

11. Miller, A (1997). *The drama of the gifted child: The search for the true self* (New York: Basic Books), p. 83.

12. Lerner, B, Raskin, R, and Davis, E B (1967). On the need to be pregnant, *International Journal of Psychoanalysis* 48: pp. 288–97.

13. Moore, B E, and Fine, B D (eds.) (1990). *Psychoanalytic terms and concepts* (New Haven: Yale University Press), p. 123.

14. Landridge, Connolly, and Sheeran (2000).

15. Stotland (1990), p. 9.

16. Landridge, Connolly, and Sheeran (2000), p. 336.

17. Potts, M, and Short, R (1999). *Ever since Adam and Eve: The evolution of human sexuality* (New York: Cambridge University Press), p. 127.

18. Schultz, R M, and Williams, C J (2002). The science of ART, *Science* 296: pp. 2188–90.

19. Dickstein, L J (1990). Effects of the new reproductive technologies on individuals and relationships, in Stotland, pp. 123–29.

20. Rosenthal, M B, and Goldfarb, J (1997). Infertility and assisted reproductive technology: An update for mental health professionals, *Harvard Review of Psychiatry* 5: pp. 159–72.

21. Downey and McKinney (1990), p. 161.

22. Gibson, F L, Ungerer, J A, McMahon, C A, Catherine, A, Leslie, G I, and Saunders, D M (2000). The mother–child relationship following in vitro fertiliza-

tion (IVF): Infant attachment, responsivity, and maternal sensitivity, *Journal of Child Psychology and Psychiatry and Allied Disciplines* 41: pp. 1015–23.

23. Landridge, Connolly, and Sheeran (2000), p. 322.

24. Corea, C (1979). *The mother machine: Reproductive technologies from artificial insemination to artificial wombs* (New York: Harper and Row).

25. Golombok, S (1999). New family forms: Children reared in solo mother families, lesbian mother families, and in families created by assisted reproduction, in *Child psychology: A handbook of contemporary issues*, ed. Batter, L, and Tamis-LeMonde, C (Philadelphia: Psychology Press), pp. 429–46.

26. Golombok (1999), p. 433.

27. Golombok, S, Murray, C, Brinsden, P, and Abdalla, H (1999). Social versus biological parenting: Family functioning and socioeconomic development of children conceived by egg or sperm donation, *Journal of Child Psychology and Psychiatry and Allied Disciplines* 40: pp. 519–27.

28. Golombok, S, MacCallum, F, and Goodman, E (2001). The "test-tube" generation: Parent–child relationships and the psychological well-being of in vitro fertilization children at adolescence, *Child Development* 72: pp. 599–608; Golombok, S, MacCallum, F, Goodman, E, and Rutter, M (2002). Families with children created by artificial insemination: A follow-up at age twelve, *Child Development* 73: pp. 952–68.

29. Gottlieb, C, Lalos, O, and Lindblad, F (2000). Disclosure of donor insemination to the child: The impact of Swedish legislation on couples' attitudes, *Human Reproduction* 15: pp. 2052–56.

30. Lindblad, F, Gottlieb, C, and Lalos, O (2000). To tell or not to tell—What parents think about telling their children that they were born following donor insemination, *Journal of Psychosomatic Obstetrics and Gynecology* 21: pp. 193–207.

31. Specific reasons to tell included that it feels natural to do so and that one should be honest and open. Parents favoring disclosure felt that children have a right to know their origins, that knowing may prove of value to the child, and that it might be detrimental for the child to find out belatedly. Parents favoring disclosure also felt that disclosure relieves tension in the family associated with keeping donor insemination a secret and that it also helps engender respect for the decision to make use of this assisted reproductive technology.

Reasons not to tell included a simple reiteration that the couple just did not want to and that it was nobody's business but theirs. More reasoned responses were concerns that learning the truth of his origins might prove harmful to the child and "to the child's relationship with or love for . . . [his or her] parents, particularly the father."

Parents who refused to answer the question about disclosure responded in ways that implied that they were against it. They saw disclosure as irrelevant or destructive to the child, commenting that the "child does not need another father" and that "one never knows anyway" whether a father is, in fact, the biological father. Reasoning the researchers adjudged inconclusive also implied that those parents were against disclosure. Such parents stated that the child resembles the mother or the father in appearance, that the family had no prior relationship with the donor, and that "social inheritance" is the most important kind of inheritance.

What the researchers refer to as "context-dependent" reasoning included that the child should reach a certain age and maturity to handle disclosure, that it might be necessary to wait for the same kind of maturation in a younger sibling(s), and that "inequality between differently conceived" siblings may be induced by telling. One couple appreciated that special circumstances may arise that would require them telling, such as inherited diseases and the need for organ donation.

32. Golombok et al. (1999).

33. Dickstein (1990), pp. 133–34.

34. Downey and McKinney (1990); Sokoloff, B Z (1987). Alternative methods of reproduction: Effects on the child, *Clinical Pediatrics* 26: pp. 11–16.

35. Turner, A J, and Coyle, A (2000). What does it mean to be a donor offspring? The identity experiences of adults conceived by donor insemination and the implications for counseling and therapy, *Human Reproduction* 15: pp. 2041–51.

36. Golombok (1999).

37. Golombok et al. (1999).

38. Scheib, J E (1997). The psychology of female choice in the context of donor insemination, *Dissertation Abstracts International: Section B: The Sciences and Engineering* 58(6-B): p. 3306; Scheib, J E, Kristiansen, A, and Wara, A (1997). A Norwegian note on "Sperm donor selection and the psychology of female mate choice," *Evolution and Human Behavior* 18: pp. 143–49.

39. Stotland (1990), p. 11.

40. Scheib, Kristiansen, and Wara (1997). We do not need to discuss here that there *are* genetic influences on character. The important point is that the women do not believe there are.

41. Foster-Frazer, K L (1990). Male donors' personality characteristics, parental relationships, and motivations to participate in artificial insemination, Ph.D. dissertation, California School of Professional Psychology at Berkeley/Alameda, pp. 3–6, 24–32; Emond, M, and Scheib, J E (1998). Why not donate sperm? A study of potential donors, *Evolution and Human Behavior* 19: pp. 313–19.

42. Barritt, J A, Willadsen, S, Brenner, C A, and Cohen, J (2001). Cytoplasmic transfer in assisted reproduction, *Human Reproduction Update* 7: pp. 426–35.

43. Kanefield, L (1999). The reparative motive in surrogate mothers, *Adoption Quarterly* 2: pp. 5–19.

44. Steadman, J H, and McCloskey, G T (1987). The prospect of surrogate mothering: Clinical concerns, *Canadian Journal of Psychiatry* 32: pp. 545–50.

45. Steadman and McCloskey (1987).

46. Steadman and McCloskey (1987).

47. Golombok, S, and MacCallum, F (2003). Practitioner review: Outcomes for parents and children following non-traditional conception: What do clinicians need to know? *Journal of Child Psychology and Psychiatry* 44: pp. 303–15.

48. Harrison, M (1990). Psychological ramifications of "surrogate" motherhood, in Stotland, pp. 97–112.

49. Salisbury, K L, LaGasse, L, and Lester, B (2003). Maternal–fetal attachment, *Journal of the American Medical Association* 289: p. 1701.

50. Einwohner, J (1998). Who becomes a surrogate: Personality characteristics, in *Gender in transition: A new frontier*, ed. Offerman-Zuckerberg, J (New York: Plenum).

51. Harrison (1990).

52. Kanefield (1999), p. 11.

53. Parker, P J (1983). Motivation of surrogate mothers: Initial findings, *American Journal of Psychiatry* 140: pp. 117–18.

54. Kanefield (1999).

55. Kanefield (1999), p. 13.

56. Harrison (1990), p. 99.

57. Harrison (1990).

58. Harrison (1990).

59. Harrison (1990).

60. Steadman and McCloskey (1987); Rowland (1987).

61. Andrews, L B (1999). *The clone age: Adventures in the new world of reproductive technology* (New York: Henry Holt), pp. 73–75.

62. Dawkins, R (1989 [1976]). *The selfish gene* (New York: Oxford University Press), p. 56.

63. Stormer, N (2003). Seeing the fetus: The role of technology and image in the maternal–fetal relationship, *Journal of the American Medical Association* 289: p. 1700.

64. Rosenthal and Goldfarb (1997).

65. Sex-linked inherited diseases appear only in offspring of a given sex. Certain forms of hemophilia are the best-known example, occurring only in the male offspring of families with a gene linked to the X chromosome. Girls, having two X chromosomes, are unlikely to show signs of the illness, as the normal chromosome dilutes the effect of the one carrying the defect, but boys, who have only one X chromosome and the tiny Y chromosome, manifest the disease.

66. Andrews (1999), pp. 140–54.

67. This is not to discount "epigenetic" risks raised in the previous chapter.

68. Das Gupta, cited in Dugger, C W (2001). Modern Asia's anomaly: The girls who don't get born, *New York Times* Week in Review, May 6: p. 4; Eckholm, E (2002). Desire for sons drives use of prenatal scans in China, *New York Times*, June 21: p. A3.

69. Mudur, G (1999). Indian medical authorities act on antenatal sex selection (news), *British Medical Journal* 319: p. 401.

70. Mackay, J (2000). *The Penguin atlas of human sexual behavior: Sexuality and sexual practice around the world* (Brighton, England: Penguin).

71. Mann, C C (2003). The first cloning superpower: Inside China's race to become the clone capital of the world, *Wired*, January: pp. 114–43.

72. Kolata, G (2001). Fertility ethics authority approves sex selection: Some experts balk at shift in position, *New York Times*, September 28: p. A16.

73. Wilson, E O (1978). *On human nature* (Cambridge: Harvard University Press), p. 40.

74. Rowland (1987).

75. Dworkin, A (1998). Sasha, in *Clones and clones: Facts and fantasies about human cloning*, ed. Nussbaum, M C, and Sunstein, C R (New York: W. W. Norton), pp. 73–77.

76. Harrison (1990), p. 111.

CHAPTER 3

1. As explained previously, if the rearing mother were also the egg or mitochondrial mother, and the child were a clone of the father, then the child would be very slightly connected to the mother genetically. The genetic link to the mother would be exclusive if the child were her clone.

2. Kass, L (1998). The wisdom of repugnance: Why we should ban the cloning of humans (originally published in the *New Republic*, 1997), in *The human cloning debate*, ed. McGee, G (Berkeley: Berkeley Hills Books), p. 162.

3. Daly, M, and Wilson, M (1998). *The Truth about Cinderella: A Darwinian view of parental love* (New Haven: Yale University Press), pp. 47–59.

4. Claxton-Oldfield, S (2000). Deconstructing the myth of the wicked stepparent, *Marriage and Family Review* 30: pp. 51–58.

5. Daly, M, and Wilson, M (1988). *Homicide* (New York: Aldine de Gruyter); Daly and Wilson (1998).

6. Daly and Wilson (1998).

7. Daly and Wilson (1998), p. 27.

8. Daly and Wilson (1998), p. 32.

9. Daly and Wilson (1998), pp. 34–35.

10. Case, A, Lin, I-F, and McLanahan, S (2000). How hungry is the selfish gene? *The Economic Journal* 110: pp. 781–804.

11. Daly and Wilson (1998), pp. 60–61.

12. Daly and Wilson (1998), p. 50.

13. Faller, K C (1990). Sexual abuse by parental caretakers: A comparison of abusers who are biological fathers in intact families, stepfathers, and noncustodial fathers, in *The incest perpetrator: A family member no one wants to treat*, ed. Horton, A L, and Johnson, B L (Newbury Park, Calif.: Sage Publications).

14. Erickson, M T (2000). The evolution of incest avoidance, Oedipus and the psychopathologies of kinship, in *Genes on the couch: Explorations in evolutionary psychology*, ed. Gilbert, P, and Bailey, K G (East Essex: Brenner-Routledge), pp. 211–31.

15. Hamilton, W D (1964). The genetical evolution of social behavior, *Journal of Theoretical Biology* 7: pp. 1–52.

16. Case, Lin, and McLanahan (2000).

17. Wilson, E O (2000). *Sociobiology: The new synthesis*, 25th anniversary edition (Cambridge: Harvard University Press); Leak, G K, and Christopher, S B (1982). Freudian psychoanalysis and sociobiology: A synthesis, *American Psychologist* 37: pp. 313–22; Tudge, C (2000). *The impact of the gene: From Mendel's peas to designer babies* (New York: Hill and Wang, a division of Straus and Giroux), pp. 172–230.

18. Daly and Wilson (1998), p. 38.

19. Becker, G S, Landes, E M, and Michael, R T (1977). An economic analysis of marital instability, *Journal of Political Economy* 85: pp. 1141–87.

20. Barash, D P (2001). *Revolutionary biology: The new gene-centered view of life* (New Brunswick, N.J.: Transaction Publishers), p. 150.

21. Bowlby, J (1969). *Attachment and loss, vol. 1: Attachment* (New York: Basic Books).

22. Barash (2001).

23. Flinn, M V (1988). Step- and genetic parent/offspring relationships in a Caribbean village, *Ethnology and Sociobiology* 9: pp. 335–69.

24. Bailey, D B, Jr., Bruer, J T, Symons, F T, and Lichtman, J W, et al. (eds.) (2001). *Critical thinking about critical periods* (Baltimore: Paul H. Brooks Publishing Co.).

25. Insel, T R (2000). Toward a neurobiology of attachment, *Review of General Psychology* 4: pp. 176–85.

26. Erickson (2000).

27. Wolf, A P (1995). *Sexual attraction and childhood association: A Chinese brief for Edward Westermark* (Stanford: Stanford University Press).

28. Barth, R P (1997). Effects of age and race of adoption versus remaining in long-term out-of-home care, *Child Welfare* 76: pp. 285–308; Howe, D, Shemmings, D, and Feast, J (2001). Age at placement and adult adopted people's experience of being adopted, *Child and Family Social Work* 6: pp. 337–49.

29. Daly and Wilson (1998), p. 32.

30. Baum, K J (2000). A comparison of lesbian and heterosexual stepfamilies: Is sex of stepparent or sex of parent the more salient factor in family adjustment? *Dissertation Abstracts International: Section B: The Sciences and Engineering* 60(9-B): p. 4874.

31. Golombok, S (1999). New family forms: Children reared in solo mother families, lesbian mother families, and in families created by assisted reproduction, in *Child psychology: A handbook of contemporary issues*, ed. Batter, L, and Tamis-LeMonde, C (Philadelphia: Psychology Press).

32. Golombok, S, Murray, C, Brinsden, P, and Abdalla, H (1999). Social versus biological parenting: Family functioning and socioeconomic development of children conceived by egg or sperm donation, *Journal of Child Psychology and Psychiatry and Allied Disciplines* 40: pp. 519–27.

33. Daly and Wilson (1998).

34. Grotevant, H D, and Kohler, J K (1999). Adoptive families, in *Parenting and child development in "nontraditional" families*, ed. Lamb, M E, et al. (Mahwah, N.J.: Lawrence Erlbaum Associates), pp. 161–90.

35. Grotevant and Kohler (1999).

CHAPTER 4

1. Hollingsworth, L D (2000). Who seeks to adopt a child? Findings from the national survey of family growth, *Adoption Quarterly* 3: pp. 1–24; Haugaard, J J (1998). Is adoption a risk factor for the development of adjustment problems? *Clinical Psychology Review* 18: pp. 47–69.

2. Grotevant, H D, and Kohler, J K (1999). Adoptive families, in *Parenting and child development in nontraditional families*, ed. Lamb, M E (Mahwah, N.J.: Lawrence Erlbaum Associates), pp. 161–90.

3. Brodzinsky, D M, Smith, D W, and Brodzinsky, A B (1998). Children's adjustment to adoption: Developmental and clinical issues, *Developmental Clinical Psychology and Psychiatry* 38: p. 5.

4. Nancy Segal has raised the idea, but to my knowledge neither she nor anyone else has developed it. See Segal, N L (1997). Behavioral aspects of intergenerational cloning: What twins tells us, *Jurimetrics* 38: pp. 57–67.

See also McGee, G, and Wilmut, I (1998). Cloning and the adoption model, in *The human cloning debate*, ed. McGee, G (Berkeley: Berkeley Hills Books). McGee and Wilmut propose an "adoption model" for cloning, but the model is of a different type than the one we will develop here. Theirs is not meant to model aspects of cloning. It utilizes adoption as a model for making ethical decisions and social and legal policy regarding cloning.

5. Levi-Strauss, C (1969). Structural analysis in linguistics and anthropology, in *Structural anthropology* (New York: Basic Books), pp. 139–61.

6. Lifton, B J (1988 [1979]). *Lost and found: The adoption experience* (New York: Harper and Row), p. 6.

7. Grotevant, H D (1997). Family processes, identity development, and behavioral outcomes for adopted adolescents, *Journal of Adolescent Research* 12: pp. 139–61.

8. Wieder, H (1977). On being told of adoption, *Psychoanalytic Quarterly* 46: pp. 1–22.

9. Haugaard (1998).

10. Epidemiological science often proceeds from finding that a group with a certain characteristic is overrepresented in a clinical population. It then sensibly asks whether this might indicate that that characteristic is a causal factor in the disorder or illness. Is the characteristic identified in the clinical population also associated with disorder or disease in a sample of the general population? Can causal factors associated with the identified characteristic be identified? Because differences among groups may be hard to distinguish in a general population sample, negative studies are only meaningful if the populations are very large, and even then, persistent overrepresentation of groups with certain characteristics in a clinical population still deserves scrutiny.

11. Brent, B C, Fan, X, Christensen, M, Grotevant, H D, and van Dulmen, M (2000). Comparison of adopted and nonadopted adolescents in a large, nationally representative survey, *Child Development* 7: pp. 1458–73.

12. Cadoret, R J (1990). Biological perspectives on adoptee adjustment, in Brodzinsky and Schecter, pp. 25–41.

13. Lewontin, R (1998). The confusion over cloning, in McGee, pp. 125–40.

14. Lifton (1988), p. 19.

15. Lifton, B J (1994). *The journey of the adopted self: A quest for wholeness* (New York: Basic Books), p. 19.

16. Brodzinsky, Smith, and Brodzinsky (1998).

17. Conversely, perhaps the spouse of a progenitor/rearing parent might feel less entitled to guide and correct the child and adolescent.

18. Cadoret (1990).

19. Lifton (1994), pp. 88–103; Lifton (1988).

20. Lifton (1988), p. 4.

21. Lifton (1988), p. 9.

22. Grotevant and Kohler (1999), pp. 162–63.

23. Bowlby, J (1969). *Attachment and loss, vol. 1: Attachment* (New York: Basic Books).

24. Ainsworth, M S (1993). Attachment as related to mother–infant interaction, *Advances in Infancy Research* 8: pp. 1–50.

25. Mackey, W C (2001). Support for the existence of an independent man-to-child affiliative bond: Fatherhood as a biocultural invention, *Psychology of Men and Masculinity* 2: pp. 51–56.

26. Winnicott, D W (1971 [1967]). Mirror-role of mother and family in child development, in *Playing and reality* (New York: Basic Books), pp. 111–18.

27. Thompson, R A (2001). Sensitive periods in attachment, in *Critical thinking about critical periods*, ed. Bailey, D B, Jr., Bruer, J T, Symons, F T, and Lichtman, J W, et al. (Baltimore: Paul H. Brooks Publishing Co.), pp. 83–106.

28. Golombok, S (1999). New family forms: Children raised in solo mother families, lesbian mother families, and in families created by assisted reproduction, in *Child psychology: A handbook of contemporary issues*, ed. Balter, L, and Tamis-LeMonde, C (Philadelphia: Psychology Press), p. 430.

29. Johnson, J J, McAndrew, F T, and Harris, P B (1991). Sociobiology and the naming of adopted and natural children, *Ethology and Sociobiology* 12: pp. 365–75; Smith, D S (1988). Research in developmental sociobiology: Parenting and family behavior, in *Sociobiological perspectives on human development*, ed. MacDonald, K B (New York: Springer-Verlag).

30. We will see that this sort of thing can actually happen between parent and child when we explore the concept of "projective identification" in chapter 5.

31. Lifton (1994), pp. 50–51.

32. Wieder, H (1977). On being told of adoption, *Psychoanalytic Quarterly* 46: pp. 1–22.

33. Sants, H J (1964). Genealogical bewilderment in children with substitute parents, *British Journal of Medical Psychology* 37: pp. 133–41. Others writing on human reproductive cloning have pointed to this issue but without reference to this psychological research.

34. Sants (1964).

35. Erikson, E (1959). Identity and the life cycle, *Psychological Issues* 1: pp. 74–88; Erikson, E (1963). *Childhood and society*, 2d edition (New York: W. W. Norton).

36. Daly and Wilson (1988), pp. 107–21, attempt to refute the Oedipal complex by citing research showing lack of gender bias in child abuse and murder until adolescence. They misunderstand this psychoanalytic concept in coming to this conclusion. It is not the adult who wants to kill the child in the oedipal phase; it is the oedipal child who wants to murder the parenting adult of the same sex. And, for the young boy, feelings of attachment and love tend to conflict with his wish to destroy his father. In addition, young children have practically no ability to inflict extreme physical harm, so we would not expect research to show them murdering their same-sex parent.

Note also, sociobiology and evolutionary psychology attribute the incest taboo to an evolutionarily driven revulsion to sexual relations between close genetic relatives based on the high risk of penalty to be paid in biological defects in offspring of incestuous pairings. See Wilson, E O (1978). *On human nature* (Cambridge: Harvard University Press), pp. 36–38.

37. Hoopes, J L (1990). Adoption and identity formation, in *The psychology of adoption*, ed. Brodzinsky, D M, and Schecter, M D (New York: Oxford University Press), pp. 144–66.

38. Silverman, M A (1985–86). Adoptive anxiety, adoptive rage, and adoptive guilt, *International Journal of Psychoanalytic Psychology* 11: pp. 301–7.

39. Lifton (1988), pp. 164–66; Greenberg, M, and Littlewood, R (1995). Post-adoption incest and phenotypic matching: Experience, personal meanings and biosocial implications, *British Journal of Medical Psychology* 68: pp. 29–44.

40. Kaplan, L (1988). The concept of the family romance, *Psychoanalytic Review* 61: pp. 169–202.

41. Lehrman, P R (1927). The fantasy of not belonging to one's family, *Archives of Neurology and Psychiatry* 18: pp. 1015–23.

42. Kaplan (1988). The family romance fantasy is the precursor of the more thorough and deeply felt disillusionment the preadolescent typically experiences concerning his or her parents (Freud, A [1949]. On certain difficulties in the pread-olescent's relation to his parents, in *The writings of Anna Freud* [1969], vol. 4 [New York: International Universities Press]).

43. Wieder, H (1977). The family romance fantasy of adopted children, *Psycho-analytic Quarterly* 46: pp. 185–201.

44. Heinemann, T V (1999). In search of the romantic family: Unconscious con-tributions to problems in foster and adoptive placement, *Journal for the Psycho-analysis of Culture and Society* 42: p. 260.

45. Freud, A (1965). Normality and psychopathology in childhood: Assess-ments of development, in *The writings of Anna Freud* [1969], vol. 6 (New York: In-ternational Universities Press).

46. Nickman, S (1985). Losses in adoption, *Psychoanalytic Study of the Child* 40: pp. 365–98.

47. Wieder (1977).

48. Lehrman (1927).

49. Josselson, R (1987). *Finding herself: Pathways to identity development in women* (San Francisco: Jossey-Bass), p. 10.

50. Josselson, R (1997). *Revising herself* (New York: Oxford University Press), p. 28.

51. Grotevant and Kohler (1999), p. 181.

52. Grotevant, H D (1992). Assigned and chosen identity components: A process perspective on their integration, in *Adolescent identity formation*, ed. Adams, G R, and Gullotta, T, and Montemayor, R (Thousand Oaks, Calif.: Sage Publications), p. 78.

53. Grotevant (1992), p. 77.

54. We could imagine a number of types of sibling relationships for a clone:

1. The clone is adopted and unrelated to the siblings with whom he is raised. He is truly chosen, in contrast to his sexually reproduced siblings.
2. The clone has a rearing parent as his progenitor. That makes him a spe-cial kind of half sibling to the sexually reproduced offspring of his rear-ing parents.
3. The clone's siblings were also cloned from the same progenitor, probably one of the rearing parents, and so are "twins" of different ages.
4. The clone's progenitor is an earlier noncloned child in the family.
    a. That child may still be in the family as a favorite child, and the family wanted to raise "him" again from infancy.

b. The earlier child may be dead, and hence the clone is a classic "replacement" child.

55. Brodzinsky, Smith, and Brodzinsky (1998), p. 92.

56. Lifton (1994), pp. 64–66.

57. Gergen, K J, and Gergen, M (1988). Narrative and self in relationship, in *Advances in experimental social psychology*, vol. 21, ed. Berkowitz, L (New York: Academic Press), pp. 17–56.

58. Lifton (1994), pp. 36–38.

59. Josselson (1987), p. 63.

60. Dr. Seuss (1959). *Happy birthday to you!* (New York: Random House).

61. Lifton (1988), p. 55.

62. Lifton (1994), p. 51.

63. Verrier, N (1987). The primal wound: A preliminary investigation into the effects of separation from birth mother on adopted children, *Pre- and Peri-natal Psychology* 2: pp. 75–85.

64. The term *false self* was coined in Winnicott, D W (1960). Ego distortions in terms of the true and false self, in *The maturational processes and the facilitating environment* (1965) (New York: International Universities Press). The concept of "'as if' personality" is put forward in Deutsch, H (1942 [1934]). Some forms of emotional disturbance and their relationship to schizophrenia, *Psychoanalytic Quarterly* 11: pp. 301–22.

65. Marcia, J (1966). Development and validation of ego-identity status, *Journal of Personal and Social Psychology* 3: pp. 551–58.

66. People writing from philosophical and legal perspectives have addressed the issue, framing it as Joel Feinberg has, in terms of a "right to an open future." We will see how the psychological literature on identity can inform debate on this alleged "right."

67. Lifton (1994), p. 71.

68. Lifton (1994), pp. 89–108.

69. Lifton (1988), p. 59.

70. Lifton (1994), p. 72.

71. Verrier (1987), p. 77.

72. Neumann, E (1973). *The child* (New York: G. P. Putnam), p. 13.

73. Lifton (1994), pp. 12–20.

74. Lifton (1994), p. 120.

75. Lifton (1994), p. 206.

76. Such a scenario gives new meaning to the title of the book by Friday, N (1977). *My mother myself: The daughter's search for identity* (New York: Delacorte).

77. Grotevant and Kohler (1999).

78. Wrobel, G M, Ayers-Lopez, S, Grotevant, H D, McRoy, R G, et al. (1996). Openness in adoption and the level of child participation, *Child Development* 67: pp. 2358–74.

79. Grotevant and Kohler (1999), p. 175.

80. Lifton (1988), pp. 108–206.

81. Lifton (1988), p. 189.

82. Lifton (1988), pp. 263–69.

83. Gaylord, C L (1976). The adoptive child's right to know, *Case and Comment* 81: p. 38.

84. Brodzinsky, Smith, and Brodzinsky (1998), pp. 79–88.

85. McGinn, M F (2000). Attachment and separation: Obstacles for adoptees, *Journal of Social Distress and the Homeless* 9: pp. 273–90.

86. Grotevant and Kohler (1999), pp. 183–84.

87. Alexander, B (2000). You again, *Wired*, February: pp. 120–35.

88. Wieder, H (1978). On when and whether to disclose about adoption, *Journal of the American Psychoanalytic Association* 26: pp. 793–811.

89. Lifton (1994), pp. 116–17.

90. Lifton (1988), p. 16.

91. Given the low rate of success, more than one egg would be employed in any cloning attempt.

92. Lifton (1988), p. 17.

93. Brinich, P M (1980). Potential effects of adoption on self and object representation, *Psychoanalytic Study of the Child* 35: pp. 107–34.

94. Schecter, M D, and Bertocci, D (1990). The meaning of the search, in Brodzinsky and Schecter.

95. Lifton (1994), p. 147.

96. Lifton (1994), p. 162.

97. Lifton (1994), p. 162.

98. Lifton (1994), pp. 211–15.

99. Lifton (1994), pp. 241–60.

100. Hollingsworth (2000).

101. Dawkins, R (1989 [1976]). *The selfish gene* (New York: Oxford University Press), pp. 101–2.

102. Barash, D P (2001). *Revolutionary biology: The new gene-centered view of life* (New Brunswick, N.J.: Transaction Publishers), p. 150.

103. Brodzinsky, Smith, and Brodzinsky (1988), p. 96; Haimes, E (1987). Now I know who I really am: Identity change and redefinitions of self in adoption, in *Self and identity: Perspectives across the lifespan*, ed. Honess, T, and Yardley, K (Boston: Routledge and Kegan Paul).

104. Brodzinsky, Smith, and Brodzinsky (1998), p. 95.

105. Clans of clones of the same progenitor might arise and prefer their own company.

106. Lifton (1988), p. 24.

107. In assessing parents, it is important to focus on their motivation for adoption, their attitudes about disclosure, and what they have to say about the adopted child's curiosity about his origins and whether the family is receiving social support in adoption (Brodzinsky, Smith, and Brodzinsky [1998], p. 90). In assessing adopted children, it is critical to explore the child's knowledge of and feelings about adoption, as well as family communication and feedback from others about it (Brodzinsky, Smith, and Brodzinsky [1998], p. 95).

CHAPTER 5

1. As explained earlier, genetic identity would be complete only in those cases in which a woman clones herself, using one of her enucleated eggs as a receptacle

for a transferred somatic cell nucleus. All other instances of cloning would dilute the genetic identity very slightly by the mitochondrial DNA of the egg donor.

2. Brodzinsky, D M, Smith, D W, and Brodzinsky, A B (1998). Children's adjustment to adoption: Developmental and clinical issues, *Developmental Clinical Psychology and Psychiatry* 38: p. 11.

3. Gaylord, C L (1976). The adoptive child's right to know, *Case and Comment* 81: pp. 38–44. For example, it would be contrary to nature for eunuchs to father a child, and so they were not permitted to adopt.

4. Grotevant, H D, and Kohler, J K (1999). Adoptive families, in *Parenting and child development in "nontraditional" families*, ed. Lamb, M E (Mahwah, N.J.: Lawrence Erlbaum Associates), pp. 161–90.

5. Modell, J, and Dambacher, N (1997). Making a "real" family: Matching cultural biologism in American adoption, *Adoption Quarterly* 1: pp. 3–33.

6. Grotevant and Kohler (1999), p. 179.

7. As a corollary, we may suppose that the more an individual offspring resembles its parents, the more likely it is to possess those traits that furthered its parents' adaptive advantages.

8. Hamilton, W D (1964). The genetical evolution of social behavior, *Journal of Theoretical Biology* 7: pp. 1–52.

9. Erickson, M T (2000). Oedipus and the psychopathologies of kinship, in *Genes on the couch: Explorations in evolutionary psychology*, ed. Gilbert, P, and Bailey, K G (East Sussex, England: Brunner-Routledge), p. 215.

10. Dawkins, R (1989 [1976]). *The selfish gene* (New York: Oxford University Press), pp. 99–100, emphasis added.

11. Barash, D P (2001). *Revolutionary biology: The new gene-centered view of life* (New Brunswick, N.J.: Transaction Publishers).

12. Christenfeld, N J, and Hill, E A (1995). Whose baby are you? (letter), *Nature* 378: p. 669; McLain, D K, Setters, D, Moulton, M P, and Pratt, A E (2000). Ascription of resemblance of newborns by parents and nonrelatives, *Evolution and Human Behavior* 21: pp. 11–23.

13. Daly, M, and Wilson, M (1982). Whom are newborn babies said to resemble? *Ethology and Sociobiology* 3: pp. 69–78; Regalski, J M, and Gaulin, S J (1993). Whom are Mexican infants said to resemble? Monitoring and fostering parental confidence in the Yucatan, *Ethology and Sociobiology* 14: pp. 97–113.

14. Daly, M, and Wilson, M (1981). Abuse and neglect of children in evolutionary perspective, in *Natural selection and social behavior*, ed. Alexander, R D, and Tinkle, D W (New York: Chiron).

15. Burch, R J, and Gallup, G, Jr. (2000). Perceptions of paternal resemblance predict family violence, *Evolution and Human Behavior* 21: pp. 429–35.

16. Barash (2001), p. 82. Some philosophers of mind might object here. I am not claiming that there are not powerful biological correlates to mental and social experience but, rather, that reducing it all to biology cannot give us a deep understanding of sentient beings.

17. The term *identification* is often used to "refer to all the mental processes by which an individual becomes like another in one or several aspects." As such, it is one type of internalization. The term *internalization* is defined as "the means by which aspects of need-gratifying relationships and functions provided for one individual by another are preserved by making them part of the self." The terms *primary identification* and *incorporation* refer to developmentally primitive levels

of internalization where there are not clear boundaries between self and other. *Secondary identification* and *introjection* describe higher developmental levels of internalization and identification where self and other are more clearly differentiated as separate from each other. See Moore, B E, and Fine, B D (1990). *Psychoanalytic terms and concepts* (New Haven: Yale University Press), pp. 102–3. For the purposes of this discussion, I will not be distinguishing among these types but will be using the term *identification* in an ordinary language way. If one wishes, one may infer the type of identification from the context.

18. Rubin, R (1984). *Maternal identity and the maternal experience* (New York: Springer Publishing Co.).

19. Parnell, R W (1961). Commentary on Sandler, J. (ed.) *Identification in children, parents and doctors*, in *Psychosomatic aspects of paediatrics*, ed. MacKeith, R, and Sandler, J (New York: Pergamon), p. 23.

20. Sandler, J (1961). Commentary on Sandler, J (ed.) *Identification in children, parents and doctors*, in MacKeith and Sandler, p. 25.

21. Platek, S M, Critton, S R, Burch, R L, Frederick, D A, Myers, T E, and Gallup, G G, Jr. (2003). How much paternal resemblance is enough? Sex differences in hypothetical investment decisions but not in the detection of resemblance, *Evolution and Human Behavior* 24: pp. 81–87.

22. Josselson, R (1987). *Finding herself: Pathways to identity development in women* (San Francisco: Jossey-Bass), p. 171, emphasis added.

23. Josselson (1987), p. 170.

24. Adam Phillips thinks that adolescents tend to be fascinated by cloning because it impinges so directly on this conflict. See Phillips, A (1998). Sameness is all, in *Clones and clones: Facts and fantasies about human cloning*, ed. Nussbaum, M C, and Sunstein, C R (New York: W. W. Norton).

25. Dawkins (1998), p. 56.

26. Tooley, M (1998). The moral status of cloning in humans, in *Human cloning*, ed. Humber, J M, and Almeder, R F (Towota, N.J.: Humana Press), pp. 67–101; Segal, N L (2002). Human cloning: Insights from twins and twin research, *Hastings Law Journal* 53: pp. 1073–84.

27. Grotevant and Kohler (1999), p. 168.

28. One could argue that the cloned offspring of a homosexual male might prove to be healthier psychologically than that of a heterosexual. Heterosexual couples can choose to sexually reproduce or, if infertile, to employ one of the several assisted reproductive technologies described in chapter 2; and lesbian couples can agree on one of them becoming pregnant through one of those technologies. However, lacking even one uterus between them, homosexual male couples have fewer alternatives for becoming biological parents. Aside from enlisting a woman to become pregnant with the sperm of one of them, cloning would be their only alternative. Because fewer alternatives exist for genetic parenthood for homosexual male couples, as a group they might be less inclined to seek cloning for reasons that would put the cloned child at psychological risk.

29. Kendler, K S, Thornton, L M, Gillman, S E, and Kessler, R C (2000). Sexual orientation in a U.S. national sample of twin and nontwin sibling pairs, *American Journal of Psychiatry* 157: pp. 1843–46.

30. Bailey, J M, and Pillard, R C (1991). A genetic study of male sexual orientation, *Archives of General Psychiatry* 48: pp. 1089–96.

31. Hershberger, S L (1997). A twin registry study of male and female sexual orientation, *Journal of Sex Research* 34: pp. 212–22.

32. Hershberger (1997); Bailey and Pillard (1991).

33. Blanchard, R, and Bogaert, A F (1997). Additive effects of older brothers and homosexual brothers in the prediction of marriage and cohabitation, *Behavior Genetics* 27: pp. 45–54.

34. Miller, A (1997). *The drama of the gifted child: The search for the true self* (New York: Basic Books), p. 6.

35. Freud, S (1914). On narcissism, in *The standard edition of the complete psychological works of Sigmund Freud,* vol. 14 (1974), ed. Strachey, J (London: Hogarth Press), pp. 69–102.

36. Kernberg, O F (1967). Borderline personality organization, *Journal of the American Psychoanalytic Association* 15: pp. 641–85; Kernberg, O F (1975). *Borderline conditions and pathological narcissism* (New York: Jason Aronson); Kohut, H (1971). *The analysis of the self: A systematic approach to the psychoanalytic treatment of narcissistic personality disorders* (New York: International Universities Press); Kohut, H (1977). *The restoration of the self* (New York: International Universities Press).

37. Pulver, S (1986). Narcissism: The term and the concept, in *Essential papers on narcissism,* ed. Morrison, A P (New York: New York University Press), pp. 91–111.

38. Pulver (1986).

39. American Psychiatric Association (2000). *Diagnostic and statistical manual, IV-TR* (Washington, D.C.: American Psychiatric Press). The criteria that best distinguish this disorder from other personality disorders are entitlement, exploitation, and excessive need for admiration and attention. See Nurnberg, H G, Martin, G A, and Pollack, S (1994). An empirical method to refine personality disorder classification using stepwise logistic regression modeling to develop diagnostic criteria and thresholds, *Comprehensive Psychiatry* 35: pp. 409–10.

40. Horowitz, M J (1989). Clinical phenomenology of narcissistic pathology, *Psychiatric Clinics of North America* 12: pp. 531–39.

41. Kernberg, O F (1989). An ego psychology object relations theory of the structure and treatment of pathologic narcissism, *Psychiatric Clinics of North America* 12: pp. 732–39.

42. Bach, S (1977). On the narcissistic state of consciousness, *International Journal of Psychoanalysis* 58: pp. 209–33.

43. Psychotherapists distinguish pathologies of self from anxiety, mood, and psychotic disorders. Where the psychopathology is primarily in the self, the problem lies mainly in the construction of the core of the personality. Narcissistic Personality Disorder and Borderline Personality Disorder are the best examples. People with disorders of the self may and frequently do also have anxiety and mood disorders, but treating those disorders without addressing the pathology of the self cannot be expected to result in a fully satisfactory recovery. For example, the depressed patient with one of these disorders, after successful treatment with antidepressant medication, may go from being depressed to feeling "empty."

44. Mazzano, J, Palacio, E F, and Zikha, G (1999). The narcissistic scenarios of parenthood, *International Journal of Psychoanalysis* 80: pp. 465–76.

45. Goldberg, A (1989). Self psychology and the narcissistic personality disorders, *Psychiatric Clinics of North America* 12: pp. 731–39.

46. Recent research distinguishes between "overt" and "covert" narcissists. Covert narcissists tend to be unhappy, but overt narcissists, in contradiction to clinical theory and experience, are truly pleased with themselves. See Rose, P (2002). The happy and unhappy faces of narcissism, *Personality and Individual Differences* 33: pp. 377–91. Let us take this study with a grain of salt, as we could not expect the paper-and-pencil assessments it utilized to tap deeper levels of consciousness. But regardless of its validity, a child parented by a narcissist, whether overt or covert, is at risk for a similarly damaging style of parenting.

47. Rinsley, D B (1980). Diagnosis and treatment of borderline and narcissistic children and adolescents, *Bulletin of the Menninger Clinic* 44: pp. 147–70.

48. Rinsley (1980).

49. Winnicott, D W (1960). Ego distortion in terms of the true and false self, in *The maturational processes and the facilitating environment* (1965) (London: Hogarth Press), pp. 140–52.

50. Miller (1997), p. 28.

51. Miller (1997), p. 9.

52. Miller (1997), pp. 21, 10.

53. "Selfobjects" are representations of people in the child's world, which the child invests with his own narcissism. The idealized parent is the paradigmatic selfobject. Selfobjects are created during normal childhood development to help maintain the child's narcissistic equilibrium and "grandiose self." With development, the child progressively develops realistic self-esteem and pride in his accomplishments as he becomes aware of his own limitations and is able to tolerate loss and disappointment. At the same time, the idealized parent becomes a more realistic "ego ideal" or desired model for the self, which is gradually internalized. See Kohut, H (1977). *The restoration of the self* (New York: International Universities Press).

54. Miller (1997), p. 30.

55. Goldberg (1989), p. 738.

56. Miller (1997), p. 31.

57. Rinsley (1989), pp. 702–3.

58. If internal unconscious psychological pressures are strong enough, then a person can be of a different age or gender than the original source of the transference feelings and still be the object of a transference enactment.

59. Kernberg, P (1989). Narcissistic Personality Disorder in childhood, *Psychiatric Clinics of North America* 12: pp. 672, 674.

60. Kernberg, P (1989), p. 676, emphasis added.

61. Goldberg (1989).

62. Goldberg (1989).

63. Narcissistic people may view themselves as "heroes" when in states of self-righteous rage—entitled, strong, and justified in destroying those whom they perceive as monsters. If they think of themselves as weak, then they will put someone else in the role of the hero, attaching themselves to that person in an idealizing transference. See Horowitz (1989), pp. 535–36. Cloning him- or herself could be an attempt by a narcissistic person to enhance "hero" status. Narcissists might wish to clone an idealized other (celebrity or famous person) so

that as its rearing parents, they could attach themselves to their hero in an idealizing transference. In both of these scenarios, the parent would not relate to the child clone as a person with a unique self, and the clone would fail to live up to expectations.

64. See Horowitz (1989); Kernberg, O F (1989); and Goldberg (1989), p. 733.

65. Paulina Kernberg offered the following as a transference interpretation to a patient manifesting a twinship transference: "Now you and I must think the same and feel the same, even look the same, because any difference would be painful as it might mean I have something you do not. Than you may feel you do not control me, or other people for that matter" ([1989], p. 691).

66. Goldberg (1989).

67. Burlingham, D (1945). The fantasy of having a twin, *Psychoanalytic Study of the Child* 1: pp. 205–10.

68. See, e.g., Poe, E A (1966). William Wilson, in *Complete stories and poems of Edgar Allan Poe* (Garden City, N.Y.: Doubleday and Co.), pp. 156–70.

69. Ogden, T H (1982). *Projective identification and psychotherapeutic technique* (New York: Jason Aronson), p. 12.

70. Crisp, P (1987). Uncontained projective identification: The vicious circle of runaway positive feedback loops, *Psychoanalytic Psychology* 4: pp. 291–99.

71. Shengold, L (1989). *Soul murder: The effects of childhood abuse and deprivation* (New Haven: Yale University Press), p. 19.

72. Shengold (1989), p. 2.

73. Rinsley (1980), p. 154.

74. Shengold, L (1978). Assault on a child's individuality: A kind of soul murder, *Psychoanalytic Quarterly* 47: pp. 419–24.

75. Kernberg, P (1989), p. 679.

76. Shengold (1978), p. 423.

77. Shengold (1978), pp. 422–23.

78. Having an aloof narcissistic mother may mitigate the effects of her narcissistic pathology, but aloofness in a parent may foster other emotional problems in the child. See Kernberg, P (1989), p. 683.

79. Bloland, S E (1999). Fame: The power and cost of fantasy, *Atlantic Monthly* 284: pp. 51–62.

80. Kernberg, O F (1975). *Borderline conditions and pathological narcissism* (New York: Jason Aronson), p. 17.

81. Hans Jonas has expressed concern that parents selecting a clone with the expectation that the child would share its progenitor's desired traits would feel greatly entitled to get what they paid for. See Jonas, H (1974). Biological engineering—A preview, in *Philosophical essays: From ancient creed to technological man*, ed. Jonas, H (Englewood Cliffs, N.J.: Prentice-Hall), pp. 141–67.

82. Kernberg (1975), p. 228.

83. Kernberg, O F (1980). Pathological narcissism in middle age, in *Internal world and external reality: Object relations theory applied* (New York: Jason Aronson), pp. 133–40.

84. In Dissociative Identity Disorder, the dissociated part or parts can assume executive control and "come out." People with this disorder typically experience "lost time" when this happens and may find evidence that they have said, done, or written things of which they have no recollection.

85. Adam Phillips uses the term *psychic clone* to describe transference representations. See Phillips (1998), pp. 88–94. I am using the term in a more limited sense, applied to dissociated aspects of the mind, representing either the self or others, but more active and autonomous than is the case in typical instances of transference.

86. Though Barry apparently did not, some such adolescents may attempt to reestablish the "pathologic binding interaction" by acting out in some way. See Kernberg, P (1989), pp. 688–89. The narcissistic progenitor/parent of a clone and the clone might be especially prone to this sort of interaction.

87. Kernberg, P (1989), p. 679.

88. Kernberg, P (1989), p. 679.

89. Richard Dawkins seems to be referring to the ego ideal when he says: "Mightn't it be wonderful to advise your junior copy on where you went wrong, and how to do it better?" ([1998], p. 55). A progenitor could take a benevolent and empathic perspective, but the advice offered a clone might be delivered more as an imperative—and experienced like one. If the progenitor views the clone as a junior "me," then his sense of parental authority and entitlement may feel close to absolute. I knew that my own son at age eight was determined not to be my "clone" when his requests for advice made it clear that I was not him and he was not me. He would preface his questions with a phrase parallel to, but importantly different from, the "If I were you . . ." preface to advice I would sometimes offer him and would ask me, "Daddy, if you were you . . . ?"

90. Kernberg, P (1989), p. 678.

91. Kernberg, P (1989), p. 688.

92. Tooley (1998), pp. 89–90.

93. Parnell (1961).

94. The idea that heightened parent–child resemblance might have problematic dimensions for a child makes sense clinically but apparently has never been researched.

CHAPTER 6

1. David Giles has provided a three-dimensional "taxonomy" of fame comprising three dimensions: type, domain, and trajectory. The fame achieved by public figures, show business "stars," accident, and through merit are all different types. Levels of fame constitute the second dimension. The fame may be domain specific: fame in the local community, national fame, and international fame. Fame trajectories constitute the third dimension. See Giles, D (1999). *Illusions of immortality: A psychology of fame and celebrity* (New York: St. Martin's Press). Sam, Jerry's father, was famous by merit and by becoming a public figure. His fame was domain specific and local, and the trajectory did not fall before his death.

2. Hopefully, public curiosity would not be as intrusive as it was for the Dion quintuplets. And first, of course, it would need to be proven that the child is, in fact, a clone by meeting demands for the required scientific evidence. See Schatten, G, Prather, R, and Wilmut, J (2003). Cloning claim is science fiction, not science (letter), *Science* 299: p. 344.

3. Singer, P, and Wells, D (1984). *The reproductive revolution: New ways of making babies* (New York: Oxford University Press), p. 161.

4. Jamison, K R (1993). *Touched with fire: Manic-depressive illness and the artistic temperament* (New York: Free Press).

5. Simonton, D K (1994). *Greatness: Who makes history and why* (New York: Guilford Press).

6. Giles (1999), p. 47.

7. Giles (1999), p. 44.

8. Bloland, S E (1999). Fame: The power and cost of fantasy, *Atlantic Monthly* 284: p. 62.

9. Bloland (1999), p. 57.

10. Lifton, B J (1998 [1979]). *Lost and found: The adoption experience* (New York: Harper and Row), p. 31.

11. Giles (1999), pp. 142–45.

12. Lifton (1998), p. 28.

13. It may be worth looking at parents of child prodigies and child stars as a model for those wishing to parent clones of the famous.

14. Knafo, D (1991). What's in a name? Psychoanalytic considerations on children of famous parents, *Psychoanalytic Psychology* 8: pp. 263–81.

15. Knafo (1991).

16. Knafo (1991).

17. Knafo (1991).

18. Such clones might have an overdetermined "institutional transference."

19. Hart, D, Maloney, J, and Damon, W (1987). The meaning and development of identity, in *Self and identity: Perspectives across the lifespan*, ed. Honess, T, and Yardley, L (New York: Routledge and Kegan Paul), p. 122.

20. Knafo (1991).

21. Bloland (1999), p. 51.

22. Bloland (1999), p. 52.

23. Freud, M (1958). *Sigmund Freud, man and father* (New York: Vanguard Press), p. 9.

24. Knafo (1991).

25. I use the term *ideal form* intentionally. For Plato, forms cannot exist in reality. The famous progenitor, while real, through fame has attained a socially metaphysical level of existence. Any clones would understandably be destined to be experienced as base physical attempts to copy the ideal metaphysical form. Any progenitor, famous or not, is likely to be selected as a progenitor because he is thought to be an ideal, though perhaps only in the mind of the rearing parents. We will examine this further in the Replacement Child Model.

26. Winnicott, D W (1960). Ego distortion in terms of the true and false self, in *The maturational processes and the facilitating environment* (1965) (London: Hogarth Press), pp. 140–52.

27. Mitchell, G, and Cronson, H D (1987). The celebrity family: A clinical perspective, *The American Journal of Family Therapy* 15: p. 239.

28. Mitchell and Cronson (1987), p. 240. This confirms work in Pittman, F S (1985). Children of the rich, *Family Process* 24: pp. 679–88.

29. Freud, S (1916). Those wrecked by success, in *The standard edition of the complete psychological works of Sigmund Freud*, vol. 14 (1974), ed. Strachey, J (London: Hogarth Press), pp. 316–31.

30. Knafo (1991).

31. Jacobson, E (1959). The "exceptions": An elaboration of Freud's character study, *Psychoanalytic Study of the Child* 14: pp. 135–54.

32. Giles (1999), pp. 81–82.

33. Knafo (1991).

34. For this very reason, identical "sibling" clones of the same progenitor might want to seek each other out, much as identical twins separated at birth may.

35. Winnicott (1960).

36. Knafo (1991).

37. Mitchell and Cronson (1987).

38. Angell, K (2000). Fruit from the literary tree: Young authors find their parents' fame a boon and a bane, *New York Times*, April 25: p. E1.

39. Angell (2000).

40. Raines, H (2000). The presidential campaign as oedipal arena (editorial), *New York Times*, June 28: p. A20.

41. Kristoff, N D (2000). A father's footsteps echo through a son's career, *New York Times*, September 11: p. A11.

42. Associated Press (2002). Sons of stars search for identity of their own, *New York Times*, June 14: p. D4.

43. Horton, D, and Wohl, R R (1956). Mass communication and parasocial interaction, *Psychiatry* 19: pp. 215–19.

44. Giles (1999), p. 95.

45. Giles (1999), pp. 142–45.

46. Nimoy, L (1975). *I am not Spock* (New York: Ballantine).

47. Giles (1999), p. 95.

48. Adler, P A, and Adler, P (1989). The gloried self: The aggrandizement and the constriction of the self, *Social Psychology Quarterly* 52: pp. 279–310.

49. Gidden, A (1991). *Modernity and self identity: Self and society in the late modern age* (Cambridge, England: Polity Press).

50. This is a facet of cloning objected to in Rifkin, J (1983). *Algeny* (New York: Viking Press).

51. Knafo (1991).

52. Sims, C (2000). Hard legacy for Japan sect leader's family: No easy time for the children of the man famous for sarin gas attack, *New York Times*, September 1: p. A3.

53. Knafo (1991).

54. Knafo (1991), p. 271.

55. Rank, O (1998 [1930]). *Psychology and the soul: A study of the origin, conceptual evolution, and nature of the soul*, trans. Richter, G C, and Lieberman, E J (Baltimore: Johns Hopkins University Press), p. 54.

56. Shengold, L (1978). Assault on a child's individuality: A kind of soul murder, *Psychoanalytic Quarterly* 47: pp. 419–24; Shengold, L (1989). *Soul murder: The effects of childhood abuse and deprivation* (New Haven: Yale University Press).

CHAPTER 7

1. Alexander, B (2001). You again, *Wired*, February: pp. 122–35.

2. Perinatal death is medically defined as loss between the twentieth week of gestation and the twenty-seventh day of life, but it is not so demarcated psychologically.

3. Leon, I G (1990). *When a baby dies: Psychotherapy for pregnancy and newborn loss* (New Haven: Yale University Press), p. 24.

4. Leon (1990), pp. 13–19.

5. Klass, D (1997). The deceased child in the psychic and social worlds of bereaved parents during the resolution of grief, *Death Studies* 21: p. 150.

6. As we have seen in the Assisted Reproductive Technologies and Arrangements Model, fetal ultrasound images may intensify these.

7. Zeenah, C H, Keener, M, and Anders, T (1986). Developing perceptions of temperament and their relation to mother and infant behavior, *Journal of Child Psychology and Psychiatry and Allied Disciplines* 27: pp. 499–512.

8. Zeenah, C H, Zeenah, P D, and Stewart, L K (1990). Parent's constructions of their infants' personalities before and after birth: A descriptive study, *Child Psychiatry and Human Development* 20: pp. 191–206.

9. Grout, L A, and Romanoff, B D (2000). The myth of the replacement child: Parents' stories and practices after perinatal death, *Death Studies* 24: pp. 93–113.

10. Kubler-Ross, E (1969). *On death and dying* (New York: MacMillan).

11. Freud, S (1917 [1915]). Mourning and melancholia, in *The standard edition of the complete psychological works of Sigmund Freud*, vol. 13 (1974), ed. Strachey, J (London: Hogarth Press), pp. 237–58.

12. Pollock, G H (1961). Mourning and adaptation, *International Journal of Psychoanalysis* 42: pp. 341–46.

13. Cain, A C, and Cain, B S (1962). On replacing a child, *Journal of the Academy of Child Psychiatry* 1: p. 452.

14. Grout and Romanoff (2000).

15. Grout and Romanoff (2000), p. 95.

16. Davis, D L, Stewart, M, and Harmon, R J (1989). Postponing pregnancy after perinatal death: Perspectives on doctor advice, *Child and Adolescent Psychiatry* 28: pp. 481–87.

17. Klass (1997), p. 164.

18. Klass (1997), p. 156.

19. Klass (1997), pp. 147–48.

20. Klass (1997), pp. 159–60.

21. Klass (1997), p. 161.

22. Klass (1997), p. 157.

23. Johnson, S E (1989). Replacement children, in *Preventive psychiatry: Early intervention and situational crisis management*, ed. Klagsbrun, S C, Kuman, G W, Clark, E J, Kutscher, A H, DeBellis, R, and Lambert, A L (Philadelphia: Charles Press), pp. 115–19.

24. Legg, C, and Sherrick, I (1976). The replacement child—A developmental tragedy: Some preliminary comments, *Child Psychiatry and Human Development* 7: p. 113.

25. Cain and Cain (1962), p. 453.

26. Cain and Cain (1962), p. 448.

27. Cain and Cain (1962).

28. Cain and Cain (1962).

29. Lundlin, T (1984). Morbidity following sudden and unexpected bereavement, *British Journal of Psychiatry* 144: pp. 84–85.

30. Johnson (1989).

31. Davis, Stewart, and Harmon (1989).

32. Johnson (1989), p. 116.

33. Johnson (1989), p. 117.

34. Grout and Romanoff (2000). For this reason we might predict that those wanting to clone a dead child would most likely be parents who lost a child beyond the perinatal period.

35. Volkan, V (1981). *Linking objects and linking phenomena: A study of the forms, symptoms, metapsychology, and therapy of complicated mourning* (New York: International Universities Press).

36. Bowlby, J (1980). *Attachment and loss, vol. 3: Loss: sadness and depression.* (New York: Basic Books); Klaus, M H, and Kennell, J H (1982). *Parent–infant bonding*, 2d edition (St. Louis: Mosby).

37. Davis, Stewart, and Harmon (1989).

38. Cain and Cain (1962), p. 444.

39. Cain and Cain (1962).

40. Cain and Cain (1962), p. 449.

41. Cain and Cain (1962).

42. Johnson (1987).

43. Johnson (1987).

44. Legg and Sherrick (1976).

45. Kogan, I (2003). On being a dead, beloved child, *Psychoanalytic Quarterly* 62: pp. 722–66.

46. Johnson (1987).

47. Cain and Cain (1962), p. 451.

48. Cain and Cain (1962), p. 451.

49. Sabbadini, A (1988). The replacement child, *Contemporary Psychoanalysis* 24: p. 531.

50. Ainsfeld, L, and Richards, A D (2000). The replacement child: Variations on a theme in history and psychoanalysis, *The Psychoanalytic Study of the Child* 55: p. 310.

51. Legg and Sherrick (1976), p. 118.

52. Legg and Sherrick (1976), p. 123.

53. Winnicott, D W (1965). The capacity to be alone, in *The maturational processes and the facilitating environment* (London: Hogarth Press), pp. 29–36.

54. Sabbadini (1988), p. 544.

55. I am indebted to Homer Curtis, M.D., for discussions regarding this case and suggesting this term.

56. Leon (1990), pp. 10–11.

57. Stack, I (1987). Prenatal psychotherapy and maternal transference to the fetus, *Infant Mental Health Journal* 54: pp. 162–67.

58. Benedek, T (1970). The psychobiology of pregnancy, in *Parenthood: Its psychology and psychopathology*, ed. Anthony, E J, and Benedek, T (Boston: Little, Brown and Co.), pp. 137–55; Raphael-Leff, I (1983). Facilitators and regulators: Two approaches to mothering, *British Journal of Medical Psychology* 59: pp. 43–53.

59. Ballou, J (1978). The significance of reconciliative themes in the psychology of pregnancy, *Bulletin of the Menninger Clinic* 42: pp. 383–413.

60. Ainsfeld and Richards (2000); Lifton, B J (1988 [1979]), *Lost and found: The adoption experience* (New York: Harper and Row), p. 36.

61. Phillips, A (1998). Sameness is all, in *Clones and clones: Facts and fantasies about human cloning*, ed. Nussbaum, M C, and Sunstein, C R (New York: W. W. Norton), pp. 88–94.

62. Brentano, F (1874). The distinction between mental and physical phenomena, in *Introduction to the philosophy of mind* (1970), ed. Morick, H (Glenview, Ill.: Scott, Foresman, and Co.), pp. 109–29.

63. Sabbadini (1988), p. 530.

64. Blumer, D (2002). The illness of Vincent Van Gogh, *American Journal of Psychiatry* 159: pp. 519–26.

65. Meissner, W W (1995). In the shadow of death, *Psychoanalytic Review* 82: pp. 551–52.

66. Meissner (1995).

67. Erikson, E (1956). The problem of ego identity, *Journal of the American Psychoanalytic Association* 4: p. 87.

68. Meissner (1995), p. 557.

69. Chen, D W (2000). See Spot run, or is that Spot's clone? Owners look to science to resurrect family pet, *New York Times*, December 9: p. B1. A step toward this goal, the cloning of a cat, has already been achieved. See Shin, T, Kraemer, D, Pryor, J, Liu, L, Rugila, J, Howe, L, Buck, S, Murphy, K, Lyons, L, and Westhusin, M (2002). A cat cloned by nuclear transplantation, *Nature* 415: p. 859.

CHAPTER 8

1. Alford, R (1988). *Naming and identity: A cross-cultural study of personal naming practices* (New Haven: HRAF Press), p. 141.

2. Hymer, S M (1985). What's in a name? *Dynamic Psychotherapy* 3: pp. 186–97.

3. Joubert, C S (1993). Personal names as a psychological variable, *Psychological Reports* 73: pp. 1123–45; Dion, K L (1983). Name, identity, and self, *Names* 3: pp. 245–57.

4. Allport, G (1961). *Pattern and growth in personality* (New York: Holt, Rinehart and Winston), p. 117.

5. Nagera, H (1967). *Vincent van Gogh: A psychological study* (London: George Allen and Unwin, Ltd.), p. 14.

6. Nagera (1967), p. 13.

7. Meissner, W W (1995). In the shadow of death, *Psychoanalytic Review* 82: pp. 535–57.

8. Sabbadini, A (1988). The replacement child, *Contemporary Psychoanalysis* 24: p. 545.

9. Smith, D S (1977). Child-naming practices and family structure change: Hingham, Massachusetts 1640–1880, the Newberry Papers in Family and Community History, Paper 76-5, Newberry Library, Chicago.

10. Smith (1977), p. 6.

11. Legg, C, and Sherick, I (1976). The replacement child—A developmental tragedy: Some preliminary comments, *Child Psychology and Human Development* 7: p. 115.

12. In the language of cultural anthropology, the context is both synchronic and diachronic.

13. Woodward, J (1988). The bereaved twin, *Acta Geneticae Medicae et Gemellologiae* 37: pp. 173–80.

14. Woodward (1988).

15. Gould, D J (1998). Dolly's fashion and Louis' passion, in *Clones and clones: Facts and fantasies about human cloning*, ed. Nussbaum, M C, and Sunstein, C R (New York: W. W. Norton), pp. 41–53.

16. Plank, R (1964). Names of twins, *Names* 12: pp. 1–5.

17. Alford (1988), p. 114. If the analogy between namesakes and clones is valid, then a corollary analogy between namesake taboos and restrictions on cloning may be as well. It would suggest a parallel between a namesake taboo without qualification or distinction and legislation completely prohibiting cloning. Namesake taboos that are not absolute would parallel the notion of legislation limiting and regulating cloning.

18. Alford (1988), p. 117.

19. Alford (1988), p. 117.

20. Alford (1988), p. 74.

21. Piaget, J (1963). *The child's conception of the world* (Patterson, N.J.: Littlefield, Adams).

22. Von Domarus, E (1944). The specific laws of logic in schizophrenia, in *Language and thought in schizophrenia: Collected papers*, ed. Kasanin, J S (Berkeley: University of California Press), pp. 104–14; Arieti, S (1974). *Interpretation of schizophrenia*, 2d edition (New York: Basic Books), pp. 229–35.

23. Freud, S (1900–1). The interpretation of dreams and On dreams, in *The standard edition of the complete psychological works of Sigmund Freud*, vol. 5 (1974), ed. Strachey, J (London: Hogarth Press).

24. See Rickey, C (2002). Alan Berliner, meet Alan Berliner, et al., *Philadelphia Inquirer*, January 4: pp. W2–W3, concerning the movie *The Sweetest Sound* and the phenomenon of "egosurfing" to find unrelated namesakes.

25. Geumple, D L (1965). Sunik: Name sharing as a factor governing Eskimo kinship terms, *Ethnology* 4: pp. 323–35; Seeman, M V (1980). Name and identity, *Canadian Journal of Psychiatry* 25: pp. 125–37.

26. Alford (1988), p. 78.

27. Alford (1988), p. 138.

28. Alford (1988), p. 133.

29. Alford (1988), p. 133; See also McAndrew, F T, King, J C, and Honoroff, L R (2002). A sociobiological analysis of namesaking patterns in 322 American families, *Journal of Applied Social Psychology* 32: pp. 851–54.

30. Alford (1988), p. 133.

31. Alford (1988), p. 141.

32. Rick brought to my attention that, especially in the southern and western United States, "Trey" is a nickname sometimes given to people bearing the suffix "III." Among others, the founder and chairman of Microsoft, Bill Gates, is a "IIId" and had "Trey" as a nickname. I have not discovered a nickname given to people with the suffix "IV."

33. Until I met Rick and found out that he was a "IVth," my only experience with numbers following a person's name had been with a temporary personal assignment. The year I was born, Steve, Stephen, and Steven were very popular names. In fact, there were four of us in my elementary school class. On the first day of school, the teacher assigned us a numerical suffix, based on the alphabetical order of our names. That made me "Steve 2." I resented not being "Steve 1" and wished for a more unique, though not unusual, name. In an analogy to cloning, multiple clones of the same progenitor, whether or not given his name with a numerical suffix, may be at risk for resenting, if not being a clone, then not being clone number one.

34. A person who is a namesake is more likely to namesake his or her child. See McAndrew, King, and Honoroff (2002).

35. Smith (1977).

36. Smith (1977).

37. Alford (1988), p. 133.

38. Taylor, R D (1974). John Doe, Jr.: A study of his distribution in space, time, and the social structure, *Social Forces* 53: p. 14.

39. Smith (1977).

40. Alford (1988), pp. 132–33.

41. Taylor (1974), p. 19, referring to a concept originated by Weber.

42. Taylor (1974).

43. Smith (1977), p. 10.

44. A clone might choose to have himself cloned, or, if his progenitor was still alive or suitably saved tissue samples were available, he might have his progenitor cloned anew. In theory, either approach would yield the same result, but so far, no one has reported cloning an animal from a clone.

45. Smith (1977); Taylor (1974).

46. Taylor (1974).

47. Alford (1987), p. 127.

48. Smith (1977), pp. 9–10, citing Massachusetts Secretary of the Commonwealth (1885). *List of persons whose names have been changed in Massachusetts* (Boston: Massachusetts Secretary of the Commonwealth).

49. Dralle, P W, and Mackiewicz, K (1981). Psychological impact of women's name change at marriage: Literature review and implications for further study, *American Journal of Family Therapy* 9: pp. 50–55. Naming patterns for alter personalities in Dissociative Identity Disorder may provide clues about the psychology of namesaking. In addition to some alters having no name and others with names quite different from the one by which the individual is typically known, frequently the names often resemble the name of the host personality, usually as diminutive or juvenile forms. Despite more than one alter having the same name, so far, I have encountered none with a "Jr." or numeral suffix.

50. Smith (1977), p. 4, argues that when it comes to naming, "non-occurrences can be as significant as actual events." The closest thing I have encountered to someone making himself a parental namesake is illustrated by Barry, whom we have met in the chapter on parent–child resemblance. His father gave his own middle name to both Barry and another son. A Catholic, Barry had to select a Confirmation name when he was twelve years old, and he chose his father's first name. In therapy, Barry said he did not know why he had done this, given his negative feelings about his father. The dissociated part of his mind, modeled after his father, clarified the mystery. This "psychic clone" of his father had already identified itself with Barry's father's name but indicated that it had also wanted Barry to receive it as a Confirmation name to try to please his father and be more like him. Perhaps it might also have been an attempt to pull ahead in a competition with his brother to be his father's favorite. Both Barry and his brother shared their father's middle name.

51. Busse, T V, Busse, K, and Busse, M (1979). Identical first names for parent and child, *The Journal of Social Psychology* 107: pp. 293–94.

52. Plank, R (1971). The use of "Jr." in relation to psychiatric treatment, *Names* 19: pp. 132–36.

53. Cameron, C (1987). The trouble with Junior: Father-naming, child abuse, and delinquency, *Sociology and Social Research* 71: pp. 200–3.

54. Cameron (1987).

55. Cameron (1987).

56. Plank (1971).

57. Cameron (1987), p. 202.

58. Zweigenhaft, Z I, Hayes, K N, and Haagen, C H (1980). The social impact of names, *Journal of Social Psychology* 110: pp. 203–10.

59. Alford (1988), p. 133.

60. Zweigenhaft and colleagues failed to include standard deviations for each group. They admit that they did not have the data to take socioeconomic status into account. That should lead to caution in interpreting their compelling and provocative results on patterns of name suffix assignment, because Alford (1988), pp. 132–33, finds that kin naming is influenced by socioeconomic class standing and occupation and that firstborn males are more likely to receive kin names. From Alford's findings, if we draw an analogy between reproductive cloning and namesaking, we might reason that people of higher socioeconomic status not only would be more able to afford cloning but would most likely want to clone kin, including themselves.

61. Zweigenhaft, Hayes, and Haagen (1980), p. 209.

62. Zweigenhaft, Hayes, and Haagen (1980), p. 209, emphasis added.

63. Zweigenhaft, Hayes, and Haagen (1980). Higher socioeconomic status, likely in many of these families, may be a confounding factor in multigenerational namesaking.

64. Lawson, E D (1984). Personal names: 100 years of social science contributions, *Names* 32: pp. 45–73.

65. Angell, K (2000). Fruit from the literary tree: Young authors find their parents' fame a boon and a bane, *New York Times*, April 25: p. E1.

66. Angell (2000), p. E1.

67. Bruni, F (2000). The other Bush: A young George, P., on the political stage, *New York Times*, April 18: p. A22.

68. Lawson (1984), p. 46.

69. If famous people were to decide to offer tissue samples for use in producing clones for adoption, then they might want to consider, as part of the agreement, requiring that the adoptive parents not make the clone a partial or complete namesake. Perhaps this would decrease the attractiveness of adopting such clones to narcissists.

70. Rubin, R (1984). *Maternal identity and maternal experience* (New York: Springer Publishing Co.), p. 136.

71. Slaves were also named by their masters in an exercise of their dominion over them. See Burnard, T (2001). Slave naming patterns: Onamastics and the taxonomy of race in eighteenth-century Jamaica, *Journal of Interdisciplinary History* 31: pp. 325–46.

72. Johnson, J J, McAndrew, F T, and Harris, P B (1991). Sociobiology and the naming of adopted and natural children, *Ethology and Sociobiology* 12: pp. 365–75.

73. Furstenberg, F F, Jr., and Talvitie, K G (1980). Children's names and paternal claims, *Journal of Family Issues* 1: pp. 31–55.

74. Johnson, McAndrew, and Harris (1991).

75. Johnson, McAndrew, and Harris (1991).

76. So far, no research has examined whether narcissistic individuals are more likely to name a child for themselves.

77. Krone, C, and Harris, C C (1988). The importance of infant gender and family resemblance with the parents' perinatal bereavement process: Establishing personhood, *Journal of Perinatal Nursing* 2: pp. 1–11.

78. Lifton, B J (1994). *Journey of the adopted self: A quest for wholeness* (New York: Basic Books), p. 206.

79. Epstein, R A (1998). A rush to caution: Cloning human beings, in Nussbaum and Sunstein, pp. 262–69, argues that a progenitor's "genesake" should have enhanced confidence in his ability to succeed because he would know that he has the same genetic makeup as his presumably successful progenitor. Such a sanguine perspective is consistent with the data on parental namesakes bearing numerical suffixes but not with what has been found for "Jr." namesakes of a parent.

80. Dawkins, R (1989 [1976]). *The selfish gene* (New York: Oxford University Press), pp. 192–201, 322–31.

81. A self-clone would be as identical genetically as a namesake would be identical memically to the self. And a clone who is also a namesake of his progenitor would be doubly identical to him. For his progenitor, such a clone would be a double self-perpetuation—identical genetically as a self-clone and identical memically as a self-namesake.

CHAPTER 9

1. These are not the only possible models of reproductive cloning, though there is more psychological theory and research relevant to these than to other possible

models. Because cloning makes it possible for single people to become biological parents, an additional conceptually appealing model is the "Single Parents by Choice Model," but there is little relevant research or clinical theory in the area at this time. Some view single women choosing to become mothers as constituting a threat to the family, and this is also a concern raised about reproductive cloning by a single person of either sex. See Rosenthal, M (1990). Single women requesting artificial insemination by donor, in *Psychiatric aspects of reproductive technology*, ed. Stotland, N L (Washington, D.C.: American Psychiatric Press). One of the few bits of research relevant to such a model is the finding that a single parent is likely to adopt a child of the same sex as him- or herself. See Haugaard, J J, Palmer, M, and Wojslawowicz, J C (1999). Single parent adoptions, *Adoption Quarterly* 2: pp. 65–74.

There may be some parallels between possible motivations of a single adult choosing to become a parent and those of a woman in a heterosexual couple who stops using birth control without her husband's knowledge or agreement to try to become pregnant. Recall that mothers within couples who unilaterally decide to have another child after losing one are more likely to have a troubled replacement child, in contrast to a couple deciding jointly to do so. See Cain, A C, and Cain, B S (1962). On replacing a child, *Journal of the American Academy of Child Psychiatry* 1: pp. 443–56; and Johnson, S E (1989). Replacement children, in *Preventative psychiatry: Early intervention and situational crisis management*, ed. Klagsbrun, S C, and Kliman, G W (Philadelphia: Charles Press), pp. 115–19. Of course, we would need to put this concern in the context of other research showing that single women seeking donor insemination or adoption desire motherhood for reasons not unlike those of married women. See Rosenthal (1990).

2. Krueger, D W (1989). *Body self and psychological self: A developmental and clinical integration of disorders of the self* (New York: Bruner/Mazel), pp. 3–31.

3. Birth order would need to be controlled for as a separate variable.

4. Myth and literature illustrate powerful wishes to murder one's double. This has been nicely reviewed in Rank, O (1971 [1925]). *The double: A psychoanalytic study*, trans. Tucker, H, Jr. (New York: New American Library); and Donniger, W (1998). Sex and the mythological clone, in *Clones and clones: Facts and fantasies about human cloning*, ed. Nussbaum, M C, and Sunstein, C R (New York: W. W. Norton), pp. 114–38.

CHAPTER 10

1. Miller, W I (1998). Sheep, joking, and the uncanny, in *Clones and clones: Facts and fantasies about human cloning*, ed. Nussbaum, M C, and Sunstein, C R (New York: W. W. Norton), pp. 78–87.

2. Brock, D W (1998). Cloning human beings: An assessment of the ethical issues pro and con, in Nussbaum and Sunstein, pp. 141–64.

3. Aubry, E J (1999). The tyranny of fashion, available at http://salon.com (accessed June 25).

4. Kamm, F (1994). Moral problems in cloning embryos, *American Philosophical Association Newsletter: Philosophy and Medicine* 94: p. 91.

5. Grotstein, J S (1997). "The sins of the fathers . . . ": Human sacrifice and the inter- and transgenerational neurosis/psychosis, *International Journal of Psychotherapy* 2: pp. 11–30.

6. Rose, N (1996). *Inventing our selves: Psychology, power, and personhood* (Cambridge: Cambridge University Press), pp. 175–76, discussing the work of Kenneth and Mary Gergen. See Gergen, K J, and Gergen, M (1988). Narrative and self in relationship, in *Advances in experimental social psychology*, vol. 21, ed. Berkowitz, L (New York: Academic Press), pp. 17–56.

7. Alexander, B (2001). You again, *Wired*, February: p. 131.

8. Dawkins, R (1989 [1976]). *The selfish gene* (New York: Oxford University Press), p. 88.

9. Dawkins (1989), pp. 238–53.

10. Rose (1996), p. 173, explaining the ideas of Emil Benveniste on the subject-creating properties of personal pronouns. See Benveniste, E (1971). *Problems in general linguistics*, trans. Meek, M E (Coral Gables: University of Miami Press).

11. Segal, N L (2000). *Entwined lives: Twins and what they tell us about human behavior* (New York: Plume [Penguin Putnam]), p. 184.

12. Kaplan, L (1998). The concept of the family romance, *Psychoanalytic Review* 61: pp. 169–202.

13. Klass, D (1997). The deceased child in the psychic and social worlds of bereaved parents during the resolution of grief, *Death Studies* 21: pp. 147–75.

14. Ammon, G (1975). Death and identity, *The Human Context* 7: pp. 94–102.

15. Turk, D J (2002). Mike or me? Self-recognition in a split-brain patient, *Nature Neuroscience* 5: pp. 841–42.

16. Green, P, and Preston, M (1981). Reinforcement of vocal correlates of auditory hallucinations by auditory feedback: A case study, *British Journal of Psychiatry* 139: pp. 204–8.

17. DeBruine, L (2002). Facial resemblance enhances trust, *Proceedings of the Royal Society of London. B* 269: pp. 1307–12.

18. Collins, J K, and LaGanza, S (1982). Self-recognition of the face: A study of adolescent narcissism, *Journal of Youth and Adolescence* 11: pp. 317–28; Collins, J K, Harper, J F, and Cassel, A J (1976). Self-body recognition in late adolescence, *Australian Psychologist* 11: pp. 153–57. To my knowledge, self-recognition has not been investigated in people with pathologic narcissism. I would hypothesize that they would have both more true and more false self-recognitions.

19. See the grouping of a number of articles under the editor's heading in *Science* (2002). Reflections on self: Immunity and beyond, *Science* 296: pp. 297–316.

20. Foucault, M (1988). *Technologies of self: A seminar with Michael Foucault*, ed. Martin, L H, Gutman, H, and Hutton, P H (Amherst: University of Massachusetts Press), p. 18.

21. Rose (1996), p. 155.

22. Rose (1996), p. 6, reviewing Charles Taylor's historical notions of the self. See Taylor, C (1989). *Sources of the self: The making of modern identity* (Cambridge: Cambridge University Press).

23. Durkheim, É (1951). *Suicide* (Glencoe, Ill.: Free Press); Funk, A G, and Wise, G M (1989). Anomie, powerlessness, and exchange: Parallel sources of deviance,

Deviant Behavior 10: pp. 53–60; Powell, E H (1988). *The design of discord: Studies on anomie* (New Brunswick, N.J.: Transaction Publishers).

24. Dawkins (1989), p. 164.

25. Fukuyama, F (2002). *Our posthuman future: Consequences of the biotechnology revolution* (New York: Farrar, Straus, and Giroux).

26. Stock, G (2002). *Redesigning humans: Our inevitable genetic future* (Boston: Houghton Mifflin).

27. Kurzweil, R (1998). *The age of spiritual machines: When computers exceed human intelligence* (New York: Viking), pp. 146–49, 205–6.

28. Kushner, D (2002). Sports fantasy is catching up with reality, *New York Times*, July 25: pp. G1, G5.

29. Stanley, A (2003). Show of awe: A thrill ride, but no blood, *New York Times*, March 23: pp. A1, B3.

30. Moore, B E, and Fine, B D (eds.) (1990). *Psychoanalytic terms and concepts* (New Haven: Yale University Press), p. 173.

31. Kluft, R P (1989). Playing for time: Temporizing techniques in the treatment of multiple personality disorder, *American Journal of Clinical Hypnosis* 32: pp. 90–98.

32. Foucault, M (1990 [1978]). *The history of sexuality, vol. 1: An introduction* (New York: Vintage Books), p. 147.

33. The reader may wish to consult the sections on religion in both the National Bioethics Advisory Commission (1997). *Cloning human beings: Report and recommendations of the National Bioethics Advisory Commission* (Rockville, Md.: National Bioethics Advisory Commission); and the President's Council on Bioethics (2002). *Human cloning and human dignity: An ethical inquiry* (Rockville, Md.: President's Council on Bioethics).

34. Rose (1996), p. 6, reviewing Taylor (1989).

35. Rank, O (1998 [1930]). *Psychology and the soul: A study of the origin, conceptual evolution and nature of the soul*, trans. Richter, C, and Lieberman, E J (Baltimore: Johns Hopkins University Press), pp. 18–19.

36. Macklin, R (1998). Personal communication.

37. Pickrell, J (2001). Experts assail plan to help childless couples (news), *Science* 291 (March 16): pp. 2061–63.

38. Krauss, C (2003). Earthlings, the prophet of clone is alive in Quebec, *New York Times*, February 24: p. A4.

CHAPTER 11

1. L'Abate, L, and L'Abate, B L (1979). The paradoxes of intimacy, *Family Therapy* 6: pp. 175–84.

2. Person, E S (1998). The sexual century, in *The sexual century* (1999), ed. Person, E S (New Haven: Yale University Press), pp. 11–30.

3. Levine, S B (1999). Male heterosexuality, in *Review of psychiatry, vol. 18: Masculinity and sexuality: Selected topics in the psychology of men*, ed. Friedman, R C, and Downey, J I (Washington, D.C.: American Psychiatric Press).

4. Greenson, R R (1968). Disidentifying with mother, *International Journal of Psychoanalysis* 49: pp. 370–74.

5. Friedman, R C, and Downey, J I (2002). *Sexual orientation and psychoanalysis: Sexual science and clinical practice* (New York: Columbia University Press), pp. 146–47.

6. Wuethrich, B (1998). Why sex? Putting theory to the test (news), *Science* 281: pp. 1980–82; Barton, N H, and Charlesworth, B (1998). Why sex and recombination? *Science* 281: pp. 1986–90.

7. Wilson, E O (1978). *On human nature* (Cambridge: Harvard University Press), p. 141.

8. Rowland, R (1987). Women as living laboratories: The new reproductive technologies, in *The trapped woman: Catch-22 in deviance and control*, ed. Figueira-McDonough, J, and Sarri, R (Newbury Park, Calif.: Sage Publications), pp. 87–88; Dworkin, A (1998). Sasha, in *Clones and clones: Facts and fantasies about human cloning*, ed. Nussbaum, M C, and Sunstein, C R (New York: W. W. Norton), pp. 73–77.

9. Person, E S (1986a). Male sexuality and power, in Person, pp. 316–32; Person, E S (1986b). The omni-available woman and lesbian sex: Two fantasy themes and their relationship to the male developmental experience, in Person, pp. 333–43.

10. Person (1986a).

11. Alexander, B (2001). You again, *Wired*, February: pp. 122–35 (see photo on p. 134).

12. Dawkins, R (1989 [1976]). *The selfish gene* (New York: Oxford University Press), p. 161.

13. Wilson (1978), pp. 124–25.

14. Kurzweil, R (1999). *The age of spiritual machines: When computers exceed human intelligence* (New York: Viking), p. 314. Kurzweil defines *virtual sex* as "sex in virtual reality incorporating a visual, auditory, and tactile environment. The sex partner can be a real or simulated person."

15. Goldberg, C (1999). Egg auction on Internet is drawing scrutiny, *New York Times*, October 28: p. A26.

16. Eskridge, W N, Jr., and Stein, E (1998). Queer clones, in Nussbaum and Sunstein, pp. 95–113; Silver, L M (1997). *Remaking Eden: Cloning and beyond in a brave new world* (New York: Avon Books), pp. 180–82.

17. Cameron, P, and Cameron, K (2002). Children of homosexual parents report childhood difficulties, *Psychological Reports* 90: pp. 71–82; Anderssen, N, Amlie, C, and Ytteroy, E A (2002). Outcomes for children with lesbian or gay parents: A review of studies from 1978 to 2000, *Scandinavian Journal of Psychology* 43: pp. 335–51.

18. Person (1998).

19. Patterson, C J (1995). Lesbian mothers, gay fathers, and their children, in *Lesbian, gay, and bisexual identities over the lifespan*, ed. D'Auselli, A R, and Patterson, C J (Oxford: Oxford University Press), pp. 262–90; Patterson, C J (1996). Lesbian mothers and their children: Findings from the Bay Area Families Study, in *Lesbians and gays in couples and families: A handbook for therapists*, ed. Laird, J, and Green, R J (San Francisco: Jossey-Bass), pp. 420–37.

20. Friedman and Downey (2002), pp. 284–307, review this history.

21. Person (1998).

22. If this argument about psychoanalytic thinking sounds familiar, I used it earlier in the book. Just as psychoanalytic ideas regarding possible psychological risks run by identical twins are not strongly supported by research studies, we still found those ideas useful in thinking about cloning.

23. Allen, F H (1940). Homosexuality in relation to the problem of human differences, *American Journal of Orthopsychiatry* 10: pp. 129–36.

24. This would require removing the male-determining Y chromosome from the nucleus of a male progenitor and replacing it with an X, maybe even one of his own. For a female it would mean removing an X chromosome and replacing it with a Y, necessarily from a man.

25. Becker, E (1973). *The denial of death* (New York: Free Press), p. 225.

26. Becker (1973), p. 240.

27. Foucault, M (1990 [1974]). *The history of sexuality, vol. 1: An introduction* (New York: Vintage Press), pp. 38–39; Smith, D (2002). On being male, female, neither, and both, *New York Times*, October 29: pp. F5, F8.

28. Kernberg, O F (1991). Sadomasochism, sexual excitement, and perversion, *Journal of the American Psychoanalytic Association* 39: p. 345.

29. Kernberg (1991).

30. Kernberg (1991), p. 346.

31. Caspi, A, McClay, J, Moffitt, T, Mill, J, Martin, J, Craig, I W, Taylor, A, and Poulton, R (2002). Role of genotype in the cycle of violence in maltreated children, *Science* 297: pp. 851–54.

32. Because low monoamine oxidase A activity is associated with impulsivity and aggressiveness, those behavioral consequences might by themselves lead a person to momentarily lose perspective regarding the humanity of others, as well as self-control. However, these individuals, disproportionately represented among people who commit the sorts of crimes noted in the research, tend to be not only aggressive and impulsive but also planfully deceitful, and they show little remorse for harming others.

The extremely nonempathic parental behaviors that constitute child abuse would seem likely to disorder, if not damage, the capacity for empathy in some people. Clinically, victims of child abuse who come to my office as adults tend to be quite concerned about others' feelings—often overly so—and they may also misread them. However, antisocial individuals, many of whom were also abused as children, rarely ask for treatment and appear to be at the other end of the empathy spectrum. As the capacity for accurate empathy could be a critical mediating psychological variable from monoamine oxidase levels to antisocial behavior, it is striking that searching for the conjunction of *monoamine oxidase* and *empathy* in PsychInfo and Medline databases yielded nothing.

33. Friedman and Downey (2002).

34. Person, E S (1978), Transvestism: New perspectives, in Person, p. 162.

35. Moore, B E, and Fine, B D (eds.) (1990), *Psychoanalytic terms and concepts* (New Haven: Yale University Press), pp. 141–42.

36. Definitions abstracted from Moore and Fine (1990).

37. Person, E (1978), pp. 161–77.

38. Freud, S (1905). Three essays on the theory of sexuality, in *The standard edition of the complete psychological works of Sigmund Freud*, vol. 7 (1974), ed. Strachey, J (London: Hogarth Press), pp. 123–243.

39. Kernberg (1991), p. 342.

40. Person (1978), pp. 171–72.

41. Becker (1973), p. 233.

42. Chasseguet-Smirgel, J (1985). *The ego ideal* (New York: W. W. Norton), p. 11.

43. Freud, S (1914). On narcissism: An introduction, in vol. 14, Strachey, p. 101.

44. Kass, L (1997). The wisdom of repugnance: Why we should ban the cloning of humans, in *The human cloning debate* (1998), ed. McGee, G (Berkeley: Berkeley Hills Books), p. 152.

45. Lasch, C (1978). *The culture of narcissism: American life in an age of diminishing expectations* (New York: W. W. Norton).

46. From analyzing vase paintings made between 570 and 470 B.C. depicting these relationships, Friedman and Downey (2002), pp. 168–69, conclude that the sexual activity, in contrast to modern pedophilic homosexual activity, did not include anal or oral sex.

47. Rank, O (1998 [1930]). *Psychology and the soul: A study of the origin, conceptual evolution, and nature of the soul*, trans. Richter, G C, and Lieberman, E J (Baltimore: Johns Hopkins University Press), pp. 30–31.

48. Becker (1973), p. 233, emphasis added.

49. Becker (1973), p. 118.

50. There are, in fact, more than one actual homunculus—sensory and motor maps of the body topographically arrayed on the brain.

51. Johnson, M, and Lakoff, G (1980). *Metaphors we live by* (Chicago: University of Chicago Press).

52. Johnson, M (1987). *The body in the mind: The bodily basis of meaning, imagination and reason* (Chicago: University of Chicago Press).

53. Chasseguet-Smirgel (1979), p. 15.

54. Person (1986a), pp. 328–29.

55. Kernberg (1991), p. 349, citing several sources.

56. Morgan, M, and Freedman, J (2000). From fear of intimacy to perversion: A clinical analysis of the film *"Sex, Lies, and Videotape," British Journal of Psychother-apy* 17: p. 91; see also Chasseguet-Smirgel (1979), pp. 17–18.

57. Chasseguet-Smirgel (1979), p. 21.

58. Kass (1997), pp. 169, 156–57.

59. Freud, S (1905). Three essays on sexuality, in vol. 7, Strachey, pp. 123–245; Laplanche, J, and Pontalis, J-B (1986). Fantasy and the origins of sexuality, in *Formations of fantasy*, Burgin, D, Donald, J, and Kaplan, C (London: Metheun), pp. 5–34; Laplanche, J (1976). *Life and death in psychoanalysis*, trans. Mehlman, J (Baltimore: Johns Hopkins University Press).

60. Mars, D (1998). A case of mother–son incest: Its consequences for development and treatment, *Journal of Clinical Psychoanalysis* 70: pp. 401–20.

61. Arkin, A M (1984). A hypothesis concerning the incest taboo, *Psychoanalytic Review* 71: pp. 375–81; Erickson, M T (1993). Rethinking Oedipus: An evolutionary perspective on incest avoidance, *American Journal of Psychiatry* 150: pp. 411–16.

62. Earlier we reviewed the fact that being a stepchild is the biggest risk factor for abuse. A rearing mother of a male clone, perhaps especially if he is a self-clone of her husband, might be more likely to behave incestuously toward him.

63. Parsons, M (2000). Sexuality and perversion a hundred years on: Discovering what Freud discovered, *International Journal of Psychoanalysis* 81: pp. 37–49.

64. Parsons (2000), p. 46.

65. Khan, M M R (1979). *Alienation in perversions* (New York: International Universities Press), p. 16.

66. Recall our earlier discussion of depersonification in chapter 5.

67. Morgan and Freedman (2000), p. 90.

68. Becker (1973), p. 249.

69. Khan (1979), p. 12.

70. Khan (1979), p. 13.

71. Khan (1979), pp. 14–16.

72. Laforge, L (1997). Review of Goldberg, A (1995). *The problem of perversion* (New Haven: Yale University Press), *International Journal of Psycho-analysis* 78: pp. 414–18.

73. Phillips, A (1998). Sameness is all, in Nussbaum and Sunstein, p. 94.

74. Dawkins (1989), p. 35.

75. Stoller, R (1975). *Perversion: The erotic form of hatred* (New York: Pantheon Press).

76. Moore and Fine (1990), p. 77.

77. Becker (1973), pp. 235–38, discussing ideas of Phyllis Greenacre. See Greenacre, P (1968). Perversions: General considerations regarding their genetic and dynamic background, *Psychoanalytic Study of the Child* 23: pp. 47–62; Greenacre, P (1953). Certain relationships between fetishism and faulty development of body image, *Psychoanalytic Study of the Child* 8: pp. 79–98; Greenacre, P (1960). Further notes on fetishism, *Psychoanalytic Study of the Child* 24: pp. 191–207; and Greenacre, P (1969). The fetish and the transitional object, *Psychoanalytic Study of the Child* 24: pp. 144–64.

78. Summers, F (1997). Review of Bach, S (1994). *Perversion and the language of love* (New York: Jason Aronson), *Journal of the American Psychoanalytic Association* 45: pp. 261–65.

79. Rank (1998), p. 30.

80. Becker (1973), p. 163.

81. Becker (1973), p. 163.

82. Rank (1998), p. 62.

83. Rank, O (1958 [1941]). *Beyond psychology* (New York: Dover Publications), pp. 63–64.

84. Hass, R G, Chaudhary, N, Klayman, E, Nussbaum, A, Pulizzi, A, and Tisan, J (2000). The relationship between the theory of evolution and the social sciences, particularly psychology, in *Evolutionary perspectives on human reproductive behavior*, ed. LeCroy, D, and Moller, P, *Annals of the New York Academy of Sciences* 907: p. 5.

85. Rank (1998), p. 108.

86. Lederberg, J (1966). Experimental genetics and human evolution, *Bulletin of the Atomic Scientists* 22: pp. 4–11.

87. Platek, S M, Burch, R L, Panyavin, I S, Wasserman, B H, and Gallup, G G, Jr. (2002). Reactions to children's faces: Resemblance affects males more than females, *Evolution and Human Behavior* 23: pp. 159–66.

88. Winnicott, D W (1960). Ego distortion in terms of true and false self, in *The maturational processes and the facilitating environment* (1965) (London: Hogarth Press), pp. 140–52.

89. Schechner, R (1971). Incest and culture: A reflection on Claude Lévi-Strauss, *Psychoanalytic Review* 58: pp. 563–70. Schechner argues that this helps to prevent incest.

90. Lisa Cahill, a Roman Catholic moral theologian quoted in the National Bioethics Advisory Commission (1997). *Cloning human beings: Report and recommendations of the National Bioethics Advisory Commission* (Rockville, Md.: National Bioethics Advisory Commission), comes to similar conclusions in her testimony but from a theological orientation.

91. Schechner (1971), p. 570.

92. Sociobiology emphasizes the biological and evolutionary penalty of incest over Lévi-Strauss's social advantages of exogamy (see Wilson [1978], pp. 36–38). However, the exchange of genes in exogamy also provides social and psychological advantages that might be just as important in preventing incest as the risk of untoward genetic and evolutionary consequences.

93. Person (1998), p. 24.

94. Robinson, P (1976). *The modernization of sex* (New York: Harper and Row).

95. Phillips (1998), pp. 91–92.

96. Phillips (1998), p. 89.

97. Chaturvedi, S K (1993). Neurosis across cultures, *International Review of Psychiatry* 5(2–3): pp. 179–91.

98. Chrzanowski, G (1986). Changing modes of neurosis and therapeutic approaches, *Contemporary Psychoanalysis* 22: pp. 241–51.

99. This idea is consistent with Lasch's statement in *The Culture of Narcissism* that "every age develops its own peculiar forms of pathology which express in exaggerated form its underlying character structure" ([1978], pp. 41–50).

100. Becker (1973), p. 240.

101. Or, given the negative connotation of the term *asexual perversion*, we also need not indict ethically any person who would pursue reproductive cloning.

102. Becker (1973), p. 244.

103. The Genetics and Public Policy Center at Johns Hopkins University with Princeton Survey Associates (2002). Public awareness about reproductive genetic technology, phone survey, December 9.

104. Person (1999).

CHAPTER 12

1. Wilson, E O (1978). *On human nature* (Cambridge: Harvard University Press), p. 196.

2. The "id" and the "ego" are the other parts of the mind in Freud's "structural" model.

3. Moore, B E, and Fine, B D (eds.) (1990). *Psychoanalytic terms and concepts* (New Haven: Yale University Press), pp. 89–90.

4. Because Freud believed that the superego develops out of efforts to resolve oedipal conflicts, he was able to explain it better for boys than for girls. He thought that fear of castration leads a boy to give up his incestuous wishes for his

mother and to identify with his father or at least his ideal of his father and his father's moral wishes. Calling this into some question is the fact that moral reasoning by boys raised in two-parent lesbian families is indistinguishable from that of boys raised in two-parent heterosexual families. See Drexler, P F (2001). Moral reasoning in sons of lesbian and heterosexual parent families: The oedipal period of development, *Gender and Psychoanalysis: An Interdisciplinary Journal* 6: pp. 19–51. However, one could argue that a boy in a two-parent lesbian family is still denied incestuous possession of either of his two mothers and might still fear castration for these desires, though not by a father. Carol Gilligan has explored moral reasoning in females and how it differs from that of males. See Gilligan, C (1982). *In a different voice* (Cambridge: Harvard University Press).

5. Piaget, J (1965). *The moral judgement of the child* (New York: Free Press).

6. Kohlberg, L (1969). Stage and sequence: The cognitive developmental approach to socialization, in *Handbook of socialization theory and research*, ed. Goslin, D A (Chicago: Rand-McNally), pp. 347–80; Gibbs, J C (1977). Kohlberg's stages of moral development: A constructive critique, *Harvard Educational Review* 47: pp. 43–61. Note that although actions of civil disobedience derive from the most advanced stage of moral reasoning, the self-righteousness that may sometimes accompany them could belie the simultaneous presence of more primitive motives and moral orientations.

7. Talmudic disputation over how to apply the many rules embraced by strictly Orthodox Jews is one example.

8. Bottum, J (2001). Against human cloning (editorial), *The Weekly Standard*, May 7: pp. 9–10.

9. U.S. Senator Orrin Hatch is a notable exception as an opponent of abortion. He sees value in therapeutic cloning to produce stem cells for medical research.

10. Hübner, K, Fuhrman, L K, Christenson, J, Kehler, J, Reinbold, R, De la Fuente, J, et al. (2003). Derivation of oocytes from mouse embryonic cells, *Science* 300: pp. 1251–56 (originally published online; available at http://10.1126/science.1083452, accessed May 1).

11. Natural law (ethics), Encarta Encyclopedia. For a historical perspective on natural law, the reader may wish to refer to Tuck, R (1979). *Natural rights theories: Their origin and development* (Cambridge: Cambridge University Press).

12. Wilson (1978), pp. 140–41, points out that theologians have employed the concept of "natural law" in relative ignorance of biology—and I would add of psychology and social sciences, as well.

13. Foucault, M (1990 [1978]). *The history of sexuality, vol. 1: An introduction* (New York: Vintage Books). Paradoxically, sometimes those condemned as engaging in "unnatural acts" try to defend themselves within the framework of natural law. For example, with male homosexuality being found to have a significant genetic determinant, many defending homosexuality try to claim that it can be as natural as heterosexuality.

14. Freud, S (1920). Beyond the pleasure principle, in *The standard edition of the complete psychological works of Sigmund Freud*, vol. 18 (1974), ed. Strachey, J (London: Hogarth Press), pp. 1–64.

15. This is consistent with what may be the "core axiom" of natural law: "What is good for a human being, and what makes for a good human being, are functions

of what a human being is" (Devine, P E [2000]. *Natural law ethics* [Westport, Conn.: Greenwood Press], p. 31).

16. Alexander, R D, Noonan, K M, and Crespi, B J (1991). The evolution of eusociality, in *The biology of the naked mole rat*, ed. Sherman, P W, Jarvis, J U M, and Alexander, R D (Princeton: Princeton University Press), pp. 1–43.

17. Of course, one individual's pursuit might undermine another's, and that is where society must try to emulate the loving family that socializes its children to take others into account in their pursuit of pleasure. The way I am applying the idea of natural law is consistent with the spirit of the American Declaration of Independence, which claims that certain "truths" are "self-evident." The most basic of these truths are human equality and "life, liberty, and the pursuit of happiness." American society and other liberal democracies embrace these truths as natural laws.

18. Such arguments, as a rule, have been ungrounded in scientific theory and data relevant by analogy to the issue and sometimes also seem nihilistic. For example, as to whether cloning people risks compromising individuality, the medical ethicist Richard Zaner argues that if we "were to become serious about valuing each person's uniqueness, a number of current social practices would have to go." He then lists a number of such practices he decries. See Zaner, R M (1998). Surprise! You're just like me! Reflections on cloning, eugenics, and other utopias, in *Human cloning*, ed. Humber, J M, and Almeder, R F (Towata, N.J.: Humana Press), p. 137. Is this a reason to refuse to pass judgment on a new practice with possibly pernicious psychological and social consequences? If one were to argue, in the spirit of Zaner, that there are a number of current practices in society that are not always congenial to optimal psychological and social development, our reply could be the same in every instance: Why add another?

The issue of individuality is closely related to the notion of replaceability, an issue we explore in the Replacement Child Model. Dan Brock briefly reviews the argument that the practice of cloning people might decrease respect for human life because individuals would be perceived as replaceable. See Brock, D W (1998). Cloning human being: An assessment of the ethical issues pro and con, in *Clones and clones: Facts and fantasies about human cloning*, ed. Nussbaum, M C, and Sunstein, C R (New York: W. W. Norton), pp. 141–64. Although it would not be logical to come to this conclusion, human beings are not completely rational. We are all influenced by unconscious primitive modes of thought. It seems likely that a clone's parents might consciously wish and perhaps unconsciously believe that the clone will be "as if" the person from whom he was cloned. Even with the best of efforts to overcome this irrational belief, a child cloned from a dead sibling, as reviewed in the chapter on the replacement child, might likely view himself as a replacement. Thus, even if the child's parents and society somehow manage to continue to value others as unique individuals in a world of cloning, *a clone might value himself less because he is one*. And when a person values himself less, it typically disturbs the way and degree to which he values others.

19. Coming from a different perspective, Francis Fukuyama also concludes that reproductive cloning would be a violation of natural law. See Fukuyama, F (2002). *Our posthuman future* (New York: Farrar, Straus, and Giroux).

20. Whether or not one believes that a human spirit preexists the physical body it will occupy, that notion may be a useful device to help one assume the perspective of the unconceived so as to advocate for their natural human rights.

21. Kaplan, A (1998). If ethics won't work here, where? in *The human cloning debate*, ed. McGee, G (Berkeley: Berkeley Hills Books), p. 84.

22. The notion of "the rights of the unconceived" directly conflicts with an argument that has been used to try to logically discredit the significance of psychological harm to a clone. The argument derives from what is called "the nonidentity problem." In essence, it claims that the clone's lack of existence, or as I would put it, "unconceived" status, prior to being cloned makes it a nonperson that eventually will owe its existence to having been cloned. And so, we should have no reason to believe that cloning is wrong. For a fuller discussion, see Brock (1998), pp. 141–64.

23. Feinberg, J (1980). The child's right to an open future, in *Whose child? Children's rights, parental authority and state power*, ed. Aiken, W, and LaFollette, H (Towata, N.J.: Rowman and Littlefield).

24. Except for adoptive parents, who typically do not get to pick and choose, parents do not get to consent to become parents of a particular child. Instead, when parents agree to try to have a child, they implicitly consent to participate in the lottery of potential outcomes of sexual reproduction and genomic recombination. With reproductive cloning, parents would be able to decide to have a child with a particular genome, but the child still has no ability to consent to be born into a particular family.

25. Page, C, and Peterman, A (2003). The right to agree, *New York Times*, January 22: p. A21.

26. Hobbes, T (1971 [1651]). *Leviathan*, Penguin Classics edition, ed. Macpherson, C B (Baltimore: Penguin Books), p. 186.

27. The Golden Rule, stated in a variety of sources in many religions, is put this way in the Talmud: "What is hateful to you, do not to your fellow man. This is the entire Law, all the rest is commentary."

28. Vogelstein and Alberts argue that we not call therapeutic cloning "cloning" at all but something else. However, language does not change by fiat or rational argument but only in the chaos of unregulated use. See Vogelstein, B, and Alberts, B (2002). Please don't call it cloning! *Science* 295: p. 1237.

29. The disease process destroys the insulin-producing cells of the pancreas. People with type I diabetes need multiple injections of insulin every day to control their blood sugar and are at risk for a wide range of medical problems resulting from cumulative damaging effects of high blood sugars. Promising research is under way to try to clone islet cells selected to be immunologically compatible for any given individual. Transplantation of such cells would give them the capacity to make insulin themselves and not have to take immunosuppressive drugs necessary to prevent the rejection of cells transplanted from a cadaver.

30. Stock, G (2002). *Redesigning humans: Our inevitable genetic future* (Boston: Houghton Mifflin).

31. Brock (1998), p. 159.

32. Harris, J (2000). Intimations of immortality, *Science* 288: p. 59.

33. Wade, N (2002). Stem cell mixing may form a human–mouse hybrid, *New York Times*, November 27: p. A21.

34. The issue of homosexuals becoming parents through cloning is specifically discussed in Eskridge, W N, Jr., and Stein, E (1998). Queer clones, in Nussbaum and Sunstein, pp. 95–113.

35. Although simply having one of each pair of chromosomes from one donor and one from another would approximate sexual reproduction, discovering and harnessing the mysteries of chromosomal recombination during meiosis would be a critical step to not only approximating but also replicating the results of sexual reproduction. Chromosomal recombination is that process whereby parts of one chromosome change places with homologous sections of the other chromosome in the pair.

36. Eskridge, W N, Jr. (2002). *Equality practice: Civil unions and the future of gay rights* (New York: Routledge).

37. Any attempt to try to anticipate the consequences of having minimally more homosexuals in society should take the sociobiological understanding of homosexuality into account. Though not having children to pass on his or her genes, the childless gay person helps family members who share some of his or her genes to adapt so that they can pass on theirs, and many of his or hers, to the next generation. See Rose, M (1981). Are there gay genes? Sociobiology and homosexuality, *Journal of Homosexuality* 6: pp. 5–34.

38. Lederberg, J (1966). Experimental genetics and human evolution, *Bulletin of the Atomic Scientists* 22: pp. 4–11; Ramsey, P (1970). *Fabricated man: The ethics of genetic control* (New Haven: Yale University Press), p. 73.

39. Jaenish, R, and Wilmut, I (2001). Don't clone humans, *Science* 291: p. 2552.

40. Landridge, D, Connolly, K, and Sheeran, P (2000). Reasons for wanting a child: A network analytic study, *Journal of Reproductive and Infant Psychology* 18: p. 336.

41. Landridge, Connolly, and Sheeran (2000), p. 336.

42. Cain, A C, and Cain, B S (1962). On replacing a child, *Journal of the Academy of Child Psychiatry* 1: p. 455.

43. The Rorschach is perhaps the best known of projective tests.

44. Phillips, A (1998). Sameness is all, in Nussbaum and Sunstein, p. 91.

45. That is sufficient reason to be against reproductive cloning and, in the end, is the one that trumps all others. However, if we were to permit cloning for any individual person or couple in particular circumstances with which we had sympathy, then doing so would give them a right denied to others. Although this would be unfair, one might argue that it would also be unfair to interfere with any effort they might make to try to compensate for the unfair consequences of fate. But allowing people with whom we feel sympathy to clone would not improve the plight of the clone. He would not be spared some of the same risks of psychological harm likely to be encountered by the clone of a pathological narcissist. In addition, allowing cloning in such "special cases" would still engender many of the risks for society as a whole that we have anticipated for unrestricted cloning.

46. Mackie, G (1997). Ending footbinding and infibulation: A convention account, *American Sociological Review* 6: pp. 999–1017.

47. Joy, B (2000). Why the future doesn't need us, *Wired*, April: pp. 238–46.

48. Wilson (1978), p. 82.

49. Wilson (1978), p. 21.

50. McKibben, B (2003). *Enough: Staying human in an engineered age* (New York: Times Books/Henry Holt); Fukuyama (2002).

51. Rank, O (1998 [1930]). *Psychology and the soul: A study of the origin, conceptual evolution, and nature of the soul*, trans. Richter, G C, and Lieberman, E J (Baltimore: Johns Hopkins University Press), p. 115.

Index

abortion: affect on cloning legislation, 251; pro-choice position on, 238; pro-life position on, 238, 239, 243; relationship to rights of the "unconceived," 245; sex-selective, 36, 37, 38, 40, 77, 164, 185

"absent but present" parent, 102

absolutist ethics, 237–39

abuse. *See* child abuse

adopted clone: adopted by narcissistic parent, 101; of famous person, 100, 152, 191; feelings/fantasies about "parents", 102–3; "sense" of progenitor's presence in, 102–3. *See also* Adoption Model

adopted sexually conceived child: attachment to child adopted past infancy, 55; family resemblance of, 73–74, 126; as parental namesake, 153, 154; "sense" of genetic parent's presence in, 102; social status of, 188. *See also* Adoption Model

adoption: as alternative to cloning, 252; commodification of embryonic clone, 184–85; statistics about, 51, 54. *See also* adopted clone; adopted sexually conceived child; Adoption Model; "adoption syndrome;" adoptive family

Adoption Model: adopted clone, 66–68 (*See also* adopted clone); capacity for intimacy by adoptee/clone, 65–66; case vignette, 49–50; "chosen child" status of adoptee/clone, 52–54, 166; compliant adoptee/clone, 62; defiant adoptee/clone, 62; disclosure to adoptee/clone, 64–65; family romance fantasy of, 59–60, 100; genealogical bewilderment of adoptee/clone, 56–57; identity formation in adoptee/clone, 60–61, 165; incest/Oedipal complex and adoptee/clone, 57–58; motherless/fatherless self of adoptee/clone, 63–64, 102–3; parental attachment to adoptee/clone, 55–56; parental bonding with adoptee/clone, 54–55; relevance by analogy to cloning, 51–52, 56, 58, 60, 64-65, 68, 69, 70; sexual self of adoptee/clone, 62–63; social aspects of adoption, 68–69; summary/conclusions, 69–70;

transference to adoptee/clone, 56.
See also integration of models
adoption syndrome, 51–52
adoptive family: difference from
stepfamily, 44; similarity to family
of a clone, 44; similarity to
stepfamily, 45–46
age of child, replacement child
replaces, 117, 120–21, 129
age difference, between
clone/progenitor, 7, 9, 12, 20, 163
Alford, Richard, 137, 139, 146, 288n60
Allen, F. H., 214, 294n23
"All the Presidents' Children," 106
alternatives, to reproductive cloning,
251–53
altruism, 76, 95, 189, 227-28. *See also*
evolutionary psychology; good
enough parent-child resemblance
and genetic relatedness; selfish
gene theory; sociobiology
ancestor worship, 136, 137, 192
Andrews, Lori, 36, 267n61
anticipated consequences, ethics of,
239–40
Arieti, Silvano, 137–38, 286n22
asexual, cloning as only
psychologically/socially, 214–15; at
cultural level, 231–32. *See also*
asexual perversion, cloning as;
sexual perversion
asexual perversion, cloning as: to
avoid mutuality in relationship,
222–24; to cope with unresolved
narcissistic concerns, 218–20; to
cope with unresolved oedipal
concerns, 220–21; as expression of
denial of death, 224–26; incest
issues in, 222; to master early
psychological trauma, 224; if only
at a social and cultural level, 232
asexual reproduction: as asexual
perversion, 217-25, 231; dissociates
sex from reproduction, 36; earliest
attempt to apply psychology to,
xiv; influence on attitude toward,
37; as not naturally occurring in

mammals, 7, 241; by
parthenogenesis, 261n10. *See also*
Assisted Reproductive
Technologies and Arrangements
Model; cloning
Ashahara, Shoko, 108
"as if" personality, 62, 299n18
"as if real" society, 199–201, 212
assisted reproductive technologies:
case vignette about infertility,
23–24; child conceived through as
"special," 35; as dissociating sex
from reproduction, 36; as highly
intentional, 36; for
homosexual/celibate man/woman,
249–50, 251–52, 264n2; overview of
desire for genetic child, 25–26;
overview of infertility, 24–25
Assisted Reproductive Technologies
and Arrangements Model:
disclosure/secrecy about child
origins, 29–31, 39, 64–65, 70, 100,
166, 169, 178, 188, 265n31–266n32;
feminist perspective on, 38–39;
relevance by analogy to cloning, 27,
28, 32, 39; summary/conclusions,
39–40. *See also* assisted reproductive
technologies; Assisted Reproductive
Technologies and Arrangements
Model, medical technologies for
genetic child/parallels to cloning;
integration of models
Assisted Reproductive Technologies
and Arrangements Model, medical
technologies for genetic
child/parallels to cloning: donor
insemination/egg donation, 27–29;
embryo adoption, 34–35; fetal
ultrasonography, 35–36; in vitro
fertilization, 26–27; selection of
sperm/egg donor by woman,
31–32; sex-selective
abortion/preselection techniques,
36–38, 40, 77, 164, 185; surrogate
motherhood, 33–35
attachment: to child adopted past
infancy, 55; of donor to clone, 67; to

embryonic/fetal clone, 34, 35–36; influence of progenitor identity on mother/clone, 55–56; maternal–fetal, in surrogacy, 33; mother–daughter, 78; mother–infant, 27, 54, 208; narcissistic, 81 (*See also* narcissistic parent; narcissistic process, in parent–child relationship); and parental grief, 112, 114, 116; parent–child, as analogous to parent–clone, 12–13; of progenitor to clone, 77; between twins, 10, 12, 14, 17, 20. *See also* bond, bonding; intimacy

attachment theory, 44–46

autonomy of self, 196

Barash, David, 44, 45, 68, 268n20, 275n16

Becker, Ernest, 219, 225, 294n25

Beethoven, Ludwig von, 135

Bible, directive against making likeness in, 199

bio-psychosocial diagnostic schema, in contemporary psychiatry, 260n11

Bipolar Affective Disorder in famous person, 99

birth father: of adopted child, 50, 67

birth mother: of adopted child, 50, 64, 67, 68; of clone, 8, 44, 68

bisexuality, 209, 230

blastocyst, 6, 238, 243

blended family, 44

blood, symbolism of, 201

bond: father–child, 153, 158; father–fetus, 227; mother–adopted embryo, 34; mother–child, as positive and powerful, 46; mother–infant, 44, 128, 208; narcissistic progenitor–clone, 31; nonprogenitor mother–clone, 44; parent–dead child, 115; parent–fetus, 35; progenitor–clone, 12, 16, 20; twinship, 10, 12, 14, 16–17, 20, 163, 263n51. *See also* attachment; mirroring

bonding: with adoptee/clone, 54–56, 154; with replacement child, 171; with stepchild, 45

Borderline Personality Disorder, 277n43

Brentano, Franz, 126, 285n62

Brock, Dan, 299n18

brothel model, of female sexuality/reproduction, 38

Bush, George H. W., 146, 151

Bush, George W., 106, 151

Cain, AC, and Cain, BS, 116, 118, 119, 120, 121, 126, 283n13

Campbell, Joseph, 100

child abuse: of clone, 43, 172, 173, 187, 224, 243; emotional, 142, 172, 173, 187, 219 (*See also* narcissistic parenting, consequences of; narcissistic processes in parent-child relationship); genetic unrelatedness as factor in, 47, 75, 93, 154; and later risk of criminal or delinquent behavior in child, 150, 186-87; of paternal namesake, 147–48, 150, 169; physical, 42–43, 142, 172, 187, 219; psychic, of twin, 11; psychological manifestations of, 33–34, 88, 91, 140, 147, 219, 224, 294n32; as result of parental narcissism, 88-89, 142, 154; sexual, 43, 88, 187, 219, 224; as social issue, 186–87; spousal, 75; of stepchild, 42–43, 45–46, 47, 75, 93, 154, 164, 177, 187

Child of the Famous Model: biological inheritance of child/clone, 98–99; child/clone of infamous person, 108; commodification of clone, 108; desire for fame, 98; desire to have famous offspring, 101; effect of death of parent/progenitor on child/clone, 108-9; effect of parasocial interactions/relationships on clone of famous person, 107; and family romance fantasy, 59; identity issues for clone, 61; psychological impact of parent fame on child/clone, 102–5; relevance by analogy to cloning, 98, 99, 100, 102, 109; social consequences of fame for famous person/child of famous person, 105–7; societal expectations of child, 98–99; summary/conclusions, 109–10; wish to have famous parents, 100. *See also* integration of models

China: foot-binding in, 232, 256; sex-specific abortion in, 37

"chosen child" status, of adoptee/clone, 36–37, 52–54, 55, 70, 166, 190, 272n54

chromosomal recombination, 7–8, 301n35

clone: biological/medical abnormality in, 8; body image of, 174–75; capacity for intimacy, 65–66, 70, 208–9; cloning of self by, 287n44; commodification of embryonic, 184–85; definition of, 8; difference from identical twin, 9; digital, 200; genetic enhancement of, 248–49; identity of (*See* identity, clonal); issue of consent for, 244–45; life cycle of, 171–74, 207; medical advantage of being, 9; natural human rights of, 242, 243, 246, 247, 300n20; self (*See* self-cloning); sexual reproduction by, 174; sibling of, 61, 105, 177–78, 189, 272n54–273n54, 282n34; suicide of, 174

cloned child syndrome, 131, 167

cloning: conception process in, 8; effect on cultural anomie, 197–98; ethics of (*See* ethics); as fashion statement, 185–86; feminist perspective on, 38, 40, 211–12, 230–31; guidelines for, 246–48; kin, 139; of pet, 130; policy guidelines on (*See* policy guidelines, for reproductive cloning); political aspects of, 179, 198–99; profit factor in, 179, 184–85; psychological consequences of, 170–75; public policy for, 246; relevance by analogy to naming (*See* naming,

relevance to cloning by analogy); as sexual perversion (*See* sexual perversion, cloning as asexual perversion); of sibling, 178; societal reaction to, 184, 187–88, 198; success rate, 8; therapeutic, x, 238, 247-48, 252, 258, 298n9; trying to attain immortality, 159, 169. *See also* cloning, wider social/cultural implications of

cloning, wider social/cultural implications of: "as if real" society, 199–201; child abuse, 186–87; desire to claim ownership, 192–93; in family romantic fantasy, 190–91; as fashionable, 185–86; genealogical bewilderment, 188–89; identity, 183–84; medicalization/commodification of reproduction, 184–85; as new social identity, 187–88; political aspects/implications of cloning, 198–99; religious concerns, 202–4; self perpetuation/denial of death, 191–92; self-recognition/preference for self-resemblance, 193–95; summary/conclusions, 204–5; as technology of self for narcissist, 183–84; wish to be emphatically understood/admired, 197–98

cloning syndrome, 51

commodification, of embryonic clone, 184–85

conjoined twins, 261n7

conjugation, 7

conscience, 57, 105, 172, 236. *See also* superego

consent issue, for unconceived child, 244–45, 300n24

context-dependent reasoning, 266n31

covert narcissist, 278n46

critical period, in attaching to child, 45–46

cryptophasia, 14

cultural anomie, effect of cloning on, 197–98

cultural implications, of cloning. *See* cloning, wider social/cultural implications of
cultural perversion, 232
cytoplasm, 6
cytoplasm, egg, 8, 32

Dali, Salvador, 135
Daly, M., 42, 45, 268n3, n5, 271n36, 275n13-14
Dawkins, Richard, 35, 68, 74–75, 78, 159, 189, 212, 228, 267n62, 280n89
default heterosexuality, 209–10
delayed twin, 261n9
dementia, self-recognition in, 193
denial of death, as motivation for cloning, 157, 191–92, 204, 247
depersonification, 83, 93, 223
differential parenting, 43, 55
differentiation phase, of separation-individuation process, 13
digital clone, 200
diploid cell, 7
disclosure/secrecy, of child origins: in adoption, 64, 100, 166, 188, 274n107; in assisted technologies/ arrangements, 29–31, 39, 64–65, 70, 100, 166, 169, 178, 188, 265n31–266n32
discrimination, against clone, 185
disidentification, of son from mother, 217
dissociative disorder, 91, 201
Dissociative Identity Disorder, 88, 129, 230, 279n84, 287n49
dizygotic twins, effect of twin death on surviving twin, 14–15
DNA: child as donor for, 8; genetic enhancement by, 248–49; mitochondrial/nuclear, 6, 8, 9, 28, 75; from two people might make cloning acceptable, 250, 256
donor insemination: choosing sex of potential child in, 37–38; disclosure of origins to child conceived by, 30, 31; donor selection by woman, 31–32; parallels to cloning, 28, 164;

parenting in family of child created by, 28–29; sex of potential child, 37–38; statistics on, 27–28
doppelgänger phenomena, 87
Dworkin, Andrea, 38

ectogenesis, 8
egg, 7, 8
egg donation: disclosure to child conceived by, 30, 31; parallels to cloning, 28, 164; parenting in family of child conceived by, 28–29
egg donor: pornographic model as, 212; woman's motivation to be, 32; woman's motivation to be, for cloning, 34, 178, 211
egg mother, of clone, 8, 9, 268n1
ego ideal, 83, 92, 218–19, 278n53; of sexual pervert, 218
embryo adoption: commodification of, 184–85; of monozygotic twin, 263n61; overview of, 34–35; as parallel to cloning, 28
embryo splitting, 6, 162
Engel, George, 14, 263n42
entitlement, parental, 53, 83, 89, 165, 176, 187. *See also* narcissistic parent; narcissistic parenting; narcissistic process, in parent-child relationship
Erikson, Erik: affect of fame on daughter of, 103; erotic desire, in cloning, 215; and family romance family, 100; as own namesake, 64, 156–57
Eskridge, Jr., William, 250, 293n16, 301n36
ethics: of absolute in abortion/cloning/reproductive choice, 237–39; and alternatives to reproductive cloning, 251–53; of anticipated consequences, 239–40; in assessment/treatment of person wanting to clone, 253–55; Golden Rule of, 245–46, 300n27; of natural rights, 240, 243–45, 256, 300n20, 300n22, 301n45; summary/ conclusions, 256–58. *See also* ethics, of

natural law; moral reasoning; policy guidelines, for reproductive cloning
ethics, of natural law: applied to cloning, 241–42; based on Golden Rule, 245–46, 300n27; as defense for "unnatural acts," 298n13; definition of, 240–41
eugenics, 186
eusocial, 242
evolutionary psychology, 45; on incest taboo, 44–45, 268n14, 269n27, 271n36, 295n61, 297n92. *See also* sociobiology
exogamy, 229, 297n92
extended phenotype, 189

false self, 62, 84, 104, 105, 109, 166, 273n64
fame: desire to achieve oneself, 99–100, 101; as immortality, 101, 109–10; taxonomy of, 280n1. *See also* Child of the Famous Model
family romance fantasy, 101, 272n42; and adopted child, 63, 100, 165, 190; and clone, 59-60, 172, 190–91; overview of, 58–60; role as a defense against incestuous wishes, 59; role in desire for famous child, 101, 167, 191
famous person: adopted clone of, 100, 191; clone of as denial of death, 192; clone of as ego ideal, 219; cloning as fashion statement, 186; motivation to parent clone of, 68, 167, 181; namesake as child of, 150–51, 159; namesake of, 140, 152; naming adopted clone of, 140, 169; role of family romance fantasy in desire for child of, 101, 167, 191. *See also* Child of the Famous Model
farming model, of female sexuality/reproduction, 38
father–child bond, 153, 158
father–fetus bond, 227
Feinberg, Joel, 273n66
feminist perspective: on cloning, 38, 40, 211–12, 230–31; on reproductive technologies, 38–39

fetal ultrasonography, 35–36, 40
fetishism, 217, 224, 225
foreclosed identity, 165, 173
Foucault, Michel, 195, 201, 215, 240–41, 291n20, 292n32, 298n13
fraternal twins, 12
Freud, Sigmund, 73, 81; affect of fame on son of, 103; anal stage of, 171, 209; on homosexuality, 213; on incest, 56; latency period of, 172; on moral reasoning, 236, 297n4–298n4; on mourning, 114, 283n11; on narcissism, 81, 277n35; oedipal period of, 57, 172; oral phase of, 170, 209; phallic phase of, 209; pleasure principle of, 241; on sexual perversity, 230
frozen embryo, 34, 35
Fukuyama, Francis, 199, 292n25, 2999n19

gametes, 7
gay man. *See* homosexual male; homosexual male couple
gedanken, xiv
gender differences: in moral reasoning, 298n4; in namesaking, 144, 148–49, 154; in psychological effect of infertility, 25; in self-cloning, 232; in sexual perversion, 217, 232
gender identification, 209, 210
genealogical bewilderment, 56–57, 70, 165, 176, 188–89, 221, 271n33
genetic offspring, motivation for desiring, 25–26
genius, self-cloning by, 219–20
genotype: definition of, 6; of identical twin, 8
germ cell, 7
gestational mother, of clone, 8, 44, 56, 63, 68
gestational surrogate, 33, 203, 264n2
Giles, David, 99, 280n1
Goldberg, Arnold, 223, 278n45, 296n72
Golden Rule, 245–46, 300n27
Golombok, Susan, 28–29, 55, 265n25, 265n27–n28

"good enough" mother, 227
"good enough" degree of parent-child resemblance and genetic relatedness, 227-30, 232-33
Gore, Al, 106

haploid, number of chromosomes, 7
Hatch, Orrin, 298n9
Hawking, Stephen, cloning, 152
hemophilia, 267n65
hermaphrodite, self-cloning akin to the image and means of reproduction of, 215, 230
heteronomous stage, of moral development, 236, 237
homosexual: as advocate for cloning, 179; passing on genes of, 250, 301n37
homosexual clone, 80, 212–13, 231
homosexual incest, 225
homosexuality: in ancient Greece, 219, 295n46; discredited ideas about applied to cloning, 213–14, 231; in some males, defended as natural, 298n13; male fear of, 80, 211; practical implications relating to cloning, 212–13
homosexual male: as progenitor, 80, 155, 178; psychological health of cloned offspring of, 176n28
homosexual male couple: assisted reproductive technologies for, 249–50, 251–52, 264n2; child reared by, 80, 213; creating clone with genes from both members of, 212–13, 231
homuncular self, 230

ICSI. *See* intracytoplasmic sperm injection
ideal form, 103, 281n25
idealizing transference, 86–87, 278n63–279n63
identical twin: basis of mutual identification of, 76; bond between, 10, 12, 14, 16–17, 20, 163, 263n51; as clone, 1; environmental effects on vulnerability of, 5, 7; grief at death of other, 163; medical advantage of being, 9; naming, 135–36; reared apart, 7, 19, 20, 262n23; sexual orientation of, 80; twinship criteria for, 6; in vitro fertilization of embryo, 261n9. *See also* Identical Twin Model; twin
Identical Twin Model: biological/psychological limitations of, 4–5; biology of twin vs. SCNT clone, 6–9; case vignette, 2–4; parent–child relationship in, 71; relevance by analogy to cloning, 5,6, 10, 16, 18, 21; societal reaction to twins, 183, 184; summary/conclusions, 19–21. *See also* identical twin; Identical Twin Model, relation to cloning; integration of models; twins
Identical Twin Model, relation to cloning: attachment, 12–13; confusing resemblance with sameness, 10–12; death of twin, 14–16; dominance/submission in twin pair, 17–18; effect of twin bond on bond with others, 16–17, 263n51; environmental effects, 9; relevance by analogy to cloning, 5, 6, 10, 16, 18; separation-individuation process, 13–14, 55; social experience of twins, 18–19
identification, definition of, 275n17
identity: assigned components of, 60-61, 133; of child/clone of famous person, 103–4; definition of, 60; false self, 62, 84, 104, 105, 109, 273n64; foreclosed, 165, 173; genealogical bewilderment in, 56–57, 70, 165, 176, 188–89, 221; of replacement child/clone, 121, 129; of twin, 73, 163; versus role confusion—Erikson's stage five, 173. *See also* Adoption Model, Child of the Famous Model; Dissociative Identity Disorder; identity, clonal; integration of models, Namesake

Model; Parent–Child Resemblance
Model; separation-individuation,
phase of development; suffix,
name; transference
identity, clonal: formation of as
parallel/opposite to adoptive
identity, 60–61, 165; genetic, 139,
274n1–275n1; as new assigned
social identity component, 187–88;
self-narrative in, 61, 189;
transference, 126, 128, 179;
understanding through naming
analogy, 138–39, 158–59
immortality: as antisexual, 225; fame
as, 101, 109–10; memic, 159, 169,
253; as motivation for cloning, 109,
131, 204, 253, 258; namesaking as
attempt at, 157; and pathological
narcissist, 176, 195; spiritual, 253.
See also denial of death, as
motivation for cloning; Namesake
Model
immune system, self-recognition in,
193
implanted animal embryo, medical
risk/survival rate of, xiii
imprinting, 45
incest: Freud on, 56; homosexual, 225;
mother–son, 222; and Oedipal
Complex, 57–58, 221–22;
preventing through exogamy,
297n92; as rebirth, 225; self-
narrative of child of, 188–89; sexual
abuse, 43, 88, 187, 219, 224; taboo
on, 57, 229, 271n36; toward adopted
child, 58, 165; toward clone, 187,
214, 222, 225, 295n62. *See also*
evolutionary psychology, on incest
taboo; family romance fantasy, role
as a defense against incestuous
wishes
incorporation, definition of,
275n17–276n17
India, sex-specific abortion in, 37
individuality of human, affect of
cloning on, 299n18
infamous person, child/clone of, 108

infertility: medical definition of, 24. *See
also* assisted reproductive
technologies; Assisted
Reproductive Technologies and
Arrangements Model
integration of models: rationale for,
162; summary/conclusions, 179–81.
See also integration of models,
application to clone; integration of
models, conceptual
strength/weakness/convergence
with other models
integration of models, application to
clone: body image issues, 174–75;
dominance/control/dependency
issues, 175; individuals
promoting/facilitating cloning,
178–79; life cycle of clone, 170–74;
mortality issues, 175; nonclone
sibling, 177–78; parents of clone,
175–77
integration of models, conceptual
strength/weakness/convergence
with other model(s): of Adoption
Model, 165–66; of Assisted
Reproductive Technologies and
Arrangements Model, 164; of Child
of the Famous Model, 167; of
Identical Twin Model, 162–64; of
Namesake Model, 168–69; of
Parent–Child Resemblance Model,
166–67; of Replacement Child
Model, 167; of Stepchild Model,
164–65
intergenerational cloning: definition
of, 7. *See also* cloning
internalization, definition of, 275n17
intimacy: adoptee/clone capacity for,
65–66, 70, 208–9; definition of, 207,
230; effect of cloning on
dissociating reproduction from,
210–11; mother–infant relationship
as basis for, 208; relationship to
sex/sexuality, 209–10; type of
intimate relationship, 208
intracytoplasmic sperm injection
(ICSI), 27

introjection, 14, 276n17
Inuit, namesaking by, 139
in vitro fertilization (IVF): of
    identical twin embryo, 261n9;
    parenting in family of child
    created by, 28; to produce genetic
    child of both parents, 26–27;
    sexing sperm for, 38, 40; surplus
    embryos as result of, 34, 35
Iroquois, namesaking by, 139
IVF. *See* in vitro fertilization (IVF)

Jacobean complex, 124
Jonas, Hans, 279n81
Josselson, Ruthellen, 60, 77,
    242n49–n50
Joy, Bill, 257, 302n47

Kaplan, Arthur, 243–44, 300n21
Kass, Leon, 41, 219, 268n2
Kernberg, Otto, 81, 215, 277n36–n41,
    279n83, 294n28
Kernberg, Paulina, 85, 278n59
kin bias/selection, 43, 74–75, 194, 226,
    228. *See also* evolutionary
    psychology; sociobiology
kin cloning, motivation for, 139
Kohlberg, Lawrence, 237, 298n6
Kohut, Heinz, 81, 84, 277n36
Kurzweil, R., 293n14

latency period, of childhood, 58
Lederberg, Joshua, xiv, 226–27, 250,
    260n6
lesbian: assisted reproductive
    technologies for, 249–50, 251–52,
    264n2; bonding to clone by, 55–56;
    in male sexual fantasy, 211; as
    progenitor, 177. *See also*
    homosexual; homosexuality
lesbian couple: child reared by, 46, 80,
    213; implications of creating clone
    with genes of both members of,
    212–13, 231
Levine, Stephen, 209–10, 292n3
Lévi-Strauss, Claude, 229, 270n5,
    297n89–n92

Lewontin, Richard, 11, 262n21
life cycle schema. *See* psychological
    theory of human development
Lifton, Betty Jean, 51, 52, 53, 54, 63,
    64–65, 69, 270n6–n15
lineal tradition of naming, 143–44
linking object, to connect to dead
    child, 120

Macklin, Ruth, ix, x, 261n7, 292n36
Mahler, Margaret, 13, 82, 171, 262n31
male child, preference for, 37, 38, 77
maternal–fetal attachment, in
    surrogacy, 33
McGee, Glenn, 259n6, 269-270n4
medical risk, for implanted animal
    embryo, xiii
memic immortality, 159, 169, 253
metaphor, 230, 295n51–n52; Miller,
    Alice, 26, 264n11
mirroring, 54–55, 84
mirror transference, 86–87
mitochondria, 6
mitochondrial DNA, 6
mitochondrial mother, of clone, 8, 9,
    268n1
mitochondrial/nuclear DNA, 6, 8, 9,
    28, 75
modeling, use in science, xiv
monoamine oxidase, 216, 294n32
monozygotic twins, 6, 261n9; effect of
    death of twin on survivor, 14–15;
    embryo adoption of, 263n61;
    sexual orientation of, 80. *See also*
    identical twins
moral reasoning, 235; Freud on, 236,
    297n4–298n4; gender difference in,
    298n4; Kohlberg on, 237; Piaget
    on, 236
mother: attitude toward child not
    genetically related, 164; birth, of
    adopted child, 50, 64, 67, 68;
    gestational, of clone, 8, 44, 68; incest
    by, 222; interaction with
    replacement child, 122; and
    intimacy, 208; mirroring by, 54–55,
    84; mitochondrial, of clone, 8, 9,

268n1; as narcissistic parent, 84, 85.
See also surrogate mother
mother–child bond, 34, 44, 46, 128, 208
mother–daughter resemblance, 77–79
mother–infant symbiosis, 13, 55
motivation, for motherhood, 113
mourning, critical ingredients in
    adequate, 114
"Mourning and Melancholia" (Freud),
    114
multideterminism, 26
Multiple Personality Disorder, 87, 88;
    Myth, defined, 52
Myths, relevant to cloning: adopted
    child as "chosen," 52, 53; the
    "double," 290n4; stork, to
    desexualize child's origins, 69, 70,
    190; the "wicked stepparent," 42,
    292n3–n4

Namesake Model: contemporary
    naming practice, other than parent,
    139–40; further research need,
    159–60; kin naming, 139, 288n60;
    lineal tradition of, 143–44;
    namesaked charitable giving to try
    to achieve immortality, 253;
    namesake as child of famous
    person, 150–51; namesake of
    nonsibling dead, 136–39, 138–38,
    192; namesaking as attempt to heal
    wounded self, 155–57, 168;
    namesaking as form of claiming
    child as own, 153; namesaking for
    those closer to death but still alive,
    139; namesaking in primitive
    society, 137–38, 168; namesaking to
    assign paternity, 153–54, 168–69;
    parent named for own child,
    151–52; partial namesake, 136, 144,
    150, 152, 156–57, 168;
    summary/conclusions, 157–59;
    trend away from dead sibling
    namesake, 134–35, 168, 169, 192;
    twin naming, 135–36. *See also*
    integration of models; Namesake
    Model, self-namesaking

Namesake Model, self-namesaking:
    child changing own first name, 146;
    decision-maker in, 146; influence on,
    143–44, 169; paternal case study
    example, 139–43; psychological
    theory/research on, 147–50, 169;
    terminating suffix by child, 145, 146;
    trend in, 145–47. *See also* suffix, name
namesake taboo, 135, 137, 286n17
naming, relevance to cloning by
    analogy: adopting/parent clone of
    famous person, 151–52; assigned
    component of identity, 133, 151,
    153; claiming adopted clone as
    own, 154–55; gender disparity in
    cloning, 149; human wishing to be
    clone of parent, 146; motivation for
    self-cloning, 156, 158; parental
    attitudes toward self-named child
    may parallel those toward self-
    clone, 150; understanding of clonal
    individual identity, 138–39, 158–59
narcissism: covert/overt, 278n46;
    forms of, 81; pathological, 71, 81,
    176, 194, 216 (*See also* Narcissistic
    Personality Disorder); wider
    social/cultural implications of, 195
narcissistic asexual pedophilia, 219
narcissistic parent: child of as
    replacement, 125–26; disclosure to
    child that he is a clone, 31;
    motivation for self-cloning, 92, 154,
    171, 219; and parent–child
    resemblance, 92–96, 166, 168;
    parenting skills of, 180;
    rationalization for self-cloning by,
    195–96; self-namesaking of clone
    by, 154, 155, 156; view of clone of
    famous person as ideal form, 103–4
narcissistic parenting: of adopted
    clone of famous person, 101, 102;
    consequences of, 83–85, 197; by
    famous person of self-clone, 102;
    Narcissistic Personality Disorder,
    81–83, 87, 277n39–n43
narcissistic processes, in parent–child
    relationship: case vignette, 90–91;

child as device to enhance self-esteem, 89–92; intense envy, 89; projective identification, 87–88; soul murder, 88–89

narcissistic transference, type of, 86–87, 38, 94, 166–67, 176

narcissist parent: adoption of clone of famous person by, 101; cloning as technology of self for, 195–97

National Bioethics Advisory Commission, x, 260n5

naturalistic notions of self, 202

natural law. *See* ethics, of natural law

natural rights, 240; of unconceived child, 243–45, 256, 300n20–n22, 301n45

natural selection, 32, 74

neurological damage, self-recognition in, 193

neurotic sexual disorder, difference from perversion, 216, 217

nominal identity, 137–38, 139

nominal realism, 137, 139

nuclear/mitochondrial DNA, 6, 8, 9, 28, 75

nucleus, of cell, 6

Oedipal complex, 57, 76, 271n36; in adoptee/clone, 57–58, 59, 165; in replacement child, 124, 128

oedipal conflict: and adoptee/clone, 57–59; asexual perversion as false solution to, 220–21; importance of resolving, to moral sensibility, 236. *See also* family romance fantasy, role as defense against incestuous wishes

omni-available woman, 211

overt narcissist, 278n46

ovum. *See* egg

Page, Christina, 245

paleologic, 137–38, 157

parasocial interactions/relationships, 107

parental expectation/entitlement, 53, 82, 87, 106, 113–14, 239, 279n81

parental grief: coping strategy for (*See* Replacement Child Model); as motivation for cloning, 111–12, 127–28, 167; for perinatal/new baby, 112, 114–16

parent–child conflict, 89, 166, 196–97

Parent–Child Resemblance Model: case vignette, 72–73, 77, 79, 90–91; consequences of narcissistic parenting, 83–85, 194; expectations/pressures for clone as replacement, 131; identification as psychological process for sexually reproduced child/clone, 76–79; kin identification/resemblance, 74–75, 194; narcissistic parenting/parent–child resemblance, 92–94; narcissistic transference/process in relationships, 86–92; parent–child sexual orientation resemblance, 79–80; physical/psychological resemblance in adoption/cloning, 73–74; relevance by analogy to cloning, 75, 80, 81, 82, 89; self-cloning of man as assurance of paternity, 74; self-recognition/preference for self resemblance, 194–95; self-resemblance/narcissism, 80–83; summary/conclusions, 94–96. *See also* integration of models; "good enough" degree of parent-child resemblance and genetic relatedness

parenthood, motivation for, 25–26, 111–12, 127

Parnell, R. W., 76, 276n19

parthenogenesis, 261n10–n11

parthenote, 238, 261n10

partial namesake, 136, 144, 150, 152, 156–57, 168

paternity, 75, 153–54, 227

pathological narcissism, self-recognition in, 194

pathologies of self, 83

pedophile, 217, 218, 295n46

peripheral sexualities, 215

personal pronoun, for
clone/progenitor, 189
perversion. *See* asexual perversion;
sexual perversion
Peterman, Amanda, 245
phenotype: definition of, 6; of identical
twin, 8–9
Phillips, Adam, 126, 223, 231, 255,
260n5, 280n85
Piaget, Jean, 137, 236, 239, 286n21,
298n5
Plank, Robert, 148, 286n16, 288n52
Plato, 281n25
pleasure principle, 241
pluripotent cell, 261n8
policy guidelines, for reproductive
cloning: ban/punish violators of
ban, 251; disapprove but legally
tolerate, 250–51; encourage without
restriction, 246; permit without
restriction, 246–47. *See also* policy
guidelines, for reproductive
cloning, circumstance/purpose for
sanctioning
policy guidelines, for reproductive
cloning, circumstance/purpose for
sanctioning: to allow infertilile
heterosexual couple to become
genetic parents, 247; to create
longer-lived/more capable
humans, 248–49; to eliminate/treat
heritable disease, 247–48; to give
deceased or dying chance to live on
genetically, 247; to permit
gay/lesbian/single heterosexual to
be genetic parent, 249–50
polymorphous perverse, immature
sexuality as, 209, 218, 230
pornography, 212, 231–32
Posttraumatic Stress Disorder, 201
President's Council on Bioethics,
260n5
prestige of decent, 144
primary identification, definition of,
275n17–276n17
primo-progenitor, 221
progenitor, definition of, xv, 8

projection, 13–14
projective testing, of person wanting
to clone, 254
pseudocyesis, 26
psychic clone, 88, 91, 231, 280n85,
288n50
psychological theory of attachment,
44–45
psychosocial stages of human
development, Erikson's theory of,
271n35; autonomy vs.
shame/doubt, 171–72; basic trust
vs. mistrust, 170–71; generativity
vs. despair, 174; identity vs. role
confusion, 60, 170, 173, 175;
industry vs. inferiority, 172–72;
initiative vs. guilt, 57, 172; integrity
vs. despair, 174; intimacy vs.
isolation, 173–74, 207

Raelian sect, 203, 211
Rank, Otto, 202, 226, 257, 258, 282n55,
290n4, 296n3
rapprochement phase, of separation-
individuation process, 13
religion: Biblical directive against
making likeness, 199, 202;
Christianity on cloning, 203; on
conceptions of self, 202; Islam on
cloning, 203; Judaism on
soul/cloning, 202, 203; opposition
to cloning, 203–4; spiritual
immortality as alternative to
cloning, 253
Replacement Child Model: affect of
cloning on respect for human life,
299n18; characteristics of
replacement child, 120–22;
disturbed pattern of interaction
between mother/replacement
child, 122; naming replacement
child, 123, 127, 130; original concept
of replacement child, 116; original
research on (Cain & Cain), 116–18,
119, 120–21, 126; relevance by
analogy to cloning, 112, 129, 130;
risk factor for replacement child,

116–20; suicide of replacement child, 127, 130; summary/conclusions, 127-132. *See also* integration of models; Replacement Child Model, broadening original concept of replacement child; Replacement Child Model, summary/conclusions, 127–32

Replacement Child Model, broadening original concept of replacement child: case vignette of different gender replacement child, 122–21; case vignette replacing dead sibling of parent, 121–26; parents unable to conceive, adopting child resembling them, 126; replacement of living child disappointing to parents, 126; transference identity of replacement child/clone, 126, 128

Replacement Child Model, summary/conclusions, 127–32; adverse psychological consequences for replacement child/clone, 128–29; broadening concept of replacement child/clone, 129–30; cloned child syndrome, 131, 167; desire for parenthood, 127; future research need, 130; implications/limitations of research, 129; replacement child resemblance to child lost, 127–29

replacement child syndrome, 167, 254

replacement pets, 130, 285n9. *See also* Replacement Child, relevance by analogy

reproduction: brothel model of, 38; farming model of, 38; medicalization/commodification of, 184–85; sexual process of, 7–8. *See also* asexual reproduction; cloning

reproductive choice, absolutist ethics on, 238, 239

sadomasochist, 217, 225

Sants, H. J., 56–57, 271n33

Schilder, Paul, xiv, 260n9

scientific model, relationship to metaphor, xiv, 260n10

SCNT. *See* somatic cell nuclear transfer

screen memory, 201

secondary identification, definition of, 276n17

Segal, Nancy, 6, 10, 16, 20–21, 262n29, 263n41

selective abortion. *See* sex-selective abortion

self-cloning: appeal to adopted person, 63, 64; as asexual perversion, 218–19, 225; as assurance of paternity, 74; and erotic desire, 215; of famous person by narcissistic parent, 102; gender difference in, 232; of homosexual person, 155; incest in, 222; motivation for, 26, 92, 156, 158, 164–65, 166–67, 176, 197, 220, 223; motivation for narcissistic parent, 92, 154, 171, 219; nonprogenitor parent of, 44, 46, 47, 75, 170, 172, 175, 176, 177; rationalization for narcissistic parent, 195–96; of true genius, 219–20. *See also* selfish gene theory

The Selfish Gene (Dawkins), 74–75, 267n62

selfish gene theory, 68, 74–75, 76, 95, 99, 189, 195, 228, 232

self object, 84, 223, 278n53

self-reflective teknonym, 151

self-resemblance, parent–child: adopted child, 73–74, 126; and narcissistic parent, 92–96, 166, 168; social/cultural implications of, 193–95. *See also* Parent–Child Resemblance Model

separation-individuation, phase of development, 13, 55, 82, 85, 121, 125, 128, 163, 167, 171, 217, 228–29

sex: dissociating reproduction from sex/intimacy, 210–11; and pornography, 212; relationship to intimacy/sexuality, 209–10; role of erotic desire in cloning, 215

sex-linked inherited disease, 267n65

sex preselection techniques, 36–38, 40, 164

sex-selective abortion, 36, 37, 38, 40, 77, 164, 185

sexuality, components of, 209

sexualization, of perversion, 223

sexual modernism, 230

sexual perversion: definition of, 216–17, 223; difference from neurotic sexual disorder, 216, 217; masochism, 217, 225; narcissistic asexual pedophilia, 219; narcissistic dimensions of, 216; pedophilia, 217, 218, 295n46; polymorphous, 209, 218, 230; sadism, 217, 225. *See also* asexual perversion, cloning as; fetishism; incest

sexual reproduction, process of, 7–8

Siamese twins. *See* conjoined twins

sibling, of clone, 61, 105, 177–78, 189, 272n54–273n54, 282n34

Single Parents by Choice Model, 290n1

slave naming patterns, 289n71

sociobiology, 43, 226. *See also* evolutionary psychology

somatic cell nuclear transfer (SCNT): cloning by, 7; identical twin as biological model for, 8–9; original idea for, xiv, 7, 260n8

soul, 195, 196, 202, 203, 219, 282n55

soul murder, 88, 279n71, n74

Spemann, Hans, xiv, 7, 260n8

sperm donor, motivation of, 32, 178

spiritual immortality, 253

stem cell/stem cell research, 238, 261n8, 298n9

Stepchild Model: adoption by stepparent may parallel non-progenitor parent to clone relationship, 47; case vignette, 41–42; factors mitigating "stepchild effect" for clone, 44–46; potential benefit of stepchild adoption, 46–47; relevance by analogy to cloning, 41, 43, 47; risk of abuse for clone, 43–44; risk of abuse for stepchild, 42–43, 45–46, 47, 75, 93, 154, 164, 177, 187; summary/conclusions, 47. *See also* integration of models

Stock, Gregory, 199, 292n26

suffix, name: as claim to status, 144; dropping of, 140–41, 142, 143, 145, 146; negative mental health consequences for "Jr.", 147, 158, 192; nickname for, 143, 287n32; political aspects of, 198; psychological interpretation of Jr./Roman numeral, 142, 148, 149–50, 158, 169, 289n79. *See* naming, relevance to cloning by analogy, assigned component of identity

suicide, 103, 104, 197: of clone, 174; of replacement child, 127, 130, 134

superego, 104, 236, 297n4–298n5. *See also* conscience

support group, for grieving parents, 115–16

surrogate mother: attachment to fetus by, 33; of clone, 33, 34, 67; gestational, 34; motivations for becoming, 33–34

survival rate, of implanted animal embryo, xiii

symbolic estate, name as, 144

Taylor, Rex, 146, 287n38

technology, role in producing "as if real" society, 199–201

technology of self, cloning as, 195–97

teknonym. *See* Namesake Model, parent named for his own child

tempered clonality, 250

theistic notions of self, 202

therapeutic cloning, 258; to eliminate/treat heritable disease, 247–48, 252; for stem cell research, x, 238, 252, 298n9

Terrence, x, 259n4

Tienari, Pekk, 17, 263n56

Tooley, Michael, 93–94, 276n26

totipotent cell, 6

transference: and choice of who to clone, 86–87, 168; clone as embodied enactment of, 126, 159, 171, 195; enactment of , 85; idealizing, 86–87, 278n63–279n63; and identity of replacement child/clone, 126, 128, 171; issues of, in parenting clone, 180–81; mirror, 86–87; narcissistic, 38, 86, 94, 166–67, 176; and replacement, 126, 177, 181; twinning/twinship, 86–87, 138, 166–67, 176, 279n65; unconscious expectations from, 180; universality of, 56, 86, 179
transvestite, 217
twin fantasy, 87
twinning reaction, 13; of replacement child, 121–22
twinning/twinship transference, 86–87, 138, 279n65
twins: attachment between, 10, 12, 14, 17, 20; conjoined, 261n7; delayed twin, 261n9; dizygotic, 14–15; fraternal, 12; identity of, 73, 163; monozygotic (*See* monozygotic twins); psychical child abuse of, 11; singletons, should be compared to, 263n5. *See also* identical twin; Identical Twin Model; monozygotic twins

twinship bond, 10, 12, 14, 16–17, 20, 163, 263n51
twinship fantasy, 191

ultrasonography, 37
unconceived child/clone, natural rights of, 243–44, 256, 300nn20&22, 301n45
utilitarian ethics, 239

van Gogh, Vincent: as replacement child, 127, 130, 134; suicide of, 134
virtual sex, 200, 212, 224, 293n14
Von Domarus's principle, 137–38, 286n22
voyeur, 217
vulnerable child syndrome, 120

Wicker, Randolfe, 189
Wilson, E. O., 257, 267n73, 268n17, 271n36
Wilson, M. *See* Daly, M.
Winnicott, D. W., 84, 262n28, 273n64, 278n49, 284n53
Woodward, Joan, 14–16, 263n43

Zaner, Richard, 299n18
Zazzo, René, 11, 14, 262n17
Zweigenhaft, Z. I., 288n60
zygote, 6

About the Author

Stephen E. Levick is in private practice in Philadelphia, where he is a clinical assistant professor of psychiatry at the University of Pennsylvania School of Medicine and on the staff of Pennsylvania Hospital. A graduate of Case Western Reserve University's undergraduate and medical schools, he received his clinical psychiatric training at Yale University School of Medicine and did a research fellowship at New York University Medical Center. Dr. Levick has published articles on normal and disturbed brain function and individual and family psychotherapy, as well as correspondence and book reviews on a variety of topics and issues. In addition to being a member of a number of professional and scientific organizations, he is a fellow in the College of Physicians of Philadelphia.